Transgenic Plants
Current Innovations and Future Trends

Edited by

C. Neal Stewart, Jr.

University of Tennessee, Knoxville,
Tennessee 37996, USA

Copyright © 2003
Horizon Scientific Press
P.O. Box 1
Wymondham
Norfolk NR18 0EH
England

www.horizonpress.com

British Library Cataloguing-in-Publication Data

A catalogue record for this book is available from the British
Library

ISBN: 1-898486-44-1

*Printed and bound in Great Britain
by Antony Rowe Ltd, Chippenham, Wiltshire SN14 6LH*

Contents

Books of Related Interest

For further information on these books contact:

Horizon Scientific Press
P.O. Box 1, Wymondham
Norfolk
NR18 0EH England

Tel: +44(0)1953-601106
Fax: +44(0)1953-603068
Email: mail@horizonpress.com
Internet: www.horizonpress.com

Our Web site has details of all our books including full chapter abstracts, book reviews, and ordering information:

Contributors

Mentewab Ayalew
Dept. Plant Sci. & Landscape Systems
University of Tennessee
Knoxville,
TN 37996-4561
USA
Email: mentewaba@yahoo.com

Chris L. Baszczynski
Pioneer Hi-Bred International, Inc.
A DuPont Company
P.O. Box 552
7250 NW 62nd Ave.
Johnston
IA 50131-0552
USA
Email:chris.baszczynski@pioneer.com

Nicole Bechtold
MEDICAGO
Usine de molécules recombinantes
1020 Route de l'Eglise
Bureau 600
Sainte-Foy (Québec)
Canada G1V 3V9

Amy M. Brunner
Department of Forest Science
Richardson Hall
Oregon State University
Corvallis
OR 97331-5752
USA

Victor B. Busov
Department of Forestry
Biltmore Hall
North Carolina State University
Raleigh
NC 27695-8002
USA

James E. Carter III
Ctr. for Mol. Biol. and Gene Therapy
School of Medicine
Loma Linda University
Loma Linda
CA 92354
USA

Nak-Won Choi
Ctr. Mol. Biology and Gene Therapy
School of Medicine
Loma Linda University
Loma Linda
CA 92354
USA

Henry Daniell
Dept. Mol. Biol. & Microbiol. and Ctr.
for Discovery of Drugs & Diagnostics
University of Central Florida
12722 Research Parkway
Orlando FL 32826-3227
USA
E-mail: daniell@mail.ucf.edu

Anil Day
School of Biological Sciences,
University of Manchester
3.614 Stopford Building
Oxford Road
Manchester
M13 9PT
UK
Email: anil.day@man.ac.uk

Barry Goldfarb
Department of Forestry
Biltmore Hall
North Carolina State University
Raleigh
NC 27695-8002
USA

William J. Gordon-Kamm
Pioneer Hi-Bred International, Inc.
A DuPont Company
P.O. Box 1004
7300 NW 62nd Ave.
Johnston
IA 50131-1004
USA
Email:
william.gordon-kamm@pioneer.com

Susan Hefle
Dept. of Food Science and Technology
Food Allergy Research and Resource
Program
University of Nebraska
USA

Muhammad Sarwar Khan
National Institute for Biotechnology &
Genetic Engineering (NIBGE),
577 Jhang Road
Faisalabad
Pakistan

William H.R. Langridge
Ctr. Mol. Biol. and Gene Therapy
School of Medicine
Loma Linda University
Loma Linda
CA 92354
USA
Email: blangridge@som.llu.edu

L. Alexander Lyznik
Pioneer Hi-Bred International, Inc.
A DuPont Company
P.O. Box 1004
7300 NW 62nd Ave.
Johnston
IA 50131-1004
USA
Email: alex.lyznik@pioneer.com

Scott A. Merkle
School of Forest Resources
University of Georgia
Athens
GA 30602
USA

Georges Pelletier
Station de Génétique et d'Amélioration
des Plantes
INRA Route de Saint-Cyr
78 026 Versailles
France

David J. Peterson
Pioneer Hi-Bred International, Inc.
A DuPont Company
P.O. Box 1004
7300 NW 62nd Ave.
Johnston
IA 50131-1004
USA
Email: dave.peterson@pioneer.com

Vipaporn Phuntumart
Biological Sciences
217 Life Science Bld.
Bowling Green State University
Bowling Green,
OH 43403
USA
Email: vipa6610@hotmail.com

Zuo-Yu Zhao
Pioneer Hi-Bred International, Inc.
A DuPont Company
P.O. Box 1004
7300 NW 62nd Ave.
Johnston
IA 50131-1004
USA
Email: zuo-yu.zhao@pioneer.com

Harold Richards, IV
Food Safety Center of Excellence
University of Tennessee
Knoxville
Tennessee 37996
USA
Email: harry@utk.edu

Cheree Rivers-Khalid
Department of Biology
California State University
San Bernardino
CA 92407
USA

C.N. Stewart, Jr.
Depart. Plant Sci. & Landscape Systems
University of Tennessee
Knoxville
Tennessee 37996
USA
Email: nealstewart@tennessee.edu

Steven H. Strauss
Department of Forest Science
Richardson Hall
Oregon State University
Corvallis
OR 97331-5752
USA
Email: steve.strauss@orst.edu

Preface

My telephone often has a reporter on the other end either wishing to talk about questionable studies involving gene flow in corn or other biosafety issues. That's good. But, while I think that it is the duty and privilege of research scientists in the field to discuss the issues for public consumption, I sometimes get somewhat weary of the routine. On occasion I want to say that the seemingly never-ending public debate over biosafety and the acceptance of this peculiar lab-based process of gene transfer into plants (plant genetic transformation) is overrated. I muse wouldn't it be fun instead, to launch into an enthusiastic lecture about the amazing technical advances that have taken place within recent history, and the discoveries and developments in genomics and transgenic technologies that are unfolding before our very eyes?

As a substitute for being a poor interviewee, this volume was compiled instead. It was a pleasure and an intellectual feast to be the editor of such a book that involved some of the world's best transgenic plant scientists. It is an understatement to say that I learned much from the process. But this fact confirms that the reader will also be in for an eye-opening experience as the pages are perused.

Just as I am beginning a new position at the University of Tennessee as the page proofs are being read and marked by the authors, I think that gene transfer is at the door of a new phase of technical advance, and yes, public acceptance. I liken this particular mode of gene transfer to the internal combustion engine, which revolutionized the transportation industry. The gas engine proved to be a new way to turn wheels, and the wheels turned so much faster. There were novel risks to be considered and new economics to be understood. At the turn of the 20th century the new technology of automobiles was not at all predictable. In the USA, the Ford Model T flooded the market in the 1910s and 1920s as the result of popular demand and favorable economics. At that time driving was indeed precarious—not so much because of speed—the worry of our day, but because there were not suitable roads and the cars often broke down. Today such issues hardly seem real as our modern carriages are controlled by 21st century electronics and extremely efficient engines. Who today would trade their personal auto for a horse and buggy? I don't see many Luddites on horseback or in carriage. Along the way to public acceptance and mass use there were bumps in the road. Recall, in the 1950s and 1960s the economic disaster of the Ford Edsel and Ralph Nader's diatribe against the Chevrolet Corvair, "Unsafe at Any Speed"? These events occurred when the automotive industry was relatively mature. They did not, however, change the course of acceptance of the technology.

When I begin to feel impatient with the bumps in the road where plant biotechnology is traversing, and when it seems like the Luddites are carrying the day in the public arena, I think about the amazing technical innovations that have caused plant transformation to come to the marketplace. These are at the centerpiece of this book. I also remind myself of all the benefits that transgenic plants can have to not only feeding the world, but also creating a better and more sustainable agriculture; also in the book. Finally, of the amazing potential that biotechnology is playing in providing solutions to environmental problems. It is very tiring to hear traditional ecologists cum environmentalists (seemingly always having eco-bumper stickers on their SUVs) painting doomsday scenarios about the environment. "Woe is us—the sky is falling." At the same time, molecular biologists, such as the ones authoring chapters in this book are inventing solutions that will play a tremendous role in providing improved environmental quality. Too bad few, if any, of these shortsighted environmental chickenlittles will ever pick up such a book to expand their technical and scientific horizons. So perhaps the thing to do is indeed answer the reporters questions about biosafety, but to also provide some information about the undeniable present and future benefits: edible vaccines and plants that detect landmines; the production of cheaper drugs, and bioremediation. We are currently at the stage of the Model T. The roads are currently rough, but soon the superhighways will come into view. The science will be ready when the path becomes wide and smooth.

C. Neal Stewart, Jr
Knoxville

From: *Transgenic Plants: Current Innovations and Future Trends*
Edited by: C. Neal Stewart, Jr.

Chapter 1

Introduction: The Future of Transgenic Plants

C. Neal Stewart, Jr. and Anil Day

Abstract

We have entered the expansion phase of agricultural biotechnology, in which genetically modified (GM) plants are playing an important role in crop germplasm improvement. In the USA, the rate of adoption of GM plants has overshadowed any other introduction of an agricultural technology. The overriding thesis of this book is that while transgenic technologies have come so far in the past 20 years, it is still in its infancy. Some individuals and organized groups have rejected GM crops because they are unable to tolerate change in agriculture, especially the rapid technological advances made possible by modern gene transfer techniques. Change is inevitable and will be driven by forces such as climate variation or population shifts that are outside of our control. Perhaps the largest impediment to an even greater rate of global adoption has been questions of biosafety. Ecological and food biosafety concerns are discussed briefly in this text, but the authors have instead focused primarily on technological advances such as plastid transformation, targeted recombination, and the overproduction of pharmaceutical proteins. These breakthroughs will usher in far-greater numbers of commercial and humanitarian opportunities, and will be undeniably attractive to most of the general public; only diehard Luddites will oppose GM technology in 20 years. While the rewards of GM research will be largely reaped by a younger generation, this book highlights genuine attempts by scientists to address some of the current concerns on the biosafety of GM crops, to breed new crops that outperform conventional crops and to confer novel properties on crops, such as the ability to cleanse toxic soils, that help to sustain a decent environment.

Introduction

Gene transfer is a rapid mode for introducing desirable traits into plants. Genetically modified (GM) crops with traits controlled by single transgenes, such as herbicide tolerance and insect resistance, are well established and grown widely, particularly in the USA and Canada. Herbicide and insect-resistant canola, cotton, maize and soybean account for a high percentage of the current acreage of GM crops. Engineered traits, such as metal tolerance, drought tolerance, and enhanced nutritional content, provide further examples of the potential of GM technologies for improving both the productivity and quality of crops. The range and scope of GM technologies is vast and includes engineering crops for non-food uses. Such GM crops include plant biosensors for monitoring environmental pollution (Kovalchuk *et al.*, 2001) and plant factories that provide a renewable and energy-efficient means of making useful industrial materials such as industrial polymers (Slater *et al.*, 1999), pharmaceuticals and vaccines (Giddings *et al.*, 2000).

The immediate future should bring an increase in plant transgenic research as academic and industrial scientists exploit the plethora of new genes discovered from the sequence analyses of completed bacterial and eukaryotic genomes. This includes the dicot *Arabidopsis thaliana* whose entire gene complement (Walbot, 2000) can be searched in publicly accessible databases. Whole genome databases will accelerate the identification of genes controlling complex traits (Moffat, 2000), such as flowering time (Suarez-Lopez *et al.*, 2001) and dormancy (Kato *et al.*, 2001), and enable their manipulation by genetic engineering. A larger gene toolkit will also enhance opportunities for multi-gene engineering to introduce multiple traits or single traits controlled by several foreign enzymes, such as a novel biochemical pathway, into crops.

Biosafety and Acceptance

There has been much public debate over the possible deleterious environmental impact of single gene traits such as engineered herbicide (Warwick *et al.*, 1999; Stewart *et al.*, 2000) and insect resistance (Shelton and Sears, 2001), which might confer a selective advantage under particular growth conditions. An increase in the diversity of GM crops should lead to wider worldwide acceptance of GM technology once the benefits of specific traits, such as enhanced nutritional content including vitamin A-enriched (Potrykus, 2001) iron-enriched rice grain (Lucca *et al.*, 2001), and plants that detoxify TNT (Hannink *et al.*, 2001) are realized. As of the date of this writing, the GM plant controversy is still boiling and agricultural companies seem to be behaving increasingly conservatively. Still, the scientists contributing to this book, who work in industry, government, and academia, supply us with a vision of technological ingenuity for the future. The basis of our unwavering enthusiasm and optimism is simple: see how far we have come in plant transformation technologies in 20 years. Yes, there are risks to transgenic plant technology, just as there are risks in any technology, but it is apparent that agricultural biotechnologies will continue to march forward. All the authors of the chapters of this book believe that transgenic plants have a strong future. While they are, in large part, investigators in basic science, their optimism spans into commercial applications in addition to fundamental research. It seems that transgenic technologies are moving at such a rapid pace that all agriculture on every continent will be affected in a substantial way. In 2002 transgenic plants are being grown mainly in the USA and Canada, with smaller areas of cultivation in Argentina, Australia, and China, to name a few countries. But as the biosafety of GM plants is addressed in a broad sense, it will, no doubt, flourish, because of its potential to improve the human state.

Current Innovations and Future Trends

The book covers a selected range of topical areas in transgenic research and is not intended to represent an exhaustive review of GM crops. This was deliberate with the intention of making the book portable. To make the book attractive to the non-specialist a diverse set of chapters is included. Each chapter is more or less self-contained so that the reader can browse through the book and start at any chapter of interest. The emphasis of this book is on scientific aspects of transgenic plants, and other important areas, such as economics and sociology, which have a direct impact on the future of transgenic plants have not been included. Ecological biosafety is covered in detail elsewhere. One of us (Stewart) will soon be devoting an entire book to this subject (2003), and has recently written a chapter in the new edition of "Plants, Genes and Crop Biotechnology" edited by Maarten Chrispeels and David Sandova (2002) that will be made available by the American Society of Plant Biologists. Besides, ecological biosafety will not drive the future of transgenic plants (but we will clarify this statement later in the chapter). Also absent are technologies that represent state-of-the-art commercial successes: input traits such as insect resistance and herbicide tolerance. Plants with these traits have been widely successful, and follow-up transgenic plants with new modes of insect resistance are in the progress of being developed (Stewart, 1999). The book does not include chapters that explicitly target topics such as gene silencing, epigenetics, nutraceuticals, pharmaceutical production, and biosensing. However, these topics are discussed in the book, and will certainly play a prominent role in shaping future directions of transgenic plant applications.

Genomics
Genomics will underpin many of the future advances in transgenic research. This volume provides a snapshot of present applications of genomics and some future paths this research area will take in the next 5 to 10 years. The genomics revolution plays a prominent role in Chapters 2, 8, and 11. In several of the chapters, the authors illuminate how genomic information can be used to make a better transgenic plant. Certainly, transgenic technology will be useful in applications of the longest-lived organisms on earth: trees. Producing transgenic trees has certain challenges. Chapter 2, written by Amy Brunner, Steve Strauss, and colleagues, describes how transgenic technologies will be useful for manipulating forest tree maturity. Indeed, it is a smaller step from *Arabidopsis* to *Populus* at the genomic level than one would initially suspect. One tremendously important area that is on the cusp of commercial success is the manipulation of genes that enhance plant disease resistance. Vipaporn Phuntumart reviews the recent progress in disease resistance biotechnology in Chapter 8. Certainly for disease resistance and many other plant-based traits, transgenic plants will be used as a tool to extract genomic information. Mentewab Ayalew describes in Chapter 11 how transgenic plants can be used in functional genomics.

Techniques
There are three nuts-and-bolts chapters in the book on how to make transgenic plants. In addition to these, there are others that describe transgenic innovations. These three chapters illustrate just how quickly this science has

progressed in 20 years. Again, focusing on trees, Scott Merkle in Chapter 3 describes advances in somatic embryogenesis and "traditional" tree transformation techniques. Chapter 4, by Georges Pelletier and Nichol Bechtold, gives an authoritative history of *in planta* genetic transformation and demonstrates some principles that successful *in planta* techniques have in common. All who do transgenic plant research look to the day that we can banish tissue culture altogether in the production of transgenic plants. In Chapter 5, Henry Daniell and Muhammad Sarwar Khan describe an expanding list of transgenes that have been expressed at very high levels upon chloroplast transformation; from agronomic traits to pharmaceuticals. While this technique is limited to a few plant taxa, it provides food for thought for the future of producing transgenic plants that can overproduce proteins at levels likely not possible with nuclear transformation. And interestingly, one of the compelling reasons given for pursuing transplastomics is transgene containment, a ecological biosafety concern.

Biosafety and Acceptance Reprise

Surprisingly, regulatory affairs, ecological and food biosafety research, as well as public perception has played, and will play, we believe, a large role in shaping the transgenic plant research landscape. Chapters 6 and 7 demonstrate that predominantly (but not totally) non-scientific issues that could affect future commercialization of transgenic plants have altered the trajectory of research programs. In one instance, there is a real problem of antibiotic resistance in medicine, in which microbes have evolved resistance to antibiotics used to treat infections. Since transgenic plants have most commonly been produced using antibiotic resistance genes for selection in tissue culture, their presence in transgenic plants has become a political and regulatory problem. The European Union in Directive 2001/18/EC has banned the use of antibiotic resistance genes that pose a risk to human and animal health in commercial transgenic plants. One of us (Anil Day), in Chapter 6, makes the point well that the scientific risk of horizontal movement from plants to microbes is low, but economic and political realities nonetheless transcend science! So in the quest to make a transgenic plant without using antibiotic resistance genes, science also has enabled a more precise mode of transformation. Chris Baszczynski and colleagues further demonstrate scientific solutions to current concerns in Chapter 7, in which they demonstrate the elegance of site-specific recombination systems. Site-specific recombination allows gene targeting and the removal of antibiotic resistance genes. More precise genetic engineering has added value since it facilitates higher gene expression by removing position effect variation resulting from gene integration at unfavorable genomic locations. While public perception,

5

politics, and to a lesser extent, (though related) economics has driven biotechnology through certain non-scientific constraints, the science has advanced and matured to enable more precision in the methodologies used to transform plants. Biosafety concerns have, in many ways, also served to act as the thin edge of a wedge to promote research in applied areas such as ecology, allergenicity, and food safety. While there have been no cases of transgenic food affecting health in a toxic or allergenic sense, there is much fear that it has the potential to be dangerous. Harry Richards and Sue Hefle address these concerns in a publicly accessible chapter (Chapter 9) on allergenicity and how transgenic foods are tested for safety.

Future Plants

Finally, transgenic plant technology has the potential to take us where no plant has gone before. One example of novel uses is illustrated in Chapter 3 written by Scott Merkle. Trees are unique in possessing very deep and extensive roots that are ideal for long-term phytoremediation of all types of toxic substances. Bill Langridge and colleagues have written an extensive account (Chapter 10) on the production of vaccines in plants. Certainly, irresistible and overtly advantageous applications such as these have the power to silence opponents of biotechnology.

An Alternative Future

In contrast with the authors of this book, detractors of transgenic plants are less optimistic about the future of transgenic plants. Their vision veers more toward each person on earth practicing organic farming and abandoning all agricultural technologies. Time will tell which position comes closer to predicting the future, but history teaches that people embrace most technologies once they are convinced they are generally safe, and they decide the technology ought to be used to improve some condition or process. The examples are numerous: automobiles, airplanes, fluoridated water, microwave ovens, and cell phones have all been disputed as unsafe. As time has passed and people became more familiar with the technologies, along with their risks and benefits, the technologies have been widely adopted. In fact, other biotechnologies, such as the overproduction of pharmaceuticals in microbes, are widely adopted because of their undeniable benefits. No doubt the safety and benefits of transgenic plants will prevail in the greater public's eyes as trillions of transgenic plants are grown every year without harm, and new applications such as plant-delivered edible vaccines come to market. We predict widespread acceptance of transgenic plant technology in the next 20 years. That is, we predict that this mode of plant improvement will be accepted, just as travel by automobile is accepted. This is not to say that there will not be unsafe or objectionable applications of the technology that will be deservedly rejected on the basis of scientific knowledge and good critical judgment. Even England's Lord Melchett, the notorious anti-GM destroyer of transgenic crops is now on the payroll of Monsanto's public relations company, Burson-Marsteller. Strange occurrences such as these prove the future is indeed hard to predict.

Conclusions

The beauty of a book such as this one is that it should help plant scientists, especially the young ones, to peer into the future by peeking between two close-together book covers. The other inevitable feature of such a book is that it will be quickly out of date! New developments in science and technology will build upon and perhaps even supplant those described here, and that is to be expected. But these future advances will be shaped by the elements of current knowledge that are expressed in this book. The fun of working in this field is that, just like the weather in England, it will be different tomorrow. And we can always hope for sunny days.

References

Giddings, G., Allison, G., Brooks, D., and Carter, A. 2000. Transgenic plants as factories for biopharmaceuticals. Nature Biotechnol. 18: 1151-1155.

Hannink, N., Rosser, S.J., French, C.E., Basran, A., Murray, J.A.H., Nicklin, S., and Bruce, N.C. 2001. Phytodetoxification of TNT by transgenic plants expressing a bacterial nitroreductase. Nature Biotechnol. 19: 1168-1172.

Kato, K., Nakamura, W., Tabiki, T., Miura, H., and Sawada, S. 2001. Detection of loci controlling seed dormancy on group 4 chromosomes of wheat and comparative mapping with rice and barley genomes. Theor. Appl. Genet. 102: 980-985.

Kovalchuk, I., Kovalchuk, O., and Hohn, B. 2001. Biomonitoring the genotoxicity of environmental factors with transgenic plants. Trends Plant Sci. 6: 306-311.

Moffat, A.S. 2000. Can genetically modified crops go 'greener'? Science.

290: 253-254.

Lucca, P., Hurrell, R., and Potrykus, I. 2001. Genetic engineering approaches to improve the bioavailability and the level of iron in rice grains. Theor. Appl. Genet. 102: 392-397.

Potrykus, I. 2001. Golden rice and beyond. Plant Physiol. 125: 1157-1161.

Shelton, A., and Sears, M. 2001. The monarch butterfly controversy: scientific interpretations of a phenomenon. Plant J. 27: 483-488.

Slater, S., Mitsky, T.A., Houmiel, K.L., Hao, M., Reiser, S.E., Taylor, N.B., Tran, M., Valentin, H.E., Rodriguez, D.J., Stone, D.A., Padgette, S.R., Kishore, G., and Gruys, K.J. 1999. Metabolic engineering of *Arabidopsis* and *Brassica* for poly(3- hydroxybutyrate-co-3-hydroxyvalerate) copolymer production. Nature Biotechnol. 17: 1011-1016.

Stewart, C.N., Jr. 1999. Insecticidal transgenes into nature: gene flow, ecological effects, relevancy and monitoring. In: Symposium Proceedings No. 72 Gene Flow and Agriculture— Relevance for Transgenic Crops. P.J.W. Lutman, ed. British Crop Protection Council, Surrey, UK. p. 179-190.

Stewart, C.N. Jr., Richards, H.A., and Halfhill, M.D. 2000. Transgenic plants and biosafety: Science, misconceptions and public perceptions. Biotechniques 29: 832-843.

Suarez-Lopez, P., Wheatley, K., Robson, F., Onouchi, H., Valverde, F., and Coupland, G. 2001. CONSTANS mediates between the circadian clock and the control of flowering in Arabidopsis. Nature. 410: 1116-1120.

Walbot, V. 2000. A green chapter in the book of life. Nature. 408: 794-795.

Warwick, S.I., Beckie, H.J., and Small, E. 1999. Transgenic crops: new weed problems for Canada? Phytoprotection. 80: 71-84.

From: *Transgenic Plants: Current Innovations and Future Trends*
Edited by: C. Neal Stewart, Jr.

Chapter 2

Controlling Maturation and Flowering for Forest Tree Domestication

Amy M. Brunner, Barry Goldfarb,
Victor B. Busov, and Steven H. Strauss

Abstract

Maturation refers to programmed age-related changes in developmental processes that occur in all organisms. In trees, where maturation occurs over years to decades, numerous morphological and physiological changes associated with this process have been described. However, there has been little progress in elucidating the mechanisms that control maturation, and only limited capability to alter maturation state for horticultural plants and forest trees. The ability to prevent the acquisition of competence to flower during maturation could enable the broad use of genetically engineered trees in plantations with acceptable ecological and social consequence. Conversely, the ability to speed the onset of flowering could allow the use of breeding methods now considered untenable in trees. In addition, the modification of cambial maturation state could allow directed improvements in wood quality, and induced reversion of mature trees to more juvenile maturation states could allow facile clonal propagation of elite, proven genotypes. New genomic information and methods are providing many fresh avenues for probing mechanisms and identifying control points. More so than any other forest tree, *Populus* possesses the genomics infrastructure, and facile transformation and clonal propagation, that could allow rapid progress in elucidating the regulatory networks that control maturation.

Introduction

All higher plants exhibit maturation or developmental phase change. The most obvious example of phase change is the transition to reproductive development, though phases can also be defined quantitatively rather than qualitatively. Maturation in vegetative characteristics varies among species, and these changes are most obvious and numerous in long-lived, woody plants. Vegetative changes that commonly occur in trees include a decline in the ability to form adventitious roots, and changes in wood, branching and leaf characteristics (Figure 1). A defining feature of phase change is that maturity in a trait is relatively stable—it cannot be easily reversed under normal growth conditions. However, juvenility is ultimately restored in the next generation. These characteristics suggest a role for epigenetic mechanisms (i.e., gene-regulating activities that do not involve changes in DNA sequence, but can be mitotically and/or meiotically inherited) in the regulation of maturation states (e.g., Poethig, 1990; Greenwood and Hutchinson, 1993; Hackett and Murray, 1993).

Rejuvenation in at least some species and traits is possible using various treatments (e.g., *in vitro* culture and serial grafting). For trees, the stability criterion distinguishes vegetative phase change from developmental changes that recur seasonally. For example, in addition to age-related changes, cottonwoods and other trees produce morphologically distinct early- and late-flush leaves each season (e.g., Eckenwalder, 1996). Wood characteristics change during a single growing season, resulting in earlywood and latewood in many species, in addition to the well-known transition from juvenile to mature wood as trees age.

Several reviews of phase change in woody plants have made a distinction between maturation and aging (e.g., Greenwood and Hutchinson, 1993). We follow this convention, as have studies of metazoan systems (reviewed in Kirkwood and Austad, 2000). Aging is considered to be a stochastic process that occurs after maturation. It may result in part from the accumulation of somatic damage and is manifest by a slow progressive decline in vigor or productivity. Some of this decline is due to increased size and complexity, particularly in trees, but separating age responses from size and complexity

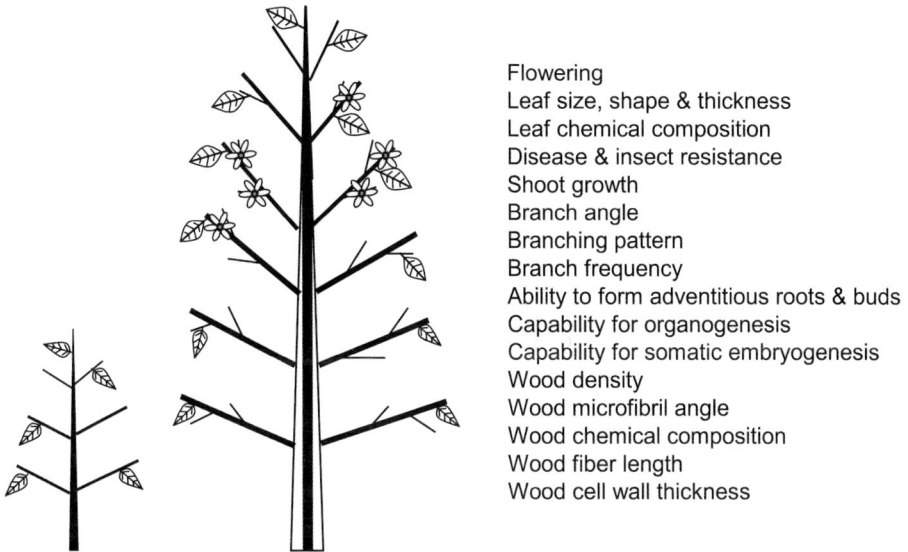

Flowering
Leaf size, shape & thickness
Leaf chemical composition
Disease & insect resistance
Shoot growth
Branch angle
Branching pattern
Branch frequency
Ability to form adventitious roots & buds
Capability for organogenesis
Capability for somatic embryogenesis
Wood density
Wood microfibril angle
Wood chemical composition
Wood fiber length
Wood cell wall thickness

Figure 1. Phenotypic changes associated with maturation in trees. Stylized juvenile (left) and mature angiosperm trees display some maturation characteristics, and additional changes are listed on the right (reviewed in Hackett and Murray 1993, Greenwood and Hutchison 1993). Mature trees show within-tree maturation gradients for some traits. Juvenile (or core) wood is depicted in black, mature wood is white.

is difficult. Whatever the underlying causes, characteristic changes in physiology and morphology occur after many years of reproductive and vegetative maturity, up to and through the 'old-growth' stage of trees (Bond, 2000). Nonetheless, though they are distinct processes, maturation and aging can alter the same traits. For example, declines in height and diameter growth with age can be the result of maturation, aging, or both (Greenwood and Hutchison, 1993).

Studies in the model annual plant *Arabidopsis* have defined at least three post-embryonic phases—a juvenile vegetative phase, an adult vegetative phase, and a reproductive phase (reviewed in Simpson *et al.*, 1999). The vegetative phases are distinguished by changes in leaf traits; most notably the distribution of trichomes, and only the adult vegetative meristem is usually competent to respond to floral induction. While *Arabidopsis* and other plants progress in a coordinated manner through vegetative maturation to the reproductive phase, the relationships between vegetative phase change and reproduction are variable among species and are particularly complex in trees. The timing of the onset of flowering in trees typically varies from a few years to many decades. While we may know when a tree first initiates flowers, in the vast majority of cases, we do not know precisely when the tree becomes competent to respond to floral inductive signals. Moreover, maturation of a particular vegetative trait may occur years before the onset of reproduction, or even after flowering occurs. For these reasons, we define juvenile vs. adult or mature trees by their reproductive status, and do not generally relate maturation in vegetative traits to specific phases (e.g., adult vegetative) of the whole organism.

We will discuss the relationships between vegetative phase change and the floral transition in the context of the whole tree, and how this may influence transgenic manipulation of maturation traits. Rather than a comprehensive review of maturation in various trees, we focus on maturation traits that are of central importance to the domestication of forest trees (see Bradshaw and Strauss, 2001). Our goal is to illustrate that the combination of forest tree genomics programs, tree-based systems for transgenic studies, and the wealth of information on regulatory genes and developmental processes in *Arabidopsis* and other annual plants, are providing powerful new tools to investigate and manipulate the maturation of trees.

Controlling Maturation: The Benefits

Breeding

Virtually all traits of economic and adaptive value change during the maturation and aging cycle in trees. Genetic improvement by breeding requires flowering for genetic recombination. The long delay before flowering and the very large size of trees when flowering occurs, preclude many options for breeding that might be considered were rapid turnover of generations feasible. This includes the introgression of a desired but rare gene into elite germplasm, especially when recessive, and the use of inbreeding/crossing to increase the efficiency of selection and better capitalize on both additive and non-additive genetic effects such as heterosis. In species threatened by exotic pests for which resistance alleles are rare, the capability for introgression could mean the difference between survival and extinction— at least from functional ecological and economic viewpoints. A means to reliably induce flowering in large numbers of very young trees, such as using an inducible transgenic system, might therefore revolutionize some kinds of tree breeding.

Gene Flow

However, if transgenes are to be used widely in forestry, it may be necessary in some places and species to restrict the ability of these genes to enter wild gene pools—either for social or ecological considerations. Thus, avoidance of flowering, which in most trees results in very wide dispersal of gametes via pollen and sometimes seed, may be required in operational plantations. This is likely to also have the desirable effect of increasing vegetative growth, especially later in life when flowering and fruit production are heavy. The simplest and most effective way to delay flowering would be to repress genes required for development of competence to flower or initiation of floral meristems, or to overexpress genes that antagonize development of these tissues. Examples of both of these kinds of genes are now known. Alternatively, strategies that do not affect reproductive maturation or meristem development could also be employed, such as via disruption of floral organs via tissue ablation or floral gene suppression (Strauss *et al.*, 1995).

Vegetative Propagation
Because of the extensive heterozygosity in tree populations and their intolerance to inbreeding (as a means of fixing desirable genotypes), large increases of genetic gain are often sought via clonal propagation of elite genotypes. However, because the genetic value of individual genotypes for yield traits cannot be reliably determined for a number of years, it is desirable to clone tissues from older trees. Alternatively, large numbers of genotypes can be maintained in a juvenile state via hedging or cryopreservation while field trials are performed. Cryopreservation is costly, can be technically challenging, and can only be applied to species for which there is a highly developed, cost-effective embryogenic propagation system in place. Hedging for long time periods, and propagation from older trees, is technically difficult. Most species show a loss of competence with age for the cellular redifferentiation required for propagation by rooted cuttings (discussed below). Epigenetic changes are also commonplace in trees, resulting in clonal propagules with undesirable mature characteristics that persist for years such as slow, branch-like growth form. If genes could be identified that enable regenerative competence to be maintained or restored, it thus would enable clonal production of juvenile-like "seedlings" from trees. Such knowledge and technology could radically increase the capability for cost-effective clonal deployment. Similarly, a system for induction of apomictic seeds in young trees would allow facile clonal propagation, and could take advantage of the already developed nursery infrastructure for seed and seedling culture. Genes related to apomixis are under intensive study in several herbaceous species (Grossniklaus *et al.*, 2001).

Wood Quality

Finally, wood quality, growth rate, pest resistance, and many other economic traits change markedly with tree age. Wood produced in young trees, and in the crown area of older trees, is referred to as juvenile wood, and differs in a large number of biochemical and structural ways from the "mature wood" produced toward the base of older trees. Nearly all of the characteristics of juvenile wood are inferior to those of mature wood for both pulp and solid wood products (Zobel and Sprague, 1998). Thus, means for accelerating the transition from juvenile to mature wood—if it could be done without compromising the rapid growth of juvenile trees—would appear highly desirable. Other targets for manipulation include crown structure; for example, more horizontal branches, common in older conifers, appear desirable for reducing knot size.

Poised To Make Major Progress

Candidate Regulatory Genes from Studies in Annual Plants

Since the sequencing of the *Arabidopsis* genome has completed, coordinated efforts are now focused on "understanding the function of all genes of a reference species within their cellular, organismal, and evolutionary context by the year 2010" (Chory *et al.*, 2000). Sequencing of the rice genome is approaching completion, and projects to construct saturation mutant lines for functional analysis are underway (reviewed in Yuan *et al.*, 2001; Hirochika, 2001). Thus, in the upcoming years, studies in *Arabidopsis* and other annual plants will provide an increasing number of candidate regulatory genes for the various traits that undergo maturation in trees (specific examples are discussed in later sections). From the standpoint of understanding tree maturation, one of the most intriguing functional genomic projects is focused on epigenetic regulation. The plant chromatin database (http://chromdb.biosci.arizona.edu) provides information on chromatin-level control of gene expression in plants, and a major project goal is to mutate and functionally analyze most of the maize and *Arabidopsis* genes that have a role in chromatin-level gene regulation.

Although studies in *Arabidopsis* and other annuals will continue to be a valuable guide, extrapolating from these studies to large, long-lived perennials has limitations. Sequencing of the *Arabidopsis* genome revealed that most genes are duplicated (*Arabidopsis* Genome Initiative, 2000). Moreover, study of the large MADS-box gene family in a variety of plant species has demonstrated that duplications specific to all taxonomic levels (e.g. orders, families, and genera) are common (e.g. Theissen *et al.*, 2000). Such duplications have resulted in genes with distinct functions, genes with highly redundant functions, and genes with distinct but overlapping functions. Although computer programs to identify orthologous genes among the different plant databases (reviewed in Yuan *et al.*, 2001), and information from comparative mapping studies will help overcome this complication, identifying tree orthologs can still be problematic. This is especially true for the large conifer genomes, where amplification and dispersal of genes to form complex families appears to have been much more prominent than in angiosperms (Kinlaw and Neale, 1997).

Furthermore, vegetative phase change affects many traits in trees that may be poorly manifest in annual plants (e.g., wood characteristics). In addition, the degree to which the genes controlling phase transitions in *Arabidopsis* play a similar role in other annual species is largely unknown. Thus, a number of genes important in tree maturation may not be predicted based on studies in annual plants. An individual gene is part of a variety of complex genetic and developmental pathways, and a gene product may perform differently depending on its molecular, physiological, or developmental context (whole organism). All the functions of a gene over multiple years in a tree, may not be inferred based on studies of orthologs in annual plants. Clearly, tree genomics projects and the ability to functionally analyze genes in trees are needed.

Tree Genomics Projects and Approaches

A number of large genomics projects are underway in forest trees. The most advanced is a large EST (expressed sequence tag) sequencing-based project at Genesis (New Zealand). They have sequenced more than one hundred thousand ESTs from a diversity of tissues in *Pinus radiata* and *Eucalyptus grandis*, and have used transgenesis in herbaceous organisms extensively to analyze function (A. Shenk, pers. comm.). However, public access to these databases is highly restricted. Smaller EST projects for eucalypts are underway in Brazil, France, and Sweden.

A large EST sequencing project is underway at North Carolina State University (Whetten *et al.*, 2001). The sources of most of the RNA for the ESTs were xylogenic tissues of *Pinus taeda*, although some shoot tip RNA was included. The project has a goal of sequencing 70,000 ESTs and, as of August 2001, 55,000 have been completed (R. Sederoff, pers. comm.; http://web.ahc.umn.edu/biodata/nsfpine and http://web.ahc.umn.edu/biodata/doepine).

Populus is the other forest taxon for which there are significant genomics projects underway. Although of economic value for diverse forest, environmental, and agronomic products, it also is considered the model forest tree for molecular genetic studies as a result of its small genome size (550 Mb), ease of clonal propagation, transformability, genome markers, and the existence of many interspecific pedigrees that facilitate QTL identification. It is therefore worthy of study in its own right as a model system for woody plants (Bradshaw, 1998). The main projects underway also focus on ESTs from xylem tissues (Mellerowicz *et al.*, 2001). The most advanced project is based in Sweden (http://Poppel.fysbot.umu.se), where a large number of scientists collaborate on diverse biological problems using the sequence and array hybridization methods developed there. Currently 49,000 EST sequences have been determined from 12 different libraries developed, mostly from aspen (*Populus tremula*), and another 7 libraries are under construction. A unigene array of 13,000 sequences has been constructed and is being used in hybridization experiments. Most of the ESTs that have been determined to date are derived from wood-forming tissues (>22,000) and leaf/apical meristems (>18,000). Other sequenced libraries derive from floral tissues (>7,000) and roots (>400). The target is 100,000 total ESTs, which they expect to produce within a year (B. Sundberg, pers. comm.).

Other genomics projects include ones in France (http://mycor.nancy.inra.fr/poplardb/index.html) and Canada. A French-based project also focuses on xylem tissues, and has a near-term goal of 20,000 ESTs and production of commercial *Populus* gene-chip that will be accessible by the scientific community. A project recently funded by Genome Canada is still being developed, but is likely to include a great deal of EST sequencing, and the construction of a detailed physical and genetic map based on BAC libraries and BAC-end-sequencing, as well as intensive QTL analysis. A parallel project is also underway there in spruce (C. Douglas, pers. comm.). Finally, it is expected the United States Department of Energy will determine the genomic sequence of *Populus* at a three to six-fold level of redundancy within the next two years (G. Tuskan and R. Dahlman, pers. comm.). This would obviously provide a quantum leap of capability for intensive genetic analysis in *Populus*.

Transgenic approaches are powerful means for functional genomic studies, and have been largely restricted to *Populus* among forest trees because of its ease of transformation and clonal propagation (e.g., Han *et al.*, 2000). Random insertional screens for genes have been carried out using *Agrobacterium*-mediated transformation and gene promoter traps, where the random insertion of a reporter gene such as GUS near to a regulatory element gives rise to cell or tissue specific expression patterns. In a preliminary study of several hundred lines, rates were observed to be similar to those of *Arabidopsis*, including a number of desired vascular and rooting-associated expression patterns (A. Groover and R. Meilan, pers. comm.). A pilot study of activation tagging (Weigel *et al.*, 2000)—where a 4X cauliflower mosaic virus 35S enhancer element was "randomly" inserted into the poplar genome via *Agrobacterium*-mediated transformation—showed a rate of recovery of mutant phenotypes of about 1%, also similar to that observed in *Arabidopsis* (unpublished data). Large-scale mutagenesis programs using these methods are likely to be successful in identifying many genes that affect woody plant development that would be missed in screens of herbaceous plants.

It is also feasible to conduct large-scale, directed transgenic programs in *Populus* where specific candidate genes related to traits of interest are suppressed or overexpressed. Transgenes producing RNAs that form duplex structures (i.e., DNA inverted repeats) appear to be highly efficient in stimulating post-transcriptional gene suppression via double-stranded RNA interference (RNAi) (Smith *et al.*, 2000). With the wealth of functional genomics information coming from studies in *Arabidopsis,* rice, and other model plant species, transgenic manipulation of tree homologs would appear to be the most direct means for developing novel biotechnological applications from genomic information.

Although stably transformed conifers have been produced from a number of species (Huang *et al*, 1991; Ellis *et al.*, 1993; Charest *et al.*, 1996; Tzfira *et al.*, 1996; Levee *et al.*, 1997; Walter *et al.*, 1998), the low frequency of transformant recovery and the time required to recover transformed plants limits the use of transformation for genomic research. These limitations appear to be more related to efficiency and speed of tissue culture regeneration system than to the introduction of DNA itself. In addition, some antibiotic selection systems that work well with angiosperms, are not very effective with gymnosperms (Merkle *et al.*, 2001). Thus, alternative selection systems could potentially improve efficiency and reduce the time until transformed plants are available. Most rapid progress will probably occur in those species for which reliable protocols exist for initiation of somatic embryogenic cultures and recovery of somatic embryos, such as *Picea abies* (Wenck *et al.*, 1999) and *Pinus radiata* (Walter *et al.*, 1999). However, transgenic approaches to functional genomics are likely to be concentrated in *Populus* for the foreseeable future.

Finally, new methods for studying genome-level methylation using arrays are under development (http://www.epigenomics.com/). Because of the expected importance of epigenetic regulation of all aspects of maturation in trees (discussed below), such methods could provide entirely new insights into the regulatory networks that govern maturation in trees.

Maturation, Meristems, Coordination, and Manipulation

The manifestation of maturation involves changes in the activity of meristems. Most of the trait transitions involve structures (e.g., leaves, branches, inflorescences) initiated by the shoot apical meristem (SAM). Because of the polar nature of shoot growth, some of these changes exhibit a spatial and temporal gradient along the axis of the shoot, such that basal regions of a mature tree may display juvenile characteristics (Figure 1). Although most adventitious root meristems are derived from vascular parenchyma cells in stems (Goldfarb *et al.*, 1998), the ability to form such roots generally declines along a gradient similar to other shoot characteristics. However, these various traits may mature at very different rates and not all show within-tree gradients.

As a product of a different meristem, the vascular cambium, wood maturation follows a different within-tree gradient; wood characteristics vary across the radius of the bole and along the height of the tree (Figure 1; reviewed in Zobel and Sprague, 1998). A young tree produces only juvenile wood, which is also called core wood, because it is formed within a given number of rings from the tree center. Wood characteristics (e.g., specific gravity, cell length) change gradually and at different rates in relation to the number of annual rings from the pith. The term "transition wood" is often used to describe intermediary phases between juvenile and mature zones. Though determinations of the mature wood zone vary depending on what particular wood trait is studied, all traits reach a relatively stable state so that the outer or mature wood is generally uniform. Because the type of wood produced is related to the age of the cambium at the point of wood formation, or distance from the pith, an older tree produces both juvenile (at the top of the bole) and mature wood (at the bottom of the bole); the overall proportion of juvenile wood decreases considerably with tree age.

Maturation in a trait involves two main regulatory steps—the initiation of phase change and the maintenance of the mature phase. Studies in a variety of plants have shown that the floral transition is regulated both by transmissible signals originating outside the SAM, and by competence of the SAM to respond to these factors (reviewed in Levy and Dean, 1998). Less is known about vegetative phase change, but studies in maize indicate that this transition is initiated by factors outside the SAM, though the adult phase may be maintained by intrinsic changes in the identity of the SAM (Irish and Karlen, 1998; Orkwiszewski and Poethig, 2000).

The long-standing hypothesis that epigenetic mechanisms (Poethig, 1990; Greenwood and Hutchinson, 1993; Hackett and Murray 1993), such as DNA methylation, have major roles in regulating phase change has not yet been proven, but accumulating evidence suggests that epigenetics is important in maturation. Early studies attempting to correlate changes in total genomic methylation with tree maturation produced inconclusive results (reviewed in Haffner *et al.*, 1991). However, changes in specific genes would not be detected by such scans, and studies in *Arabidopsis* and other eukaryotes show a high degree of specificity in epigenetic regulation.

Studies in plants have shown the importance of DNA methylation in regulating gene expression, and have begun to link methylation to chromatin remodeling enzymes (reviewed in Habu *et al.*, 2001). This connection has already been clearly demonstrated in other eukaryotic systems (reviewed in Muller and Leutz, 2001), and it appears likely that reversible modifications of chromatin, such as DNA methylation and histone acetylation, act in a combinatorial fashion to provide multiple dimensions of transcriptional control in plants. For example, the *ddm1* mutation in *Arabidopsis* induces a severe reduction in genomic 5-methylcytosine level and various developmental abnormalities (Vongs *et al.*, 1993). *DDM1* does not encode a methyltransferase, but rather a protein homologous to SWI2/SNF2, a class of ATPases, which function in multiprotein chromatin remodeling complexes that regulate transcription (Jeddeloh *et al.*, 1999). Changes in chromatin structure and DNA methylation of a maize anthocyanin regulatory gene were shown to coincide with the shift from juvenile to adult growth (Hoekenga *et al.*, 2000), and reduction in the level of RPD3-type histone deacetylases in *Arabidopsis* delayed transition to the reproductive phase (Wu *et al.*, 2000; Tian and Chen, 2001).

The mechanistic relationships between phase change in various vegetative features and reproductive competence are unknown, but there are indications that changes, in a least a subset of vegetative traits and the floral transition, are often regulated independently (reviewed in Lawson and Poethig, 1995; Greenwood, 1995; Jones, 1999). In *Arabidopsis,* a loss-of-function mutation in a cyclophilin 40 gene resulted in accelerated transition to the adult vegetative phase, but did not change the timing of the floral transition (Berardini *et al*, 2001). While mutations in some flowering-time genes alter both vegetative and reproductive phase changes, others appear specific to the floral transition (reviewed in Simpson *et al.*, 1999).

Independence of vegetative and reproductive phase change increases developmental plasticity, and thus, the potential for heterochronic evolution (discussed in Lawson and Poethig, 1995; Diggle, 1999). Recent studies in *Eucalyptus globulus* have shown that the transition from juvenile to adult leaves and the onset of flowering are under independent quantitative genetic control, and quantitative genetic analysis also suggests that the timing of this vegetative phase change is an adaptive trait (Jordan *et al.*, 1999; 2000). There are a number of alternative models that may explain the complex and prolonged maturation in trees (see Greenwood, 1995; Hackett and Murray,

1997). Nonetheless, the proposal (e.g., Poethig, 1990) that phase changes in different traits are controlled by independent pathways that share a few common regulatory elements appears to be generally consistent with our current knowledge. This model allows for coordination in the maturation of two or more traits, but maturation in one trait is not a prerequisite for maturation in a second trait. It is important to note that if there is a universally applicable model, there are still likely to be taxon-specific deviations. For example, in some species, the floral transition may be dependent on maturation in one vegetative trait, but be independent of other vegetative phase changes.

Short-distance communication via the plasmodesmata, and long-distance signaling via the phloem play a role in specific phase changes (e.g., Bernier *et al.*, 1993), and could potentially provide a mechanism for coordinating the maturation of various traits. Long-distance movement of signaling molecules such as phytohormones and sucrose have long been recognized. Sucrose signaling has an important role in the floral transition (reviewed in Bernier *et al.*, 1993), and also in wood development (Uggla *et al.*, 2001). Auxin has a major role in wood development (Sundberg *et al.*, 2000), branching (e.g., Chatfield *et al.*, 2000), and adventitious root formation (Hackett, 1988), and integrates developmental processes throughout the plant (Berleth and Sachs, 2001). An important aspect of distance signaling is that while a molecule may be widely transported via phloem, there are mechanisms for locally modulating the activity of the molecule. Plasmodesmata alter their transport capacity both temporally and spatially in different regions of the plant (reviewed in Zambryski and Crawford, 2000). In addition, the activity of transported phytohormones may be modulated locally via mechanisms such as degradation and the formation and hydrolysis of inactive conjugates. For example, the *Arabidopsis* gene *SUPERSHOOT*, which is strongly expressed in leaf axils, appears to suppress axillary meristem initiation by locally attenuating cytokinin levels (Tantikanjana *et al.*, 2001).

Recent studies have shown that mRNAs are also transported via the phloem and that a translocated mRNA can be functional (Ruiz-Medrano *et al.*, 1999; Kim *et al.*, 2001). Moreover, the delivery and exit of phloem transcripts appears to be selective. Another potential mechanism for coordinating maturation processes via long-distance signaling is RNA-mediated gene silencing (reviewed in Matzke *et al.*, 2001). Epigenetic states of genes can be transferred from one part of a plant to another via a mobile silencing signal; however, this mechanism has not been shown to play a role in normal plant development.

The mechanisms underlying maturation in various traits, how they are coordinated, and whether or not phase change in different traits is independently regulated influence our ability to manipulate maturation traits for tree improvement. Independent regulation of individual traits or suites of traits is desirable, because it allows greater flexibility. For example, this may allow genetic engineering of trees to delay reproductive phase change, and at the same time accelerate wood maturation. Conversely, it may be desirable to maintain juvenility in two or more traits, such as adventitious rooting ability, reproductive capability, and shoot growth. In such cases, targeting one common regulatory gene for transgenic manipulation rather than several different trait-specific genes may be the preferred approach. As discussed above, maturation in traits may be genetically separable, and at the same time, coordinated to various degrees at the whole plant level. Thus, genetic manipulation of one maturation trait could have indirect effects on other maturation traits and plant processes. As a whole, these possibilities highlight the need for research including transgenic manipulation in trees (rather than annual plant models) and testing of transgenic trees over multiple years. Also of importance is the development of facile conditional gene expression/suppression systems. For some traits, it may be expedient for a particular phase to be maintained, but the alternate phase to be inducible at particular times. The prevention of the floral transition is desirable when trees are grown in production plantations, but accelerated flowering is needed to advance tree breeding.

The Switch from Vegetative to Reproductive Development

Physiological and genetic studies in a variety of plants indicate that the transition to flowering is under multifactorial control, involving perception of environmental cues and internal signals related to the developmental state of the plant, changes in gene expression, and mobile signals that must be transported to the SAM (Bernier *et al.*, 1993). Different factor(s) of this regulatory network are predicted to become limiting factor(s) in different species or genotypes, or in a given genotype grown under different environmental conditions.

Most of the recent advances in unraveling the genetic networks that interact to control flowering have come from studies of the facultative long-day plant, *Arabidopsis* (reviewed in Simpson *et al.*, 1999; Araki 2001). Genetic analyses of mutants and natural variation in ecotypes have identified at least 80 loci that affect flowering time in *Arabidopsis* (Levy and Dean, 1998). Based on mutant responses to environmental cues and analyses of genetic epistasis, at least four pathways regulate flowering time in *Arabidopsis* (Figure 2; reviewed in Simpson *et al.*, 1999). Plants measure day length by integrating signals from photoreceptors and an endogenous circadian clock, and long days promote flowering via this photoperiod pathway. Extended periods of cold temperature promote flowering in many ecotypes via the vernalization pathway. Genes in the autonomous pathway probably respond to an internal 'developmental clock'. Under short-day photoperiods, flowering depends on a gibberellic acid (GA) signal transduction pathway. Ultimately, the interplay among flowering pathways activates floral meristem identity genes and the competency of SAM to respond to floral induction signals.

Studies in *Arabidopsis* have also revealed that quantitative regulation of gene expression and redundancy are important features of the flowering pathway network. Additional characteristics of this regulatory network are that it includes both suppressors and promoters of the floral transition, that related genes may have opposite effects, and that regulation involves transcriptional, post-transcriptional, and epigenetic mechanisms. These pathways include genes, such as photoreceptors, that regulate a wide variety of plant responses as well as genes that appear specific to the floral transition. In addition, downstream genes that integrate multiple flowering pathways have been identified. Specific examples that illustrate these features are discussed below.

TERMINAL FLOWER 1 (TFL1) is a floral repressor that has a role in all growth phase transitions (Ratcliffe *et al.*, 1998), and also acts to maintain identity of the indeterminate inflorescence meristem (reviewed in Pidkowich *et al.*, 1999). While *tfl1* mutants progress more rapidly through all phases and apical meristems are converted to terminal floral meristems, all phases in *35S::TFL* plants are greatly extended, resulting in much larger plants with highly branched inflorescences. TFL1 belongs to the family of phosphatidylethanolamine-binding proteins (PEBP; Bradley *et al.*, 1997) that appear to act as regulators of kinase signaling pathways (Banfield and Brady, 2000). Six genes encode PEBP proteins in *Arabidopsis* and another of these, *FT*, promotes the floral transition (Kardailsky *et al.*, 1999; Kobayashi *et al.*, 1999).

FT and the MADS-box gene, *SUPRESSOR OF OVEREXPRESSION OF CONSTANS1 (SOC1),* integrate flowering pathways (Samach *et al.*, 2000; Onouchi *et al.*, 2000; Lee *et al.*, 2000). Both genes are downstream targets of the long-day promotion pathway, and GAs also positively regulate *SOC1* expression (Borner *et al.*, 2000). *SOC1* and *FT* are also regulated by the autonomous and vernalization pathways via another MADS-box gene, the floral repressor *FLC* (Sheldon *et al.*, 1999; Michaels and Amasino, 1999). The level of *FLC* activity is proportional to flowering time and to the magnitude of the vernalization response. Moreover, the coding sequence of *FLC* alleles from early and late flowering ecotypes are identical, indicating that alleles differ in some aspect of their regulation (Sheldon *et al.*, 2000). Autonomous pathway genes that repress *FLC* expression, include two genes,

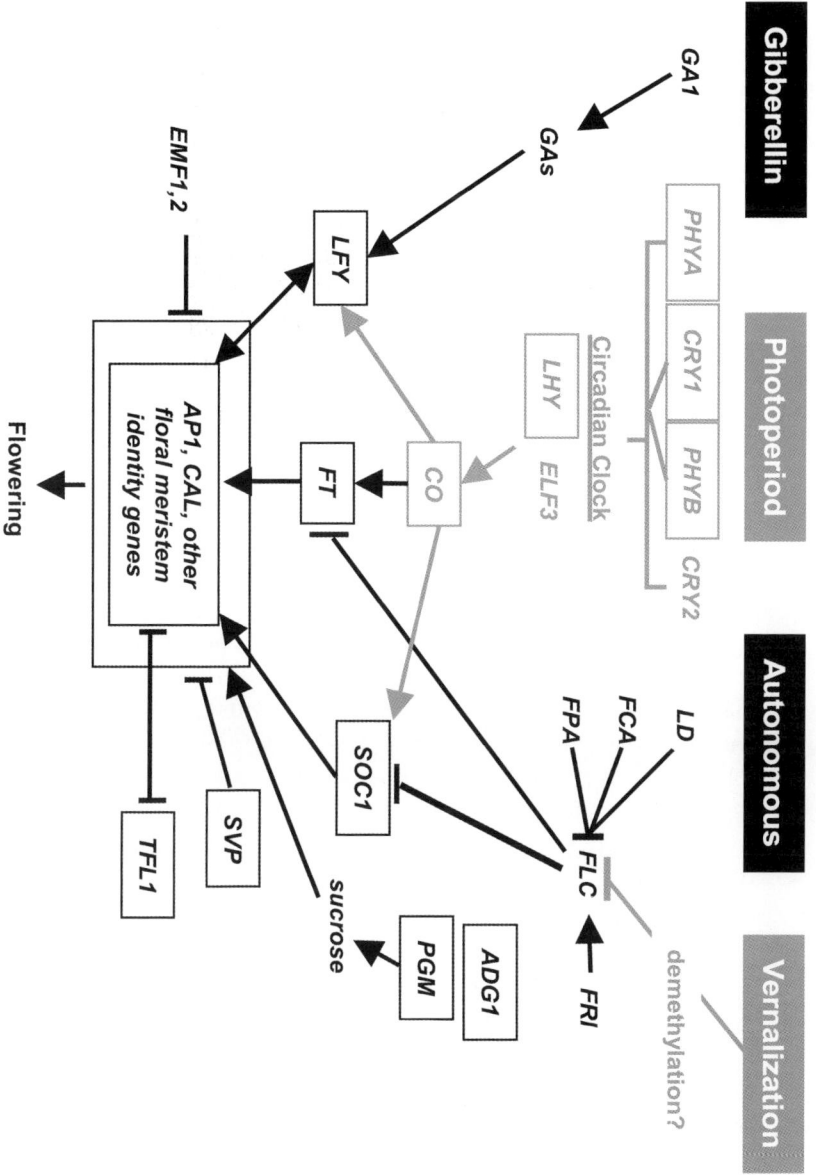

Figure 2. Model of pathways regulating the floral transition in *Arabidopsis*. The model is derived from Araki, 2001; Blazquez, 2001; and Simpson *et al.*, 1999, and does not include all interactions and genes. Arrows indicate positive regulation and short solid lines indicate repression. Some genes have not been placed in a particular pathway. Boxed genes indicate putative poplar orthologs that have been isolated via direct cloning (Brunner *et al.*, 2001, Rottmann *et al.*, 2000, Howe *et al.*, 1998) or searches of the poplar EST database (http://Poppel.fysbot.umu.se/index.html; Tandre *et al.*, 2001).

FCA and *FPA*, which encode proteins containing RNA recognition domains, and *LUMINIDEPENDENS (LD)*, which encodes a homeobox protein that may also bind RNA (MacKnight *et al.*, 1997; Auckerman *et al.*, 1999; Schomburg *et al.*, 2001). Thus, these genes may regulate *FLC* transcript levels post-transcriptionally.

A characteristic of the vernalized state is its mitotic, but not meiotic, stability, indicating epigenetic regulation. Similarly, the vernalization-induced reduction in *FLC* transcript levels occurs in all tissues, and is mitotically stable, but high levels of *FLC* transcript are restored in the next generation (Sheldon *et al.*, 2000; Michaels and Amasino, 1999). Evidence for epigenetic regulation via DNA methylation is indirect. Both chemical demethylation and antisense-*METHYLTRANSFERASE1*- induced demethylation accelerated flowering in vernalization-responsive plants, and *FLC* transcripts were also reduced (reviewed in Finnegan *et al.*, 2000). The late-flowering phenotype of dominant *fwa* mutants is the result of hypomethylation and the corresponding ectopic expression of *FWA* (Soppe *et al.*, 2000). In wild-type plants, *FWA* is expressed only in siliques and germinating seedlings, and its lack of expression in mature plants is associated with extensive methylation of two direct repeats in its 5' region, suggesting that *FWA* may promote establishment of the vegetative phase.

Still unknown is the extent to which genes that regulate the floral transition in *Arabidopsis* also regulate this transition in other annual plants. Only a few flowering-time genes have been cloned from other species, but it appears that the function of at least some genes is generally conserved among divergent species. For example, a major quantitative trait locus, *HD1*, controlling the photoperiodic response in rice, a short-day plant, was recently shown to be a homolog of the transcription factor *CONSTANS (CO)* (Yano *et al.*, 2000). *CO* acts downstream of the photoreceptors and clock-associated proteins in the long-day flowering promotion pathway (Putterill *et al.*, 1995). While GA signaling is an important flowering promotion pathway in *Arabidopsis*, it has a minimal effect in pea, another facultative long-day plant (Weller *et al.*, 1997). Moreover, GA inhibits flowering in the long-day plant, *Fuchisia hybrida* (King and Ben-Tal, 2001). Interestingly, GA appeared to inhibit the floral-promoting effect of sucrose by reducing sucrose content in the shoot apex, and also was associated with reduced import of assimilate from leaves. While GAs can induce precocious flowering in some conifers, they appear to have little effect in most angiosperm trees, and conversely, GA-inhibitors have been effective floral promoters in some angiosperm trees (reviewed in Meilan, 1997). Combined with the complex carbohydrate and sink-source relations in trees, the likely roles of sucrose and long-distance signaling in the floral transition and vegetative maturation (discussed in previous section), these results are especially intriguing, and underline the great physiological and evolutionary diversity in control of flowering.

Another possible connection to changing sink-source relationships and transmission of signals via plasmodesmata and phloem comes from studies of a key regulator of the floral transition in maize, *INDETERMINATE1 (ID1)* (Colasanti and Sundaresan, 2000). *ID1* provides molecular genetic support for the florigen hypothesis—that a flowering signal is produced in the leaves and translocated to the SAM (Colasanti *et al.*, 1998). Day-neutral varieties of maize make the transition to flowering after initiating a particular number of leaves; *id1* mutants produce many more leaves, and eventually only produce aberrant inflorescences. *ID1* expression increases as plants approach the floral transition, and it is expressed in immature leaves, but not in the SAM. Within the leaves, *ID1* expression decreases as they emerge and become photosynthetically active (i.e., as the leaves transition from sink to source tissues) (Colasanti and Sundaresan, 2000). Moreover, plasmodesmata alter their within-leaf permeability along developmental gradients during the sink-to-source transition (reviewed in Pickard and Beachy, 1999).

While overexpression or suppression of a single gene can dramatically alter the time to flowering in annual plants, can such manipulations have a similar effect in trees without causing unwanted side-effects? The finding that overexpression of the *Arabidopsis* floral meristem identity gene *LEAFY* (*LFY*) induced the formation of flowers in transgenic poplar shortly after transformation generated much excitement, and indicated that flowering in trees might be usefully manipulated (Weigel and Nilsson, 1995). However, these flowers were not entirely normal, trees were dwarfed and highly branched, and additional studies showed that *LFY*'s ability to induce early flowering in poplar was highly dependent on genotype (Rottmann *et al.*, 2000). In contrast to poplar, overexpression of either *LFY* or *APETALA1* accelerated normal flowering and fruit production in a citrus cultivar (Pena *et al.*, 2001). Developmental differences between subtropical evergreen citrus, and temperate deciduous poplar were suggested as possible reasons for the different responses. Another possibility is that unlike forest trees, fruit trees are likely to have undergone selection for early and intense flowering, and thus, may be more amenable to induction by these genes.

Normally, *LFY* is activated by both long-day and GA floral promotion pathways (Blazquez and Weigel, 2000). Constitutive *LFY* expression accelerates flowering in *Arabidopsis*, but only after plants begin to produce adult vegetative phase leaves, indicating that the shoot first has to acquire competence to respond to *LFY* (Weigel and Nilsson, 1995). Functional comparisons between *LFY* and the *Populus trichocarpa LFY* homolog (*PTFL*) suggest that the regulation of floral genes in poplar is influenced by maturation (Rottmann *et al.*, 2000). Although *LFY* induced precocious flowering

Table 1. Examples of candidate reproductive and adventitious rooting maturation genes

Gene	Source Organism	Encoded Protein	Function	Reference
ID1	Maize	Zinc-finger	Regulates a leaf-generated signal for the floral transition.	Colasanti *et al.*, 1998
FLC	*Arabidopsis*	MADS domain transcription factor	Strong dosage-dependent repressor of flowering; downregulated by autonomous pathway genes and vernalization; possibly regulated *via* epigenetic mechanisms.	Sheldon *et al.*, 1999; Michaels & Amasino 1999
SVP	*Arabidopsis*	MADS domain	Dosage-dependent repressor of flowering, largely independent of photoperiod and vernalization.	Hartmann *et al.*, 2000
SOC1/ AGL20	*Arabidopsis*	MADS domain	Promotes the floral transition, integrates signals from the autonomous, vernalization, photoperiod, and GA pathways.	Samach *et al.*, 2000; Lee *et al.*, 2000; Borner *et al.*, 2000
FT	*Arabidopsis*	PEBP	Promotes the floral transition, integrates signals from the photoperiod and autonomous pathways.	Kardailsky *et al.*, 1999; Kobayashi *et al.*, 1999
TFL1	*Arabidopsis*	PEBP	Delays the floral transition, appears to regulate the length of all phases.	Ratcliffe *et al.*, 1998
EMF1	*Arabidopsis*	Novel	Mutants flower extremely early with essentially no vegetative phase.	Aubert *et al.*, 2001
FCA	*Arabidopsis*	RNA-binding domains	Promotes flowering independent of photoperiod and vernalization.	Macknight *et al.*, 1997
FWA	*Arabidopsis*	Homeodomain	Expression regulated by DNA methylation; possibly promotes establishment of vegetative phase.	Soppe *et al.*, 2000

Gene	Organism	Type	Description	Reference
PIN1/ EIR1/ AGR1	*Arabidopsis*	Transmembrane protein	Important for polar auxin transport, which affects lateral root initiation.	Galweiler *et al.*, 1998; Luschnig *et al.*, 1998; Chen *et al.*, 1998; Bennett *et al.*, 1998; Casimiro *et al.*, 2001
AtIAA3/ AtIAA7/ AtIAA17	*Arabidopsis*	Transcription factors	Protein-protein or DNA protein binding; activators and/or repressors; can stimulate or inhibit lateral and adventitious root formation.	Nagpal *et al.*, 2000; Tian & Reed, 1999; Rouse *et al.*, 1998
AIR1	*Arabidopsis*	Subtilisin Protease	Secreted in extracellular space; induced by auxin; expressed in lateral root primordia; up regulated by NAC1.	Neuteboom *et al.*, 1999
AIR3	*Arabidopsis*	Proline/Glycine Rich Protein	Secreted in extracellular space; induced by auxin during lateral root formation.	Neuteboom *et al.*, 1999
NAC1	*Arabidopsis*	Transcription Factor	Transcription factor; transduces the auxin signal for lateral root formation; regulates *AIR1*.	Xie *et al.*, 2000
CyclinB1;1	*Arabidopsis*	Cell Cycle Regulation	Induced by auxin; expressed at the site of lateral root primordia formation.	Dubrovsky *et al.*, 2000
ACT7	*Arabidopsis*	Actin Cytoskeleton	Induced by auxin; expressed at the site of lateral root emergence.	Kandasamy *et al.*, 2001
MtN21	*Medicago*	Transmembrane protein	Expressed during nodule formation; pine homolog down regulated in mature shoots.	Gamas *et al.*, 1996; Busov *et al.*, 2001
SQN	*Arabidopsis*	Cyclophilin40	Promotes juvenile leaf morphology.	Berardini *et al.*, 2001

infrequently in most poplar genotypes, it induced vegetative alterations, such as increased branching, more often. On the other hand, *PTLF* induced early flowering in *Arabidopsis*, but only very rarely did so in poplar. However, some *35S::PTLF* transgenics began to show increased branching after a few years of growth. These differences between the *PTLF* and *LFY* transgenics suggest regulatory factors that constrain *PTLF* activity may be involved in maintenance of juvenility in poplar, and that heterologous *LFY* was less affected by these factors.

While modulating the expression level of *LFY* or its orthologs may not be a generally applicable way to manipulate flowering time in forest trees, there are an increasing number of candidate genes that may fulfill this goal (Table 1). Intuitively, genes that alter the competency of the SAM to respond to floral induction signals and/or genes whose functions are largely independent of environmental cues appear to be the strongest transgenic candidates. However, the alteration of gene regulation is an important evolutionary mechanism (Doebley and Lukens, 1998). Thus, a flowering-time gene that is regulated by photoperiod in a long-day plant, could exhibit constitutive or growth-regulated expression in a day-neutral plant. A fruitful approach to selecting candidate transgenes is to investigate tree genes homologous to the well-studied annual plant developmental genes.

However, informative expression studies of putative maturation genes in trees are not simple because of the long developmental time, and complex seasonal cycle of growth and differentiation. For example, some of the genes that regulate flowering time in annual plants may regulate the seasonal flowering time of mature trees or the within-crown distribution of flowers, but have little or no affect on the acquisition of competence to flower over years (i.e., the reproductive maturation process). To broach these constraints we have collected various tissues at different seasonal times from the upper crown of one female and one male *Populus trichocarpa* x *P. deltoides* genotype (Figure 3A). Ramets derived from juvenile trees of each clone were represented in a continuous age gradient of one to six years (i.e., they had been through one to six growing seasons when we began our collections). For both genotypes, inflorescences were first initiated at age four. We used RNAs extracted from these various tissues to determine whether changes in gene expression level could be correlated with reproductive maturation.

A number of putative poplar orthologs of annual plant flowering time genes have been identified (Figure 2). As an example, preliminary quantitative expression studies of *P. trichocarpa ID1-LIKE5* (*PTID1L5*) are shown in Figure 3 B. In contrast to maize, where leaves initiate, emerge and develop to maturity in one growing season, leaf development in poplar spans two growing seasons and is interrupted by dormancy. In addition, it is unknown when flowering signals occur, or when meristems in the leaf axils are committed to forming an inflorescence rather than a vegetative meristem. Compared to juvenile ramets, *PTIDL5* was markedly upregulated in newly expanding spring shoots (SAM, leaves, internode) from mature ramets that would soon initiate inflorescences. Somewhat surprisingly, expression was highest in vegetative buds collected in the preceding fall from mature ramets, suggesting the possibility that signals for flowering may occur months before the initiation of flowers. Expanding this approach to study of a large number of genes via EST microarray hybridization might identify many additional genes, and networks of regulatory interactions, that take part in regulating vegetative to floral phase transition. The function of select genes or combinations of genes could then be studied via RNAi suppression or overexpression in transgenic trees.

An Example of Vegetative Phase Change: Competence for Adventitious Root Formation

The perennial vine *Hedera helix* (L) (English ivy) exhibits distinct juvenile (rooting) and mature (non-rooting) phases (Geneve *et al.*, 1988). Alternatively, for many other species, the decline in rooting ability occurs gradually, along with a progressive change in other traits (Wareing, 1959). This is the typical situation in many gymnosperms, such as loblolly pine (*Pinus taeda* L), hybrid larch (*Larix* spp.), radiata pine (*Pinus radiata* D. Don), and coast redwood (*Sequoia sempervirens* D. Don [Endl] (e.g., Peer and Greenwood, 2001). Many angiosperm tree species, such as *Quercus robur*, *Eucalyptus nitens*, *E. grandis*, *Fagus sylvatica*, and *Persea americana*, also exhibit a decline in rooting ability with increasing age (e.g., Maile and Nieuwenhuis, 1996). While the effects of maturation on rooting ability appear to be the same for gymnosperms and angiosperms, it is less clear whether the mechanisms of maturation are the same. There have been frequent reports of rejuvenation in angiosperms (e.g. Brand and Lineberger, 1992), but relatively few in gymnosperms (Monteuuis, 1987; Huang *et al.*, 1992). Thus, the changes in developmental competency associated with maturation appear to be more reversible in angiosperms than in gymnosperms.

Although many empirical studies have been conducted to define the effects of maturation in vegetative tissues (reviewed in Hackett and Murray, 1993), the underlying causes have proven more difficult to elucidate. Despite considerable recent progress in the study of adventitious rooting, many of the biochemical and genetic steps leading to the initiation and development of *de novo* root meristems are not yet known (Altman and Waisel, 1997). However, several groups have identified and studied the expression of tree

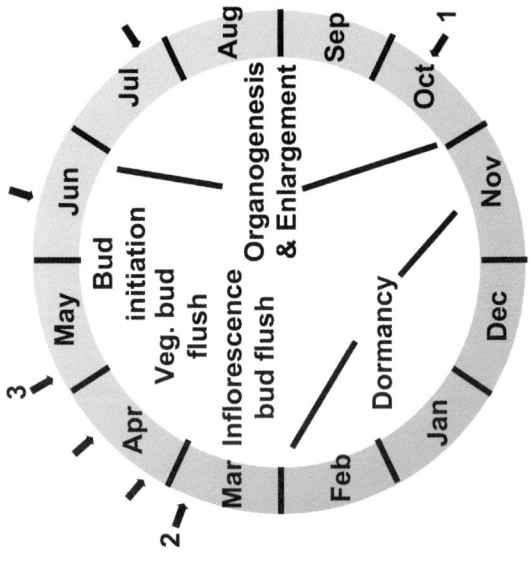

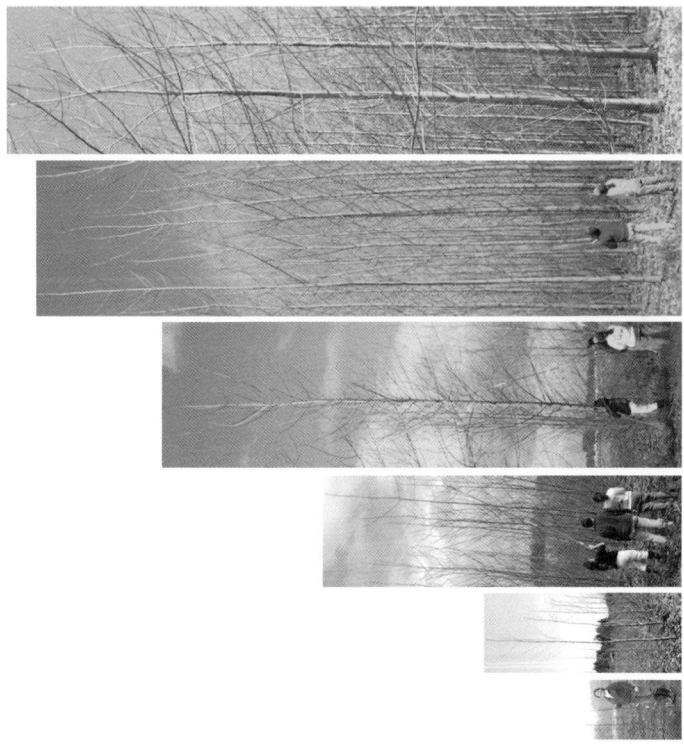

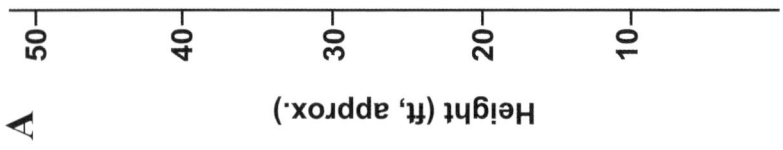

Flowering

Height (ft, approx.)

50
40
30
20
10

A

Age

1   2   3   4   5   6

Jul
Aug
Sep
Oct   1
Nov
Dec
Jan
Feb
Mar
Apr   3
May
Jun

Bud initiation
Veg. bud flush
Inflorescence bud flush
Organogenesis & Enlargement
Dormancy

2

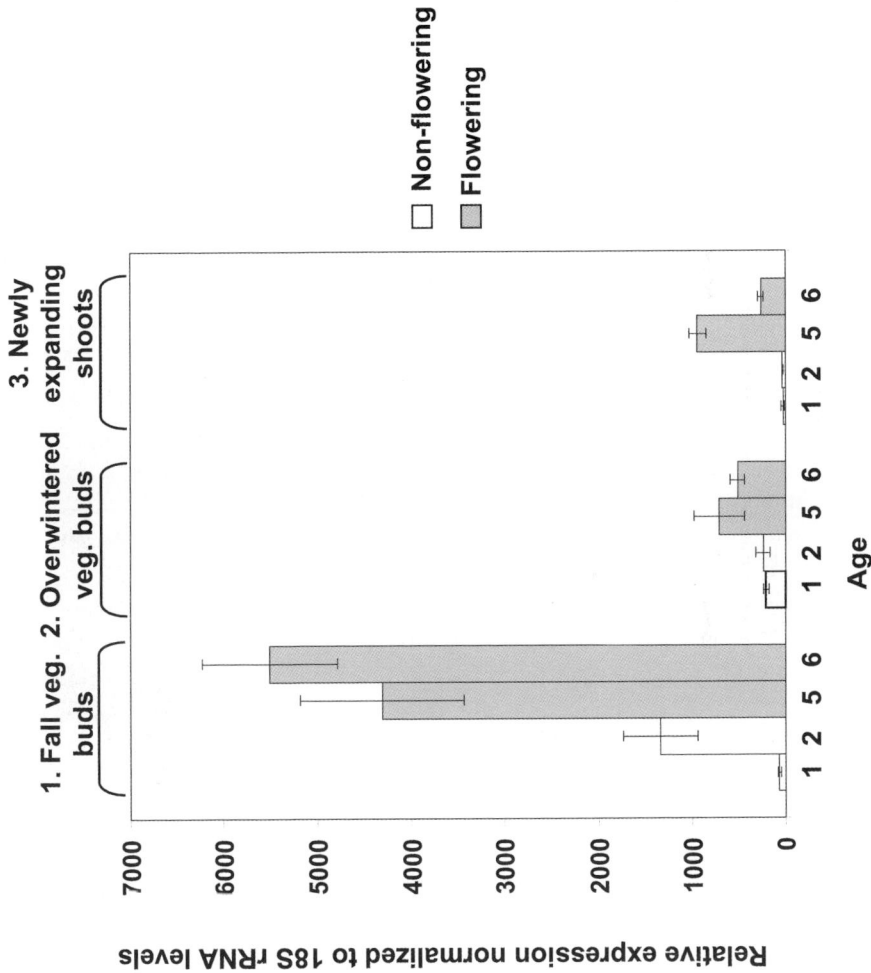

Figure 3. Variation in gene expression with age and seasonal cycle in a poplar genotype for a poplar homolog of the maize flowering time gene *INDETERMINATE1*. (A), An age gradient of a female poplar clone in Spring 2001 (left), and a typical seasonal cycle of *Populus trichocarpa x deltoides* clones in NW Oregon showing times (arrows) and types of tissue collections. Time of collection of tissues used for experiments shown in (B) are indicated by numbers. (B), RNA expression levels of *PTID1L5* measured by real-time reverse-transcriptase PCR and standardized based on ribosomal RNA expression levels for each tissue sample. Bars show one standard deviation based on three replicate measurements.

genes that could be involved. Using an *in vitro* apple (*Malus domestica*) stem disc system that previously has been useful for determining hormonal relations during root initiation (Van der Krieken *et al.*, 1992), Butler and Gallagher (2000) identified an *ARRO-1* gene that codes for a novel 2-oxoacid-dependent dioxygenase. This gene is up-regulated during root initiation and is induced by the root-stimulating auxins, IAA and IBA, but not by 2,4-D. While causing many other auxin effects, 2,4-D does not result in root initiation in this system. Further work with the apple disc system is underway to use microarrays to identify additional root formation-specific genes (Konings *et al.*, 2000).

Ermel *et al.* (2000) used cotyledon explants of walnut (*Juglans regia* L.) undergoing adventitious root formation to determine expression of genes during the different stages. Two walnut chalcone synthase genes (*CHS1* and *4*) and a transcript hybridizing with the *Arabidopsis LRP-1* (*LATERAl ROOT PRIMORDIUM-1*) (Smith and Federoff, 1995) gene were expressed during the early stages of primordium formation prior to meristem development. These genes were not expressed during the prior stage of callus-like cell division, which may correspond to the dedifferentiation step.

In gymnosperms, several groups have used hypocotyl cuttings in which root initiation is rapid and predictable (Diaz-Sala *et al.*, 1996; Goldfarb *et al.*, 1998) to identify and study gene expression. Lindroth *et al.* (2001) found localized expression of an S-adenosylmethionine synthetase gene in emerging adventitious root primordia of *Pinus contorta*. In *Pinus taeda*, expansin expression was induced by auxin treatment in hypocotyl and epicotyl cuttings undergoing root formation (Hutchison *et al.*, 1999). Auxin treatment also resulted in rapid and auxin-specific expression of five members of the Aux/IAA family of auxin-response genes in loblolly pine hypocotyl and stem cuttings (Goldfarb *et al.*, 1997).

In addition to research on trees, a greater understanding of a closely related process, lateral root development, may provide clues to aspects of adventitious root formation. *Arabidopsis* mutants have been instrumental in identifying candidate genes for involvement in lateral root formation. Increasingly, portions of the auxin transport and signal transduction pathways are becoming understood (reviewed in Leyser and Berleth, 1999; Bennett *et al.*, 1998). Several gain of function mutations in the Aux/IAA gene family of auxin regulated transcriptional regulators result in either a proliferation or inhibition of lateral or adventitious roots (Rouse *et al.*, 1998; Tian and Reed, 1999; Nagpal *et al.*, 2000). In addition, a mutation in *NAC1*, a member of a family of plant-specific transcriptional regulators, affects lateral root formation and perturbs expression of *AIR3* (Xie *et al.*, 2000), a downstream gene associated with the emergence of lateral roots (Neuteboom *et al.*, 1999). The complete determination of the lateral root pathway will be an important step for scientists researching adventitious rooting and how it is influenced by maturation. However, because roots of mature trees continue to form lateral roots, but stems from mature trees lose the ability to form adventitious roots, maturation must be affecting steps in the adventitious pathway that are unique. Unlike lateral roots, which arise directly from cell divisions of root pericycle cells (Casimiro *et al.*, 2001), most adventitious root meristems organize only after preliminary cell divisions from vascular parenchyma cells in stems (Goldfarb *et al.*, 1998). Thus, it is possible that dedifferentiation occurs during these initial cell divisions. If so, it is tempting to speculate that maturation affects the adventitious root formation process by altering the ability of these cells to dedifferentiate.

Compiling a complete list of genes that could be regulated by maturation during adventitious root formation will be expedited by functional genomics approaches. Large numbers of relatively unselected sequences can be imprinted on microarrays and screened with probes made from mRNA of different treatments or tissues. For example, in a recent study, over 3000 ESTs were screened with RNA prepared from loblolly pine cuttings with and without auxin treatment. The screen turned up numerous genes whose expression level appeared to differ with auxin treatment. Nine sequences were selected for confirmation with northern analysis and all nine were more highly expressed in cuttings treated with auxin. Interestingly, there were also differences in expression level of some of the sequences between juvenile and mature cuttings (Busov *et al.*, 2001). This type of approach offers great potential for achieving a comprehensive understanding of adventitious root initiation and how it is affected by maturation of the donor plant.

Conclusions

Maturation is a highly complex, multidimensional, and precisely regulated process when considered at the phenotypic or genomic level. It is therefore not surprising that approaches that consider single genes, single traits, or gross genomic measurements, have failed to provide significant insights into its control. The development of genomic methods that can assess expression of entire gene networks over maturation gradients, and ultimately epigenetic state-changes, may provide the quantum leap of technology needed for progress. This work is best focused on one or a very few tree genera for which the entire suite of genomic technology, including transformation to allow rigorous analysis of gene function, can be applied. *Populus* would appear to be the obvious choice for an angiosperm forest tree. A good gymnosperm candidate is spruce, because genomic studies are planned, and tissue culture and transformation systems are available. The large EST collections and arrays for other important tree species, especially pine and eucalypts, will also be a powerful resource for studying maturation if they are put in public domain. This would enable changes in networks of gene expression across maturation gradients to be studied in species where the ability to manipulate maturation might have major economic consequences. If support for genomic analysis of maturation in trees continues, the upcoming decade is expected to see some substantial progress in describing what maturation in trees is at a biological system level, as well as improved ability to manipulate it via transgenic and molecular breeding approaches.

References

Altman, A., and Waisel, Y. eds. 1997. Biology of Root Formation and Development. Plenum Press. New York and London. pp. 376.

The *Arabidopsis* Genome Initiative. 2000. Analysis of the genome sequence of the flowering plant *Arabidopsis thaliana*. Nature. 408: 796-815.

Araki, T. 2001. Transition from vegetative to reproductive phase. Curr. Opin. Plant Biol. 4: 63-68.

Aubert, D., Chen, L., Moon, Y.-H., Martin, D., Castle, L.A., Yang, C.-H., and Sung, Z.R. 2001. EMF1, a novel protein involved in the control of shoot architecture and flowering in *Arabidopsis*. Plant Cell 13: 1865-1875.

Auckerman, M.J., Lee, I., Weigel, D., and Amasino, R.M. 1999. The *Arabidopsis* flowering-time gene *LUMINIDEPENDENS* is expressed primarily in regions of cell proliferation and encoded a nuclear protein that regulates *LEAFY* expression. Plant J. 18: 195-203.

Banfield, M.J. and Brady, R.L. 2000. The structure of Antirrhinum CENTRORADIALIS protein (CEN) suggests a role as a kinase regulator. J. Mol. Biol. 297: 1159-1170.

Bennett, M.J., Marchant, A., May, S.T., and Swarup, R. 1998. Going the distance with auxin: unravelling the molecular basis of auxin transport. Philosophical Transactions of the Royal Society of London Biological Sciences. 353: 1511-1515.

Berardini, T.Z., Bollman, K., Sun, H., and Poethig, R.S. 2001. Regulation of vegetative phase change in *Arabidopsis thaliana* by cyclophilin 40. Science. 291: 2405-2407.

Berleth, T., and Sachs, T. 2001. Plant morphogenesis: long-distance coordination and local patterning. Curr. Opin. Plant Biol. 4: 57-62.

Bernier, G., Havelange, A., Houssa, C. Petitjean, A., and Lejeune, P. 1993. Physiological signals that induce flowering. Plant Cell. 5: 1147-1155.

Blazquez, M.A. 2001. Flower development pathways. J. Cell Science. 113: 3547-3548.

Blazquez, M.A., and Weigel, D. 2000. Integration of floral inductive signals in *Arabidopsis*. Nature. 404: 889-892.

Bond, B.J. 2000. Age-related changes in photosynthesis of woody plants. Trends Plant Sci. 5: 349-353.

Borner, R., Kampmann, G., Chandler, J., Gleibner, R., Wisman, E., Apel, K., and Melzer, S. 2000. A MADS domain gene involved in the transition to flowering in *Arabidopsis*. Plant J. 24: 591-599.

Bradley, D., Ratcliffe, O., Vincent, C., Carpenter, R., and Coen, E. 1997. Inflorescence commitment and architecture in *Arabidopsis*. Science. 275: 80-82.

Bradshaw Jr., H.D., and Strauss, S.H. 2001. Breeding strategies for the 21st Century: Domestication of poplar. In: Poplar Culture in North America, Part 2. D.I. Divkmann, J.G. Isebrands, J.E. Eckenwalder, and J. Richardson, eds. NRC Research Press, National Research Council of Canada, Ottawa, ON, Canada. p. 383-394.

Bradshaw, H.D. 1998. Case history in genetics of long-lived plants: molecular approaches to domestication of a fast-growing forest tree: *Populus*. In: Molecular Dissection of Complex Traits. A.H. Paterson, ed. CRC Press, NY. p. 219-228.

Brand, M.H., and Lineberger, R.D. 1992. *In vitro* rejuvenation of *Betula* (Betulaceae): Morphological evaluation. Am. J. Bot. 79: 618-625.

Brunner, A.M., Dye, S.J., Skinner, J.S., Ma, C., Meilan, R., and Strauss, S.H. 2001. Maturation and flowering in *Populus*: gene expression changes correlated with reproductive competency and genes for manipulating flowering. Abstract. In: IUFRO / Molecular Biology of Forest Trees. 22-27 July 2001 Stevenson, WA, USA.

Busov, V.B., Whetten, R., Sederoff, R., Spiker, S., Lanz-Garcia, C., and Goldfarb, B. 2001. Maturation effects on auxin-induced genes from loblolly pine (*Pinus taeda* L.). Abstract. In: IUFRO / Molecular Biology of Forest Trees. 22-27 July 2001 Stevenson, WA, USA.

Butler, E.D., and Gallagher, T.F. 2000. Characterization of auxin-induced *ARRO-1* expression in the primary root of *Malus domestica*. J. Exp. Bot. 51: 1765-1766.

Casimiro, I., Marchant, A., Bhalerao, R.P., Beeckman, T., Dhooge, S., Swarup, R., Graham, N., Inze, D., Sandberg, G., Casero, P.J., and Bennett, M. 2001. Auxin transport promotes *Arabidopsis* lateral root initiation. Plant Cell. 13: 843-852.

Charest, P.J., Devantier, Y., and Lachance, D. 1996. Stable genetic transformation of *Picea mariana* (black spruce) via particle bombardment. *In Vitro* Cell Dev. Biol. Plant. 32: 91-99.

Chen, R., Hilson, P., Sedbrook, J., Rosen, E., Caspar, T., and Masson, P.H. 1998. The *Arabidopsis thaliana AGRAVITROPIC 1* gene encodes a component of the polar-auxin-transport efflux carrier. Proc. Natl. Acad. Sci. USA. 95: 15112-15117.

Chatfield, S.P., Stirnberg, P., Forde, B.G., and Leyser, O. 2000. The hormonal regulation of axillary bud growth in *Arabidopsis*. Plant J. 24: 159-169.

Chory, J., Ecker, J.R., Briggs, S., Caboche, M., Coruzzi, G.M., Cook, D., Dangl, J., Grant, S., Guerinot, M.L., Henikoff, A., Martienssen, R., Okada, K., Raikhel, N.V., Somerville, C.R., and Weigel, D. 2000. National Science Foundation-sponsored workshop report: "the 2010 project". Plant Physiol. 123: 423-425.

Colasanti, J., and Sundaresan, V. 2000. 'Florigen' enters the molecular age: long-distance signals that cause plants to flower. Trends Biochem. 25: 236-240.

Colasanti, J., Yuan, Z., and Sundaresan, V. 1998. The *indeterminate* gene encodes a zinc finger protein and regulates a leaf-generated signal required for the transition to flowering in maize. Cell. 93: 593-603.

Diaz-Sala, C., Hutchison, K., Goldfarb, B., and Greenwood, M. 1996. Maturation-related loss in rooting competence by loblolly pine stem cuttings: the role of auxin transport, metabolism and tissue sensitivity. Physiol. Plant. 97: 481-490.

Diggle, P.K. 1999. Heteroblasty and the evolution of flowering phenologies. Int. J. Plant. Sci. 160(6 Suppl): S123-S134.

Doebley, J., and Lukens, L. 1998. Transcriptional regulators and the evolution of plant form. Plant Cell. 10: 1075-1082.

Dubrovsky, J.G., Doerner, P.W., Colon-Carmona, A., and Rost, T.L. 2000. Pericycle cell proliferation and lateral root initiation in *Arabidopsis*. Plant Physiol. 124: 1648-1657.

Eckenwalder, J.E. 1996. Systematics and evolution of Populus. In: Biology of *Populus*. R.F. Stettler, H.D. Bradshaw, P.E. Heilman, and T.M. Hinckely, eds. NRC Research Press, National Research Council of Canada, Ottawa, ON, Canada. p. 7-32.

Ellis, D.D., McCabe, D.E., McInnis, S., Ramachandran, R., Russell, D.R., Wallace, K.M., Martinell, B.J., Roberts, D.R., Raffa, K.F., and McCown, B.H. 1993. Stable transformation of *Picea glauca* by particle acceleration. Bio/Technology. 11: 84-89.

Ermel, F.F., Vizoso, S., Charpentier, J.P., Jay-Allemand, C., Catesson, A.M., and Couee, I. 2000. Mechanisms of primordium formation during adventitious root development from walnut cotyledon explants. Planta. 211: 563-574.

Finnegan, E.J., Peacock, W.J., and Dennis, E.S. 2000. DNA methylation, a key regulator of plant development and other processes. Curr. Opin. Gen. Dev. 10: 217-223.

Galweiler, L., Guan, C., Miller, A., Wisman, E., Mendgen, K., Yephremov, A., and Palme, K. 1998. Regulation of auxin transport by *AtPIN1* in *Arabidopsis* vascular tissue. Science. 282: 2226-30.

Gammas, P., Niebel, F.C., Lescure, N., and Cullimore, J.V. 1996. Use of subtractive hybridization approach to identify new *Medicago truncatula* genes induced during root nodule development. Molecular Plant Microbe Interactions. 9: 233-242.

Goldfarb, B., Hackett, W.P., Furnier, G.R., Mohn, C.A., and Plietzsch, A. 1998. Adventitious root initiation in hypocotyl and epicotyl cuttings of eastern white pine (*Pinus strobus* L.). Physiol. Plant. 102: 513-522.

Goldfarb, B., Lian, Z., Lanz-Garcia, C., and Whetten, R. 1997. Auxin-induced gene expression during rooting of loblolly pine stem cuttings. In: Biology of Root Formation and Development. A. Altman and Y. Waisal, eds. Plenum Press, New York. p. 163-167.

Geneve, R.L., Hackett, W.P., and Swanson, B.T. 1988. Adventitious root initiation in de-bladed petioles from the juvenile and mature phase of English ivy. J. Am. Soc. Hort. Sci. 113: 630-635.

Greenwood, M.S. 1995. Juvenility and maturation in conifers: current concepts. Tree Physiology. 15: 433-438.

Greenwood, M., and Hutchison, K.W. 1993. Maturation as a developmental process. In: Clonal Forestry I: Genetics and Biotechnology. M.R. Ahuja and W.J. Libby, eds. Springer-Verlag, Berlin. p. 14-33.

Grossniklaus, U., Nogler, G.A., and van Dijk, P.J. 2001. How to avoid sex: the genetic control of gametophytic apomixis. Plant Cell. 13: 1491-1498.

Habu, Y., Kakutani, T., and Paszkowski, J. 2001. Epigenetic developmental mechanisms in plants: molecules and targets of plant epigenetic regulation. Curr. Opin. Gen. Dev. 11: 215-220.

Hackett, W.P., and Murray, J.R. 1997. Approaches to understanding maturation or phase change. In: Biotechnology of Ornamental Plants. R.L. Geneve, J.E. Preece, and S.A. Merkle, eds. CAB International, Wallingford. p. 73-86.

Hackett, W.P., and Murray, J.R. 1993. Maturation and rejuvenation in woody species. In: Micropropagation of Woody Plants. M.R. Ahuja, ed. Kluwer, Netherlands. p. 93-105.

Hackett, W.P. 1988. Donor plant maturation and adventitious root formation. In: Adventitious Root Formation in Cuttings. T.D. Davis, B.E. Hassig, N. Sankhla, eds. Dioscorides Press. Portland. Oregon. p. 11-28.

Haffner, V., Enjalric, F., Lardet, L. and Carron, M.P. 1991. Maturation of woody plants: a review of metabolic and genomic aspects. Ann. Sci. For. 48: 615-630.

Han, K.-H., Meilan, R., Ma, C., and Strauss, S.H. 2000. An *Agrobacterium* transformation protocol effective in a variety of cottonwood hybrids (genus *Populus*). Plant Cell Reports. 19: 315-320.

Hartmann, U., Hohmann, S., Nettesheim, K., Wisman, E., Saedler, H., and Huijser, P. 2000. Molecular cloning of *SVP*: a negative regulator of the floral transition in *Arabidopsis*. Plant J. 21: 351-360.

Hirochika, H. 2001. Retrotransposons of rice: their regulation and use for functional genomics. Curr. Opin. Plant Biol. 4: 118-122.

Hoekenga, O.A., Muszynski, M.G., and Cone, K.C. 2000. Developmental patterns of chromatin structure and DNA methylation responsible for epigenetic expression of a maize regulatory gene. Genetics. 155: 1889-1902.

Howe, G.T., Bucciaglia, P.A., Hackett, W.P., Furnier, G.R., Cordonnier-Pratt, M.-M., and Gardener, G. 1998. Evidence that the Phytochrome gene family in black cottonwood has one *PHYA* locus and two *PHYB* loci but lacks members of the *PHY/F* and *PHYE* subfamilies. Mol. Biol. Evol. 15: 160-175.

Huang L.-C., Lius S., Huang B.-L., Murashige T., Mahdi E.F.M., and Van Gundy R. 1992. Rejuvenation of *Sequoia sempervirens* by repeated grafting of shoot tips onto juvenile rootstocks *in vitro*. Plant Physiol. 98: 166-173.

Huang, Y., Diner, A.M., and Karnosky, D.F. 1991. *Agrobacterium rhizogenes* mediated genetic transformation and regeneration of a conifer: *Larix decidua*. *In Vitro* Cell and Dev. Biol. 27: 201-207.

Hutchison, K.W., Singer, P.B., McInnis, S., Diaz-Sala, C., and Greenwood, M.S. 1999. Expansins are conserved in conifers and expressed in hypocotyls in response to exogenous auxin. Plant Physiol. 120: 827-832.

Irish, E.E., and Karlen, S. 1998. Restoration of juvenility in maize shoots by meristem culture. Int. J. Plant Sci. 159: 695-701.

Jeddeloh, J.A., Stokes, T.L., and Richards, E.J. 1999. Maintenance of genomic methylation requires a SWI2/SNF2-like protein. Nat. Genet. 22: 94-97.

Jones, C.S. 1999. An essay on juvenility, phase change and heteroblasty in seed plants. Int. J. Plant Sci. 160 (6 Suppl.): S105-S111.

Jordan, G.J., Potts, B.M., and Wiltshire, R.J.E. 1999. Strong, independent, quantitative genetic control of the timing of vegetative phase change and first flowering in *Eucalyptus globulus* ssp. *globulus* (Tasmanian blue gum). Heredity. 83: 179-187.

Jordan, G.J., Potts, B.M., Chalmers, P., and Wiltshire, R.J.E. 2000. Quantitative genetic evidence that the timing of vegetative phase change in *Eucalyptus globulus* ssp. *globulus* is an adaptive trait. Aust. J. Bot. 48: 561-567.

Kandasamy, M.K., Gilliland, L.U., McKinney, E.C., and Meagher, R.B. 2001. One plant actin isovariant, ACT7, is induced by auxin and required for normal callus formation. Plant Cell. 13: 1541-1554.

Kardailsky, I., Shukla, V.K., Ahn, J.H., Dagenais, N., Christensen, S.K., Nguyen, J.T., Chory, J., Harrison, M.J., and Weigel, D. 1999. Activation tagging of the floral inducer *FT.* Science. 286: 1962-1965.

Kim, M., Canio, W., Kessler, S., and Sinha, N. 2001. Developmental changes due to long-distance movement of a homeobox fusion transcript in tomato. Science. 293: 287-289.

King, R.W., and Ben-Tal, Y. 2001. A florigenic effect of sucrose in *Fuchsia hybrida* is blocked by gibberellin-induced assimilate competition. Plant Physiol. 125: 488-496.

Kinlaw, C.S., and Neale, D.B. 1997. Complex gene families in pine genomes. Trends Plant Sci. 2: 356-359.

Kirkwood, T.B., and S.N. Austad. 2000. Why do we age? Nature. 408: 233-238.

Kobayashi, Y., Kaya, H., Goto, K., Iwabuchi, M., and Araki, T. 1999. A pair of related genes with antagonistic roles in mediating flowering signals. Science. 286: 1960-1962.

Konings, M.C.J.M., van der Krieken, W., and van der Geest, A.H.M. 2000. Analysis of gene expression during adventitious root formation in apple: use of cDNA microarrays to monitor root formation and outgrowth potential of planting stock. (Abstract) In: Third International Symposium on Adventitious Root Formation. 27 June – 1 July 2000, Veldhoven, The Netherlands

Lawson, E.J.R., and Poethig, R.S. 1995. Shoot development in plants: time for a change. Trends Genet. 11: 263-268.

Lee, H., Suh, S.-S., Park, E., Cho, E., Ahn, J.H., Kim, S.-G., Lee, J.-S., Kwon, Y. M., and Lee, I. 2000. The *AGAMOUS-LIKE 20* MADS domain protein intergrates floral inductive pathways in *Arabidopsis.* Genes Dev. 14: 2366-2376.

Levee, V., Lelu, M.-A., Jouanin, L., Cornu, D., and Pilate, G. 1997. *Agrobacterium tumefaciens*-mediated transformation of hybrid larch (*Larix kaempferi* x *L. decidua*) and transgenic plant regeneration. Plant Cell Rep. 16: 680-685.

Levy, Y.Y., and Dean, C. 1998. The transition to flowering. Plant Cell. 10: 1973-1989.

Leyser O., and Berleth T. 1999. A molecular basis for auxin action. Cell Dev. Biol. 10: 131-137.

Lindroth, A.M., Saarikoski, P., Flygh, G., Clapham, D., Gronroos, R., Thelander, M., Ronne, H., and von Arnold, S. 2001. Two S-adenosylmethionine synthetase-encoding genes differentially expressed during adventitious root development in *Pinus contorta*. Plant Mol. Biol. 46: 335-346.

Luschnig, C., Gaxiola, R.A., Grisafi, P., and Fink, G.R. 1998. EIR1, a root-specific protein involved in auxin transport, is required for gravitropism in *Arabidopsis thaliana*. Genes Dev. 12: 2175-2187.

Macknight, R., Bancroft, I. Page, T. Lister, C. Schmidt, R., Love, K., Westphal, L., Murphy, G., Sherson, S., Cobbett, C., and Dean, C. 1997. *FCA,* a gene controlling flowering time in *Arabidopsis*, encodes a protein containing RNA-binding domains. Cell. 89: 737-745.

Maile, N., and Niewenhuis, M. 1996. Vegetative propagation of *Eucalyptus nitens* using stem cuttings. South African Forestry Journal. 175: 29-34.

Matzke, M., Matzke, A.J.M., and Kooter, J.M. 2001. RNA: guiding gene silencing. Science. 293: 1080-1083.

Meilan, R. 1997. Floral induction in woody angiosperms. New Forests. 14: 179-202.

Mellerowicz, E.J., Baucher, M., Sundberg, B., and Boerjan, W. 2001. Unraveling cell wall formation in the woody dicot stem. Plant Mol. Biol. 47: 239-274.

Merkle, S.A., Smith, D.R., Battle, P.J., Montello, and P.M., Meagher, R.B. 2001. Goodbye to antibiotic selection: transformation of slash and loblolly pines with mercuric ion resistance genes. Abstract In: IUFRO / Molecular Biology of Forest Trees. 22-27 July 2001 Stevenson, WA, USA.

Michaels, S.D., and Amasino, R.M. 1999. *FLOWERING LOCUS C* encodes a novel MADS domain protein that acts as a repressor of flowering. Plant Cell. 11: 949-956.

Monteuuis, O., 1987. *In vitro* meristem culture of mature and juvenile *Sequoiadendron giganteum*. Tree Physiology. 3: 265-272.

Muller, C., and Leutz., A. 2001. Chromatin remodeling in development and differentiation. Curr. Opin. Genet. Dev. 11: 167-174.

Nagpal, P., Walker, L.M., Young, J.C., Sonowala, A., Timpte, C., Estelle, M., and Reed, J.W. 2000. *AXR2* encodes a member of the Aux/IAA protein family. Plant Physiol. 123: 563-573.

Neuteboom, L.W., Ng, J.M., Kuyper, M., Clijdesdale, O.R., Hooykaas, P.J., and van der Zaal, B.J. 1999. Isolation and characterization of cDNA clones corresponding with mRNAs that accumulate during auxin-induced lateral root formation. Plant Mol. Biol. 39: 273-78.

Onouchi, H., Igeno, M.I., Perilleux, C., Graves, K., and Coupland, G. 2000. Mutagenesis of plants overexpressing *CONSTANS* demonstrates novel interactions among *Arabidopsis* flowering-time genes. Plant Cell. 12: 885-900.

Orkwiszewski, J.A.J., and Poethig, R.S. 2000. Phase identity of the maize leaf is determined after leaf initiation. Proc. Natl. Acad. Sci. USA. 97: 10631-10636.

Pena, L., Martin-Trillo, M. Juarez, J., Pina, J., Navarro, L., and Martinez-Zapater, J.M. 2001. Constitutive expression of *Arabidopsis LEAFY* or *APETALA1* genes in citrus reduces their generation time. Nature Biotech. 19: 263-267.

Peer, K.R., and Greenwood, M.S. 2001. Maturation, topophysis and other factors in relation to rooting in *Larix*. Tree Physiol. 21: 267-272.

Pickard, B.G., and Beachy, R.N. 1999. Intercellular connections are developmentally controlled to help move molecules through the plant. Cell. 98: 5-8.

Pidkowich, M.S., Klenz, J.E., and Haughn, G.W. 1999. The making of a flower: control of floral meristem identity in *Arabidopsis*. Trends Plant Sci. 4: 64-70.

Poethig, R.S. 1990. Phase change and the regulation of shoot morphogenesis in plants. Science. 250: 923-930.

Putterill, J., Robson, F., Lee, K., Simon, R., and Coupland, G. 1995. The *CONSTANS* gene of *Arabidopsis* promotes flowering and encodes a protein showing similarities to zinc finger transcription factors. Cell. 80: 847-857.

Ratcliffe, O.J., Amaya, I., Vincent, C.A., Rothstein, S., Carpenter, R., Coen, E.S., and Bradley, D.J. 1998. A common mechanism controls the life cycle and architecture of plants. Development. 125: 1609-1615.

Rottmann, W.H., Meilan, R. Sheppard, L.A., Brunner, A.M., Skinner, J.S., Ma, C., Cheng, S., Jouanin, L., Pilate, G., and Strauss, S.H. 2000. Diverse effects of overexpression of *LEAFY* and *PTLF*, a poplar (*Populus*) homolog of *LEAFY/FLORICAULA*, in transgenic poplar and *Arabidopsis*. Plant J. 22: 235-246.

Rouse, D., Mackay, P., Stirnberg, P., Estelle, M., and Leyser, O. 1998. Changes in auxin response from mutations in an AUX/IAA gene. Science. 279: 1371-1373.

Ruiz-Medrano, R., Xoconostle-Cazares, B., and Lucas, W.J. 1999. Phloem long-distance transport of *CmNACP* mRNA: implications for supracellular regulation in plants. Development. 126: 4405-4419.

Samach, A., Onouchi, H., Gold, S.E., Ditta, G.S., Schwarz-Sommer, Z., Yanofsky, M.F., and Coupland, G. 2000. Distinct roles of *CONSTANS* target genes in reproductive development of *Arabidopsis*. Science. 288: 1613-1616.

Schomburg, F.M., Patton, D.A., Meinke, D.W., and Amasino, R.M. 2001. *FPA*, a gene involved in floral induction in *Arabidopsis* encodes a protein containing RNA-recognition motifs. Plant Cell. 13: 1-11.

Sheldon, C.C., Finnegan, E.J., Rouse, D.T., Tadege, M., Bagnall, D.J., Helliwell, C.A., Peacock, W.J., and Dennis, E.S. 2000. The control of flowering by vernalization. Curr. Opin. Plant Biol. 3: 418-422.

Sheldon, C.C., Burn, J.E., Perez, P.P., Metzger, J., Edwards, J.A., Peacock, W.J., and Dennis, E.S. 1999. The *FLF* MADS box gene: a repressor of flowering in *Arabidopsis* regulated by vernalization and methylation. Plant Cell 11: 445-458.

Simpson, G.G., Gendall, A.R., and Dean, C. 1999. When to switch to flowering. Ann. Rev. Cell Dev. Biol. 99: 519-550.

Smith, N.A., Surinder, S.P., Wand, M.-B., Stoutjesdijk, P.A., Green, A.G., and Waterhouse, P.M. 2000. Total silencing by intron-spliced hairpin RNAs. Nature. 407: 319-320.

Smith, D.L., Federoff, N.V. 1995. *LRP1*, a gene expressed in lateral and adventitious root primordia of *Arabidopsis*. Plant Cell. 7: 735-745.

Soppe, W.J.J., Jacobsen, S.E., Alonso-Blanco, C., Jackson, J.P., Kakutani, T., Koornneef, M., and Peeters, A.J.M. 2000. The late flowering phenotype of *fwa* mutants is caused by gain-of-function epigenetic alleles of a homeodomain gene. Mol. Cell. 6: 791-802.

Strauss, S.H., Rottmann, W.H., Brunner, A.M., and Sheppard, L.A. 1995. Genetic engineering of reproductive sterility in forest trees. Molec. Breed. 1: 5-26.

Sundberg, B., Uggla, C., and Tuominen, H. 2000. Cambial growth and auxin gradients. In: Cell and Molecular Biology of Wood Formation. R. Savidge, J. Barnett, and R. Napier, eds. BIOS Scientific Publishers LTD, Oxford.

Tandre, K., Brunner, A.M., Skinner, J.S., Campaa, L., Strauss, S.H., and Nilsson, O. 2001. A functional genomics approach to the control of flowering in *Populus*. Abstract. In: IUFRO / Molecular Biology of Forest Trees. 22-27 July 2001 Stevenson, WA, USA.

Tantikanjana, T., Yong, J.W.H., Letham, D.S., Griffith, M., Hussain, M., Ljung, K., Sandberg, G., and Sundaresan, V. 2001. Control of axillary bud initiation and shoot architecture in *Arabidopsis* through the *SUPERSHOOT* gene. Gene Dev. 15: 1577-1588.

Theissen, G., Becker, A., Di Rosa, A., Kanno, A., Kim, J.T., Munster, T., Winter, K.-U., and Saedler, H. 2000. A short history of MADS-box genes in plants. Plant Mol. Biol. 42: 115-149.

Tian, L., and Chen, Z.J. 2001. Blocking histone deacetylation in *Arabidopsis* induces pleiotropic effects on plant gene regulation and development. Proc. Natl. Acad. Sci. USA. 98: 200-205.

Tian, Q., and Reed, J.W. 1999. Control of auxin-regulated root development by the *Arabidopsis thaliana* SHY2/IAA3 gene. Development. 126: 711-721.

Tzfira, T., Yarnitzky, O., Vainstein, A., and Altman, A. 1996. *Agrobacterium rhizogenes*-mediated DNA transfer in *Pinus halepensis* Mill. Plant Cell Rep. 16: 26-31.

Uggla, C., Magel, E., Moritz, T., and Sundberg, B. 2001. Function and dynamics of auxin and carbohydrates during earlywood/latewood transition in Scots pine. Plant Physiol. 125: 2029-2039.

Van der Krieken, W.M., Breteler, H., and Visser, M.H.M. 1992. Uptake and metabolism of indolebutyric acid furing root formation on *Malus* microcutting. Acta Botanica Neerlandica. 41: 435-442.

Vongs, A., Kakutani, T., Martienssen, R., and Richards, E.J. 1993. *Arabidopsis thaliana* DNA methylation mutants. Science. 260: 1926-1928.

Walter, C., Grace, L.J., Donaldson, S.S., Moody, J., Gemmell, J.E., Van der Maas, S., Kvaalen, H., and Lonneborg, A. 1999. An efficient biolistic transformation protocol for *Picea abies* (L) Karst embryogenic tissue and regeneration of transgenic plants. Can. J. For. Res. 29: 1539-1546.

Walter, C., Grace, L.J., Wagner, A., White, D.W.R., Walden, A.R., Donaldson, S.S., Hinton, H., Gardner, R.C., and Smith, D.R. 1998. Stable transformation and regeneration of transgenic plants of *Pinus radiata* D. Don. Plant Cell Rep. 17: 460-468.

Wareing, P.F. 1959. Problems of juvenility and flowering in trees. Journal of the Linnaeus Society (London). 56: 282-289.

Weigel, D., and Nilsson, O. 1995. A developmental switch sufficient for flower initiation in diverse plants. Nature. 377: 495-500.

Weigel, D., Ahn, J.H., Blazquez, J., Borevitz, Christensen, S.K., Fankhauser, C., Ferrandiz, C., Kardailsky, I., Neff, M.M., Nguyen, J.T., Sato, S., Wang, Z., Xia, Y., Dixon, R.A., Harrison, M.J., Lab, C., Yanofsky, M.F., and Chory, J. 2000. Activation tagging in *Arabidopsis*. Plant Physiol. 122: 1003-1013.

Weller, J.L., Reid, J.B., Taylot, S.A., and Murfet, I.C. 1997. The genetic control of flowering in pea. Trends Plant Sci. 2: 412-418.

Wenck, A.R., Quinn, M., Whetten, R.W., Pullman, G., and Sederoff, R. 1999. High-efficiency *Agrobacterium*-mediated transformation of Norway spruce (*Picea abies*) and loblolly pine (*Pinus taeda*). Plant Mol. Biol. 39: 407-416.

Whetten, R.W., Sun, Y-H., Zhang, Y., and Sederoff, R. 2001. Functional genomics and cell wall biosynthesis in loblolly pine. Plant Mol. Biol. 47: 275-291.

Wu, K.K., Malik, K., Tian, L., Brown, D., and Miki, B. 2000. Functional analysis of a RPD3 histone deactylase homologue in *Arabidopsis thaliana*. Plant Mol. Biol. 44: 167-176.

Xie Q, Frugis G, Colgan D, and Chua N-H. 2000. *Arabidopsis* NAC1 transduces auxin signal downstream of TIR1 to promote lateral root development. Genes Dev. 14: 3024-3036.

Yano, M., Katayose, Y., Ashikari, M., Yamanouchi, U., Monna, L., Fuse, T., Baba, T., Yamamoto, K., Umehara, Y., Nagamura, Y., and Sasaki, T. 2000. *Hd1*, a major photoperiod sensitivity QTL in rice, is closely related to the *Arabidopsis* flowering time gene *CONSTANS*. Plant Cell. 12: 2473–2483.

Yuan, Q., Quakenbush, J., Sultana, R., Pertea, M., Salzberg, S.L., and Buell, C.R. 2001. Rice bioinformatics. Analysis of rice sequence data and leveraging data to other plant species. Plant Physiol. 125: 1166-1174.

Zambryski, P., and Crawford, K. 2000. Plasmodesmata: gatekeepers for cell-to-cell transport of developmental signals in plants. Ann. Rev. Cell Dev. Biol. 16: 393-421.

Zobel, B.J., and Sprague, J.R. 1998. Juvenile Wood in Forest Trees. Springer-Verlag, Berlin.

From: *Transgenic Plants: Current Innovations and Future Trends*
Edited by: C. Neal Stewart, Jr.

Chapter 3

Transgenic Trees: Advances in Somatic Embryogenesis, Transformation and Engineering with Phytoremediation Genes

Scott A. Merkle

Abstract

Forest trees have great potential to be genetically manipulated, not only for improved fiber production, but also for generation of novel goods and services, by applying the same biotechnological tools that have been used with agronomic crop species. Unlike crop species, however, forest trees are undomesticated and thus present unique problems with regard to genetic manipulation. *In vitro* propagation, in particular somatic embryogenesis, and gene transfer are two of the tools widely predicted to have substantial impact on forest productivity in the coming decades. Embryogenic culture systems have been developed for most commercially-important forest trees and have already been employed extensively as targets for gene transfer, most commonly via microprojectile-mediated transformation (biolistics). Trees expressing genes for herbicide resistance and modified wood have been produced. In addition, transgenic trees have been proposed for new products and services that have not been previously associated with forestry, such as remediation of contaminated soils and water. Trees engineered with heavy metal detoxification genes are currently being tested for their ability to handle these pollutants. However, in order for these advances in *in vitro* propagation and genetic manipulation to be translated into economically viable plantations, a number of significant technical bottlenecks must be overcome, with regard to both transformation and somatic seedling production. In addition, recent reports indicate that alternative approaches to embryogenic cultures and biolistics may actually be superior for propagation and gene transfer with some forest trees. Progress is being made in surmounting these problems and novel approaches to improving genetic manipulation of the top commercial forest species are currently under development.

Introduction

The first trees regenerated from tissue culture in the 1960s were produced via adventitious shoots (Winton, 1968; Wolter, 1968). *In vitro* regeneration of these *Populus* spp. trees was later followed by plantlet production from cultures of commercially important conifers, also via adventitious buds (Sommer *et al.*, 1975). For the past 15 years, however, research efforts in forest industry and forest research institutions have focused on developing *in vitro* propagation via somatic embryogenesis for the top commercial forest species, which are for the most part, conifers. The first report of production of a transgenic forest tree also employed an adventitious shoot-producing system. Emulating a highly successful transformation system originally developed with tobacco, poplar leaf disks were infected with *Agrobacterium tumefaciens*, followed by regeneration of transgenic trees via adventitious shoots (Fillatti *et al.,* 1987). Again, however, while this approach has proven highly successful with poplars and some other hardwood trees, embryogenic cultures became the primary vehicles for gene transfer over the past decade for the top commercial forest trees. Given this background, it would appear that somatic embryogenesis must have significant advantages over other *in vitro* propagation systems, both for mass clonal propagation and for gene transfer purposes. These apparent advantages, frequently cited by those working with embryogenic cultures of forest trees and other plants (e.g., Parrott *et al.*, 1991), include the potential for high-volume propagation using liquid culture and the fact that the propagules produced are analogous to seed embryos. This latter feature obviates the need for separate elongation, rooting or even acclimatization steps required by the other in vitro propagation systems, and gives somatic embryos the potential to be handled as artificial seeds. These advantages theoretically should translate to savings in labor over micropropagation via axillary enhancement or adventitious shoots. Similarly, with regard to gene transfer, since theoretically every cell of an embryogenic culture should have the ability to proliferate into a new population of cells capable of producing plantlets, these cultures would appear to present excellent targets for genetic engineering. Biolistic gene transfer, in particular, appears to be especially suitable for use with embryogenic cultures, providing an alternative to *Agrobacterium*-mediated gene transfer, to which some forest species, primarily conifers, had until recently proven recalcitrant. Thus, while the earliest report of a transgenic conifer employed *Agrobacterium rhizogenes* inoculation of *Larix decidua* seedlings, followed by regeneration of transgenic adventitious shoots from phytohormone-autotrophic calluses (Huang *et al.*, 1991), almost all reports of regeneration of transgenic conifers that followed over the past decade employed microprojectile-mediated gene transfer.

Given these advantages and the fact that embryogenic cultures have been generated for most commercially important conifers and hardwoods throughout the world, one might expect to have already seen establishment of intensively-managed forests composed of elite and/or genetically-engineered clones propagated via somatic embryogenesis. However, this vision has yet to be realized, due to the fact that even the best of the systems described in the literature have lacked commercial viability. Although each system is characterized by its own problems, two barriers preventing commercial application appear repeatedly: (1) genotypic variation, which often limits high-frequency plant production and/or transformability to a low number of clones, and (2) inability to apply the system for propagation of proven genotypes (since most starting material for the cultures is derived from seeds or seedlings). One other problem for which evidence has recently been reported in embryogenic cultures of forest trees is that of somaclonal variation—genetic variation arising during culture (e.g., Tremblay *et al.*, 1999). Thus the trees derived from a given embryogenic culture line may not all be "true-to-type." Despite these bottlenecks, some encouraging progress has been made in the past few years, both in the area of improving somatic embryogenesis for propagation of the most commercially-important species, and in producing potentially valuable transgenic clones. In addition, fresh approaches to *in vitro* propagation and gene transfer employing adventitious shoots and *Agrobacterium*-mediated gene transfer may offer useful alternatives to the biolistics/embryogenic culture paradigm for transforming some important species. Finally, genes are now being tested in forest trees that may result in products and services for which trees have never previously been employed, such as phytoremediation. Here I will briefly review the background and recent advances in each of these areas.

Recent Advances in Forest Tree Somatic Embryogenesis

Since the first reports of somatic embryogenesis in a coniferous tree (*Picea abies*) were published (Hakman *et al.*, 1985), the Norway spruce model has been extended to virtually all of the important species of the Pinaceae, with some modifications to optimize the model for use with *Pinus, Pseudotsuga* and *Abies*. While induction and proliferation of these cultures has become routine, the problem mentioned earlier with regard to genetic variation in induction and embryo production characterizes most of these systems. In addition, low frequencies of somatic seedling production have particularly been a problem for some of the top commercial pines. Much of the recent work in conifer embryogenesis has been conducted by scientists based at forest products companies, and is thus mostly disclosed only in patents, but some recent progress has been reported in the scientific literature. Research with spruces and pines, in particular, has focused on improving somatic embryo quality, which should lead to higher plantlet production rates. Starting from protocols already patented by industry researchers, treatments with abscisic acid, polyethylene glycol and maltose were confirmed to promote the highest production of mature *Pinus taeda* (loblolly pine) somatic embryos (Li *et al.*, 1997; 1998a). Gelling agent concentration was shown to have an impact on both initiation of pine embryogenic cultures and on maturation of somatic embryos. While a relatively low concentration (2 mg/l) of gellan gum was optimal for initiation of embryogenic *P. taeda* cultures (Li *et al.*, 1998b), a fivefold higher level of gellan gum aided maturation and germination of somatic embryos of *Pinus strobus* (Klimaszewska and Smith, 1997) *Pinus sylvestris* and *Pinus pinaster* (Lelu *et al.*, 1999). Another surprising finding with *P. sylvestris* cultures was that somatic embryos developed and matured spontaneously with no exposure to exogenous plant growth regulators. Lower than routinely-used PGR concentrations (2.2 µM each of 2,4-D and BA) in the initiation medium also resulted not only in enhanced embryogenic culture initiation frequency for *P. strobus*, but in higher subsequent somatic embryo production (Klimaszewska *et al.*, 2001).

One scale-up approach widely predicted to have an impact on clonal propagation of forest trees via somatic embryogenesis was the application of bioreactor technology. While stirred-tank and air-lift bioreactors have been tested with embryogenic conifer cultures (Tautorus *et al.*, 1992; 1994; Ingram and Mavituna, 2000), somatic seedling production figures from them have not been published. As these models of bioreactors were originally developed for bacteria and yeast cultures, perhaps it is no surprise that they have not proven highly suitable for production of all types of plant somatic embryos. As has been the case with other plant species, bioreactor designs in which the plant tissues remain submerged in liquid medium may actually prove more suitable for production of secondary metabolites from suspension cultures of trees cells, such as taxol from *Taxus* spp. (Son *et al.*, 2000).

Alternative bioreactor configurations that avoid full-time submersion of the embryogenic material have shown promise with multiple woody species. The temporary immersion bioreactor known as the RITA was used to produce synchronous populations of *Hevea brasiliensis* (Etienne *et al.*, 1997) and *Coffea arabica* (Etienne-Barry *et al.*, 1999) somatic embryos that converted to plantlets at high frequencies. No reports of commercial forest trees produced using this device are currently available. The most promising results for potential scale-up of any commercially-important forest species via somatic embryogenesis have been those reported for white spruce (*Picea glauca*) using flat-bed bioreactors (Attree *et al.*, 1994). In this system, somatic embryos are supported within a culture chamber on a flat absorbent pad above the surface of a liquid culture medium. A fresh supply of medium is continuously pumped into one end of the chamber, while spent medium flows out by gravity from the opposite end. The authors reported that over 6300 white spruce embryos could be produced at a time with this system, with an ultimate conversion rate of 92%. This flat bed bioreactor technology has been further developed and scaled-up by scientists at Cellfor in British Columbia, Canada. Along with other, proprietary technologies for desiccation, and for sowing and germination, this bioreactor configuration is currently being employed by Cellfor for large-scale production of somatic seedlings of various commercially-important conifers, including white pine weevil-resistant Sitka spruce (*Picea sitchensis*) clones (El-Kassaby *et al.*, 2001).

Most of the embryogenic culture systems reported for forest trees to date have relied on genetically unproven, immature tissues (i.e. from seeds or seedlings) as explant material. In fact, most reports of somatic embryogenesis in tree species in reality describe "embryo cloning," in which a zygotic embryo is induced to replicate itself indefinitely. This type of embryogenesis has been described as coming from "pre-embryogenic determined cells" (Sharp *et al.*, 1980) and represents a relatively straightforward process whereby the explanted cells, which are already close to the embryonic state, are induced to maintain a program of repetitive embryo production. To some extent, the problem of somatic embryogenesis relying on genetically unproven material may be handled by long term storage of culture material (e.g., by cryopreservation) while trees derived from them are tested in the field. Embryogenic cultures have, in fact, proven to be excellent candidates for cryostorage, and reliable and relatively inexpensive protocols have been developed for embryogenic cultures of both coniferous and hardwood species (e.g., Kartha *et al.*, 1988; Hargreaves and Smith, 1992; Vendrame *et al.*, 2001). However, the full power of *in vitro* clonal propagation for forest tree improvement could be realized if the extra steps required by the use of immature material could be eliminated. Somatic embryogenesis from tissues of mature trees has been reported for some forest species via the other route described by Sharp *et al.* (1980), i.e., through "induced embryogenic determined cells." This pathway is more difficult to induce, since the starting material consists of differentiated cells that must undergo major epigenetic changes to initiate somatic embryo production. Inducing this re-determination is especially challenging for woody perennials, since many of the genes associated with embryogenesis in the starting material may have been turned off years or even decades previously.

Nevertheless, the past few years have seen encouraging progress with regard to the propagation of proven genotypes via somatic embryogenesis from tissues of mature trees. Embryogenic cultures of *Pinus radiata* (radiata pine; Smith, 1999) and *Picea abies* (Norway spruce; Pâques *et al.*, 1997) were initiated from trees up to 20 and 25 years old, respectively. Although details of the protocols were not reported, both studies presented evidence that the material underwent rejuvenation in the process. For hardwood species, floral and inflorescence tissues, in particular, have proven to be useful explants for initiation of embryogenic cultures from mature trees. Embryogenic cultures have been initiated from pistils of basket willow (*Salix viminalis* L; Grönroos *et al.*, 1989), anther filaments of *Aesculus hippocastanum* (Jorgensen, 1989; Kiss *et al.*, 1992), petals, staminodes and filaments of *Theobroma cacao* (Lopez-Baez *et al.*, 1993; Alemanno *et al.*, 1996), female inflorescences of *Elaeis guineensis* (Teixeira *et al.*, 1994), anthers of *Quercus petraea* (Jorgensen, 1991), and male catkins of *Quercus bicolor, Quercus rubra* and *Quercus velutina* (Gingas, 1991; Wann, 1994). Both staminate and pistillate inflorescences excised from dormant buds of *Liquidambar styraciflua* (sweetgum) proved capable of initiating embryogenic cultures (Figure 1), even following frozen storage of the buds for up to 2 months (Merkle and Battle, 2000).

While encouraging, most of the recent progress with somatic embryogenesis is still the result of empirical experimentation, which has long been a hallmark of plant tissue culture research. Thus, it is significant that the past few years have seen the application of genomic approaches to improve regeneration from forest tree cultures in a more systematic manner, by understanding the developmental changes occurring *in vitro* at the gene expression level. Over 400 expressed sequence tags (ESTs) identified by

Figure 1. Repetitively embryogenic culture derived from a staminate inflorescence explant of sweetgum (*Liquidambar styraciflua*).

differential display as having altered expression patterns during loblolly pine zygotic embryo development were used to produce macroarrays. These macroarrays were used to monitor the expression of genes during somatic embryo development under varied culture conditions to look for conditions that would more closely mirror the gene expression patterns in developing zygotic embryos. (Cairney *et al.*, 1999). Such DNA arrays are a potentially powerful tool for rapid optimization of *in vitro* culturing techniques for production of high-quality somatic embryos.

Advances in Gene Transfer

As discussed earlier, while production of the first transgenic conifers employed neither embryogenic cultures nor biolistics, the combination of these two tools has been responsible for the majority of the progress in conifer transformation over the past decade. In the first of these reports, developing somatic embryos of *Picea glauca* were bombarded with a vector carrying *nptII* and a *Bacillus thuringiensis* (Bt) endotoxin gene, followed by selection on very low levels of kanamycin (Ellis *et al.*, 1993). While the frequency of escapes (colonies of cells surviving antibiotic selection that later proved not to be transgenic) was very high, some of the embryogenic callus colonies that proliferated from the bombarded embryos continued growth following transfer to higher kanamycin concentrations and were shown to be stably transformed. Somatic seedlings were regenerated from the transformed callus. Later studies with *Picea mariana, Picea abies,* and *Larix laricina* showed that suspension cultured embryogenic conifer cells also made suitable target material for biolistic gene transfer (Charest *et al.*, 1996; Clapham *et al.* 2000; Brukin *et al.*, 2000; Klimaszewska *et al.* 1997). The Clapham *et al.* report is notable in it use of a low-cost particle inflow gun, selection of transclones using the *bar* gene, which confers resistance to the herbicide Basta (phosphinothricin) and the fact that a high percentage of the Basta-resistant transclones also expressed the *gusA* gene that was co-bombarded with the *bar* gene.

Despite progress with other conifers, transformation of pine species proved especially problematic and reports of stable transformation of members of this genus were absent until Walter *et al.* (1998) reported regeneration of transformed *P. radiata* plantlets from microprojectile-bombarded embryogenic cultures. Using the *uid*A gene encoding β-glucuronidase (GUS) under the control of either a double CaMV 35S promoter or an artificial Emu promoter (Last *et al.,* 1991), more than 150 transgenic radiata pine plantlets were produced from 20 independent transformation events, using four different embryogenic clones. More recently, the same lab reported the first production of a pine engineered with a herbicide resistance gene (Bishop-Hurley *et al.*, 2001). *Pinus radiata* and *P. abies* embryogenic cultures were co-bombarded with vectors carrying the *bar* and *nptII* genes and selected using geneticin. As with the Clapham *et al.* (2000) report, co-transformation rates were very high. Trees regenerated from the transgenic cultures that were sprayed with the herbicide Buster (glufosinate) continued to grow with minor or no damage while non-transgenic controls died within 8 weeks of spraying (Figure 2).

Despite encouraging results with biolistics, concerns about high copy numbers and complex integration patterns of transgenes, both of which have been associated with transgene silencing (Finnegan and McElroy, 1994), have prompted researchers to continue testing *Agrobacterium*-Ti plasmid-mediated transformation for coniferous species. Recently, *Agrobacterium*-mediated transformation was achieved in both *Pinus* and *Picea*. By adding extra copies of virulence genes (*virG, virB*) to disarmed strains of *Agrobacterium tumefaciens*, transformation efficiencies for embryogenic *P. abies* cultures co-cultivated with *A. tumefaciens* were increased 1000-fold, as determined by GUS expression. A 10-fold increase in transient GUS expression was obtained using the same approach with embryogenic *P. taeda*

Figure 2. Transgenic and control radiata pine (*Pinus radiata*) plants 4 weeks following spraying with the herbicide Buster (glufosinate). Control plants (2 plants on the right) were killed, whereas transgenic plants (2 plants on left) showed only browning at tips of needles. From Bishop-Hurley *et al.* (2001). Photo courtesy of Dr. Christian Walter, New Zealand Forest Research Institute.

cultures (Wenck *et al.*, 1999). Co-cultivation of *P. strobus* embryogenic tissue with *A. tumefaciens* carrying a 35S-35S-AMV *gus: nptII* fusion also resulted in the regeneration of stably transformed somatic embryos (Levee *et al.*, 1999). Even more recently, the ability to regenerate adventitious shoots from *P. taeda* callus (Tang *et al.*, 1998) opened the way for *Agrobacterium*-mediated transformation of this species. Mature zygotic embryos from seven *P. taeda* families were co-cultivated with *A. tumefaciens* strain GV3101 carrying a plasmid with genes for GUS and hygromycin resistance (Figure 3). Ninety transgenic plants were regenerated from hygromycin-resistant callus (Figure 4) derived from 3 families (Tang *et al.*, 2001), thus demonstrating the return to prominence not only of *A. tumefaciens* for conifer transformation, but of adventitious shoot producing cultures as a regeneration system for transgenic conifer trees.

As with conifers, the first report of stably transformed hardwood trees is over a decade old, yet additional progress in transformation of hardwood trees has remained concentrated in a few genera. *Agrobacterium*-mediated

Figure 3. *Agrobacterium*-mediated transformation of *Pinus taeda*. GUS expression in loblolly pine zygotic embryos inoculated with *A. tumefaciens* strain GV3101. Photo courtesy of Dr. Wei Tang, Department of Forestry, North Carolina State University.

protocols have been developed for most *Populus* species and hybrids (Han *et al.*, 2000) Transformation of one of the top commercial hardwood genera, *Eucalyptus*, has been reported by two groups, via co-cultivation of leaf explants from *in vitro* grown plants (Mullins *et al.*, 1997) or seedling hypocotyls (Ho *et al.*, 1998) with *A. tumefaciens*.

Since transformation protocols are most highly developed for hardwoods, it is among hardwoods, specifically *Populus* species, that the most striking breakthroughs in modification of wood quality traits by genetic engineering of steps in the lignin biosynthesis pathway have been reported. Transgenic hybrid poplars in which cinnamyl alcohol dehydrogenase (CAD) activity was reduced through the expression of an antisense CAD gene showed a modest reduction in Klason lignin and superior pulping characteristics (Lapierre *et al.*, 1999). The CAD downregulated poplars also displayed a red coloration, mainly in the outer xylem. A 90% lower caffeic acid O-methyltransferase (COMT) activity did not change lignin content but dramatically increased the frequency of guaiacyl units and resistant biphenyl linkages in lignin, thereby lowering the efficiency of kraft pulping. Homologous sense suppression of CAOMT altered lignin composition in transgenic quaking aspen (*Populus tremuloides*) (Tsai *et al.*, 1998). Antisense inhibition of the 4-coumarate: CoA ligase (4CL) gene in transgenic quaking aspen led to a reduction in lignin content of up to 45%, and an increased growth rate. An increase in cellulose content up to 15% was also observed in the transgenic lines (Hu *et al.*, 1999).

Transgenic Trees and Phytoremediation

One promising potential novel application of transgenic trees, which is now receiving growing attention, is phytoremediation, the use of plants to stabilize, reduce or detoxify pollutants. Some non-engineered forest trees have already been tested for their ability to clean pollutants from soil and water. Newman *et al.* (1997) reported that *Populus trichocarpa* x *P. deltoides* hybrids took up trichloroethylene and degraded it to carbon dioxide and other non-toxic metabolites. *Populus deltoides* x *P. nigra* hybrid rooted cuttings were demonstrated to take up, hydrolyze and dealkylate the pesticide atrazine to less toxic metabolites (Burken and Schnoor, 1997). Willow (genus *Salix*) trees are being tested extensively in Europe for "filtering" municipal wastewater. Stands of these trees may be able to function as purification plants while at the same time producing fuel wood (Perttu and Kowalic, 1997; Riddell-Black *et al.*, 1995).

Augmenting trees that have already shown potential to handle pollutants with foreign genes for breakdown of organic pollutants or detoxification of

Figure 4. *Agrobacterium*-mediated transformation of *Pinus taeda*. Transgenic shoots derived from hygromycin-resistant calluses from inoculated loblolly pine embryos. Photo courtesy of Dr. Wei Tang, Department of Forestry, North Carolina State University.

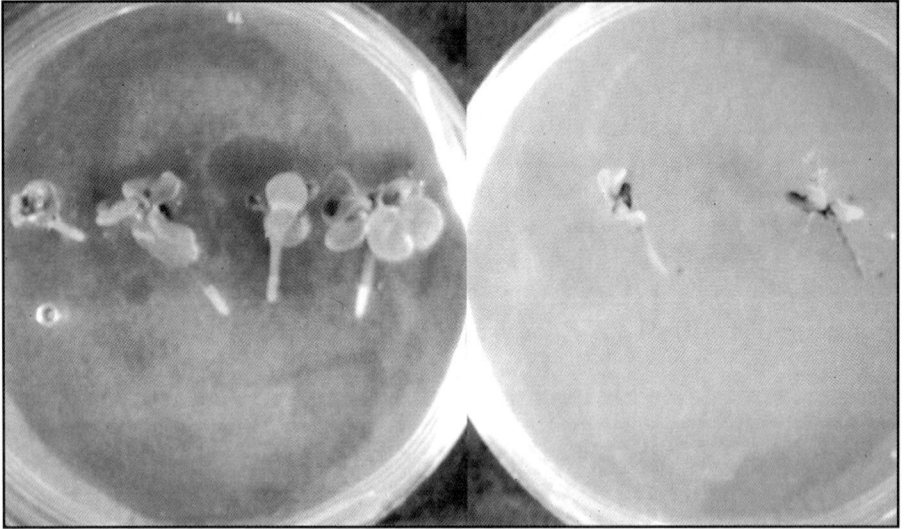

Figure 5. Transgenic *merA* and non-transformed control yellow-poplar somatic embryos after 14 days on germination medium with 50 μM HgCl$_2$. Embryos transformed with *merA* (on left) converted mercuric ion to the less toxic metal while non-transformed embryos (on right) were killed. From Rugh *et al.* (1998).

heavy metals may enhance their potential for phytoremediation applications. *Liriodendron tulipifera* (yellow-poplar) trees transformed with a modified bacterial mercuric ion reductase gene (*merA*) via microprojectile bombardment of embryogenic cultures grew well on normally toxic levels of ionic mercury *in vitro* by reducing it to less toxic elemental mercury (Figure 5; Rugh *et al.*, 1998). Eastern cottonwood trees transformed via *A. tumefaciens* with modified *merA* genes evolved 2-4 times the amounts of elemental mercury vapor relative to wild-type plantlets when exposed to a solution of ionic mercury. In a preliminary test, *merA* cottonwood plants regenerated from the cultures and potted in mercuric ion-contaminated soil showed little inhibition of growth, while non-transformed controls died within 48 hours (Che *et al.*, 2001). No doubt, as additional genes conferring the potential to break down or detoxify pollutants are identified, transgenic forest trees will become an even more important component in phytoremediation research.

Implications

Genetic improvement of forest trees has traditionally lagged behind the advances made with agronomic and horticultural crops, as the result of a number of features inherent to this category of woody plants. The traits most often cited as barriers to more rapid improvement include long life cycles, large size, and highly heterozygous genomes. In addition, most of the traits most commonly targeted for improvement (growth rate, form) are quantitatively inherited. However, research progress over the past decade has demonstrated that the powerful tools of biotechnology can be applied to manipulate forest trees in a manner parallel to what has been accomplished with agronomic crops. Indeed, due to the problems associated with breeding trees by traditional methods, the ability to clonally multiply superior genotypes *in vitro* and engineer trees for such traits as insect and pathogen resistance and altered wood quality may ultimately have an even greater impact on forest trees than on crop species. The bottlenecks preventing the research advances summarized here from translating into commercial success are substantial. Somatic embryogenesis, promoted for some time as the key to mass propagation of superior and genetically engineered genotypes, has yet to reach its potential, and challenging barriers remain to be overcome before it can be applied for manipulating the top commercial species on an economically realistic scale. The application of embryogenic and organogenic regeneration systems for engineering with genes having potentially large impacts on forest productivity (as well as for non-traditional uses such as phytoremediation) is only beginning to be tested. Even as these technical barriers are overcome, there remain important ecological, social and ethical considerations to be worked out with regard to deployment of clonal and transgenic plantations. Thus, it appears that this aspect of forest biotechnology may remain on the verge of accomplishing great gains for forestry for at least a few more years.

References

Alemanno, L., Berthouly, M., and Michaux-Ferriere, N. 1996. Histology of somatic embryogenesis from floral tissues of cocoa. Plant Cell Tiss. Org. Cult. 46: 187-194.

Attree, S.M., Pomeroy, M.K., and Fowke, L.C. 1994. Production of vigorous, desiccation tolerant white spruce (*Picea glauca* [Moench] Voss) synthetic seeds in a bioreactor. Plant Cell Rep. 13: 601-606.

Bishop-Hurley, S.L., Zabkiewicz, R.J., Grace, L., Gardner, R.C., Wagner, A., and Walter, C. 2001. Conifer genetic engineering: transgenic *Pinus radiata* (D.Don) and *Picea abies* (Kaarst) plants are resistant to the herbicide Buster. Plant Cell Rep. 20: 235-243.

Brukin, V., Clapham, D., Elfstrand, M., and von Arnold, S. 2000. Basta tolerance as a selectable and screening marker for transgenic plants of Norway spruce. Plant Cell Rep. 19: 899-903.

Burken, J.G., and J.L. Schnoor. 1997. Phytoremediation: uptake of atrazine and the role of root exudates. J. Environ. Eng. 122: 958-963.

Cairney, J., Xu, N.F., Pullman, G.S., Ciavatta, V.T., and Johns, B. 1999. Natural and somatic embryo development in loblolly pine - Gene expression studies using differential display and DNA arrays. Appl. Biochem. Biotech. 77-79: 5-17.

Charest P.J., Devantier, Y., and Lachance, D. 1996. Stable genetic transformation of *Picea mariana* (black spruce) via particle bombardment. *In Vitro* Cell. Dev. Biol.- Plant 32: 91-99.

Che, D.S., Meagher, R.B., Lima, A., Heaton, A.C.P., and Merkle, S.A. 2001. Expression of mercuric ion reductase in eastern cottonwood confers mercuric ion reduction and resistance [Abstract]. In: Proceedings of the 26th Biennial Southern Forest Tree Improvement Conference, June 26-29, 2001, Athens, GA . p. 193.

Clapham, D., Demel, P., Elfstrand, M., Koop, H.U., Sabala, I., and von Arnold, S. 2000. Gene transfer by particle bombardment to embryogenic cultures of *Picea abies* and the production of transgenic plantlets. Scandinavian J. For. Res. 15: 151-160.

El-Kassaby, Y.A., King, J., Ying, C.C., Yanchuk, A., Alfaro, R.I., and Leal, I. 2001. Somatic embryogenesis as a delivery system for specialty products with reference to resistant Sitka spruce to the white pine weevil. In: Proceedings of the 26th Biennial Southern Forest Tree Improvement Conference, June 26-29, 2001, Athens, GA. p. 154-168.

Ellis, D.D., McCabe, D.E., McInnis, S., Ramachandran, R., Russell, D.R., Wallace, K.M., Martinell, B.J., Roberts, D.R., Raffa, K.F., and McCown, B.H. 1993. Stable transformation of *Picea glauca* by particle acceleration. Bio/Technology. 11: 84-89.

Etienne, H., Lartaud, M., Michaux-Ferriere, N., Carron, M.P., Berthouly, M., and Teisson, C. 1997. Improvement of somatic embryogenesis in *Hevea brasiliensis* (Mull. Arg.) using the temporary immersion technique. *In Vitro* Cell. Dev. Biol.-Plant. 33: 81-87.

Etienne-Barry, D., Bertrand, B., Vasquez, N., and Etienne, H. 1999. Direct sowing of *Coffea arabica* somatic embryos mass-produced in a bioreactor and regeneration of plants. Plant Cell Rep. 19: 111-117.

Fillatti, J.J., Sellmer, J., McCown, B., Haissig, B., and Comai, L. 1987. *Agrobacterium* mediated transformation and regeneration of *Populus*. Mol. Gen. Genet. 206: 192-199.

Finnegan, J., and McElroy, D. 1994. Transgene inactivation: plants fight back! Bio/Technology. 12: 883-888.

Gingas, V.M. 1991. Asexual embryogenesis and plant regeneration from male catkins of *Quercus*. HortScience. 26: 1217-1218.

Grönroos, L., von Arnold, S., and Eriksson, T. 1989. Callus production and somatic embryogenesis from floral explants of basket willow (*Salix viminalis* L.). J. Plant Physiol. 134: 558-566.

Hakman, I., Fowke, L.C., Von Arnold, S., Eriksson,T.1985. The development of somatic embryos in tissue cultures.initiated from immature embryos of *Picea abies* (Norway spruce). Plant Sci. 38: 53-59.

Han, K.H., Meilan, R., Ma, C., and Strauss, S.H. 2000. An *Agrobacterium tumefaciens* transformation protocol effective on a variety of cottonwood hybrids (genus *Populus*). Plant Cell Rep. 19: 315-320.

Hargreaves, C. and Smith, D.R. 1992. Cryopreservation of *Pinus radiata* embryogenic tissue. Comb. Proc. Intl. Plant Prop. Soc. 42: 327-333.

Ho, C.K., Chang, S.H., Tsay, J.Y., Tsai, C.J., Chiang, V.L., and Chen, Z.Z. 1998. *Agrobacterium tumefaciens*-mediated transformation *of Eucalyptus camaldulensis* and production of transgenic plants. Plant Cell Rep. 17: 675-680.

Huang,Y.H., A.M. Diner and D.F. Karnosky. 1991. *Agrobacterium rhizogenes*-mediated genetic transformation and regeneration of a conifer: *Larix decidua. In Vitro* Cell. Dev. Biol. 27P: 201-207.

Hu, W.J., Harding, S.A., Lung, J., Popko, J.L., Ralph, J., Stokke, D.D., Tsai, C.J., and Chiang, V.L. 1999. Repression of lignin biosynthesis promotes cellulose accumulation and growth in transgenic trees. Nat. Biotech. 17: 808-812.

Ingram, B., and Mavituna, F. 2000. Effect of bioreactor configuration on the growth and maturation of *Picea sitchensis* somatic embryo cultures. Plant Cell Tiss. Org. Cult. 61: 87-96.

Jorgensen, J. 1989. Somatic embryogenesis in *Aesculus hippocastanum* L. by culture of filament callus, J. Plant Physiol. 135: 240-241.

Jorgensen, J. 1991. Somatic embryogenesis in *Aesculus hippocastanum* and *Quercus petraea* from old trees (10 to 140 years), In: Woody Plant Biotechnology. M.R. Ahuja, ed. Plenum Press, New York. p. 351-352.

Kartha, K.K., Fowke, L.C., Leung, N.L., Caswell, K.L., and Hakman, I. 1988. Induction of somatic embryos and plantlets from cryopreserved cell cultures of white spruce (*Picea glauca*). Plant Physiol. 132: 529-539.

Kiss, J., Heszky, L.E., Kiss, E., and Gyulai, G. 1992. High efficiency adventive embryogenesis on somatic embryos of anther, filament and immature proembryo origin in horse-chestnut (*Aesculus hippocastanum* L.), Plant Cell Tiss. Org. Cult. 30: 59-64.

Klimaszewska, K., and Smith, D.R. 1997. Maturation of somatic embryos of *Pinus strobus* is promoted by a high concentration of gellan gum. Physiol Plant. 100: 949-957.

Klimaszewska, K., Park, Y.-S., Overton, C., MacEacheron, I., and Bonga. J.M. 2001. Optimized somatic embryogenesis in *Pinus strobus* L. *In Vitro* Cell. Dev. Biol.- Plant. 37: 392-399.

Klimaszewska, K., Devantier, Y., Lachance, D., Lelu, M.A., and Charest, P.J. 1997. *Larix laricina* (tamarack): Somatic embryogenesis and genetic transformation. Can. J. For. Res. 27: 538-550.

Lapierre C., Pollet, B., Petit-Conil, M., Toval, G. Romero, J., Pilate, G., Leple, J.C., Boerjan, W., Ferret, V., De Nadai, V., Jouanin, L. 1999. Structural alterations of lignins in transgenic poplars with depressed cinnamyl alcohol dehydrogenase or caffeic acid O-methyltransferase activity have an opposite impact on the efficiency of industrial kraft pulping. Plant Physiol. 119: 153-163.

Last , D.I., Brettell, R.I.S., Chamberlain, D.A., Chaduhury, A.M., Larkin, P.J., Marsh, E.L., Peacock, W.J., Dennis, E.S. 1991. pEmu - an improved promoter for gene expression in cereal cells. Theor. Appl. Genet. 81: 581-588.

Lelu, M.A., Bastien, C., Drugeault, A., Gouez, M.L., and Klimaszewska, K. 1999. Somatic embryogenesis and plantlet development in *Pinus sylvestris* and *Pinus pinaster* on medium with and without growth regulators. Physiol. Plant. 105: 719-728.

Levee, V., Garin, E., Klimaszewska, K., and Seguin, A. 1999. Stable genetic transformation of white pine (*Pinus strobus* L.) after cocultivation of embryogenic tissues with *Agrobacterium tumefaciens*. Mol. Breed. 5: 429-440.

Li, X.Y., Huang, F.H., and Gbur, E.E. 1997. Polyethylene glycol-promoted development of somatic embryos of loblolly pine (*Pinus taeda* L.). *In Vitro* Cell. Dev. Biol.-Plant 33: 184-189.

Li, X.Y., Huang, F.H., and Gbur, E.E. 1998. Effect of basal medium, growth regulators and Phytagel concentration on initiation of embryogenic cultures from immature zygotic embryos of loblolly pine (*Pinus taeda* L.). Plant Cell Rep. 17: 298-301.

Li, X.Y., Huang, F.H., Murphy, J.B., and Gbur, E.E. 1998. Polyethylene glycol and maltose enhance somatic embryo maturation in loblolly pine (*Pinus taeda* L.). *In Vitro* Cell. Dev. Biol.-Plant 34: 22-26.

Lopez-Baez, O., Bollon, H., Eskes, A., and Petiard, V. 1993. Somatic embryogenesis and plant regeneration from flower parts of cocoa *Theobroma cacao* L., C.R. Acad. Sci. Paris. 316: 579-584.

Merkle, S.A., and Battle, P.J. 2000. Enhancement of embryogenic culture initiation from tissues of mature sweetgum trees. Plant Cell Rep. 19: 268-273.

Mullins, K.V., Llewellyn, D.J., Hartney, V.J., Strauss, S., and Dennis, E.S. 1997. Regeneration and transformation of *Eucalyptus camaldulensis.* Plant Cell Rep. 16: 787-791.

Newman, L.A., Strand, S.E., Choe, N., Duffy, J., Ekuan, G., Ruszau, M., Shurtleff, B.B., Wilmoth, J., Heilman, P., and Gordon, M.P. 1997. Uptake and biotransformation of trichloroethylene by hybrid poplars. Environ. Sci. Technol. 31: 1062-1067.

Pâques, M., Bercetche, J., Harvengt, L. 1997. Somatic embryogenesis: a way to recover clones from needles of selected Norway spruce [Abstract]. Program and Abstracts, Joint Meeting of the IUFRO Working Parties. 2.04-07 and 2.04-06, Somatic Cell Genetics and Molecular Genetics of Trees. August 12-16, 1997. Quebec City, Quebec, Canada.

Parrott, W.A., Merkle, S.A., and Williams, E.G. 1991. Somatic embryogenesis: potential for use in propagation and gene transfer systems. In: Advanced Methods in Plant Breeding and Biotechnology D.R. Murray, ed. CAB International, Wallingford, Oxon (U.K.). p. 158-200.

Perttu, K.L., and Kowalik, P.J. 1997. *Salix* vegetation filter for purification of waters and soils. Biomass and Bioenergy. 12: 9-19.

Riddell-Black, D., Rowlands, C., and Snelson, A. 1995. Heavy metal uptake from sewage sludge amended soil by *Salix* and *Populus* species grown for fuel. Fourteenth Annual Symposium on Current Topics in Plant Biochemistry, Physiology, and Molecular Biology. April 19-22, 1995. Columbia, MO. p. 51-52.

Rugh, C.L., Senecoff, J.F., Meagher, R.B., and Merkle, S.A. 1998. Development of transgenic yellow poplar for mercury phytoremediation. Nat. Biotech. 10: 925-928.

Sharp, W.R. , Söndahl, M.R. , Caldas, L.S. , and Maraffa, S.B. 1980. The physiology of *in vitro* asexual embryogenesis. In: Horticultural Reviews, Vol. 2. J. Janick, ed. AVI Publishing Co, Inc., Westport, CN. p. 268-310.

Smith, D.R. 1999. Successful rejuvenation of radiata pine. In: Proceedings of the 25th Biennial Southern Forest Tree Improvement Conference, July 11-14, 1999, New Orleans, LA . p. 158-167.

Sommer, H.E., Brown, C.L., and Kormanick, P.P. 1975. Differentiation of plantlets of longleaf pine (*Pinus palustris* Mill.) tissue cultured *in vitro.* Botanical Gazette. 136: 196-200.

Son, S. H., Choi, S. M., Lee, Y. H., Choi, K. B., Yun, S. R., Kim, J. K., Park, H. J., Kwon, O. W., Noh, E. W., Seon, J. H., Park, Y. G. 2000. Large-scale growth and taxane production in cell cultures of *Taxus cuspidata* (Japanese yew) using a novel bioreactor. Plant Cell Rep. 19: 628-633.

Tang, W., Ouyang, F., and Guo, Z.C. 1998. Plant regeneration through organogenesis from callus induced from mature zygotic embryos of loblolly pine. Plant Cell Rep. 17: 557-560.

Tang, W., Sederoff, R., and Whetten, R. 2001. Regeneration of transgenic loblolly pine (*Pinus taeda* L.) from zygotic embryos transformed with *Agrobacterium tumefaciens*. Planta 213: 981-989.

Tautorus, T.E., Lulsdorf, M.M., Kikcio, S.I., and Dunstan, D.I. 1992. Bioreactor culture of *Picea mariana* Mill. (black spruce) and the species complex *Picea glauca-engelmannii* (interior spruce) somatic embryos. Growth parameters. Appl. Microbiol. Biotechnol. 38: 46-51.

Tautorus, T.E., Kikcio, S.I., Lulsdorf, M.M., and Dunstan, D.I. 1994. Nutrient utilization during bioreactor culture, and maturation of somatic embryo cultures of *Picea mariana* and *Picea glauca-engelmannii*. *In Vitro* Cell. Dev. Biol.- Plant. 30: 58-63.

Teixeira, J.B., Söndahl, M.R., and Kirby, E.G. 1994. Somatic embryogenesis from immature inflorescences of oil palm. Plant Cell Rep. 13: 247-250.

Tremblay, L., Levasseur, C., and Tremblay, F.M. 1999. Frequency of somaclonal variation in black spruce (*Picea mariana*, Pinaceae) and white spruce (*P. glauca*, Pinaceae) derived from somatic embryogenesis and identification of some factors involved in genetic instability. Am. J. Bot. 86: 1373-1381.

Tsai, C.J., Popko, J.L., Mielke, M.R., Hu, W.J., Podila, G.K., and Chiang, V.L. 1998. Suppression of O- methyltransferase gene by homologous sense transgene in quaking aspen causes red-brown wood phenotypes. Plant Physiol. 117: 101-112.

Vendrame, W.A., Holliday, C.P., Montello, P.M., Smith, D.R., and Merkle, S.A. 2001. Cryopreservation of yellow-poplar (*Liriodendron tulipifera*) and sweetgum (*Liquidambar* spp.) embryogenic cultures. New Forests. 21: 283-292.

Walter, C., Grace, L.J., Wagner, A., White, D.W.R., Walden, A.R., Donaldson, S.S., Hinton, H., Gardner, R.C., and Smith, D.R. 1998. Stable transformation and regeneration of transgenic plants of *Pinus radiata* D. Don. Plant Cell Rep. 17: 460-468.

Wann, S.R. 1994. Cloning famous trees. In: Proceedings of the Second International Symposium on Applications of Biotechnology to Tree Culture, Protection, and Utilization, October 2-6, 1994. C.H. Michler, M.R. Becwar, D. Cullen, W.L. Nance, R.R. Sederoff and J.M. Slavicek, eds. Bloomington, MN. p. 21.

Wenck, A.R., Quinn, M., Whetten, R.W., Pullman, G., and Sederoff, R. 1999. High-efficiency *Agrobacterium*- mediated transformation of Norway spruce (*Picea abies*) and loblolly pine (*Pinus taeda*). Plant Mol. Biol. 39: 407-416.

Winton, L. 1968. Plantlets from aspen tissue cultures. Science. 160: 1234-1235.

Wolter, K.E.1968. Root and shoot initiation in aspen callus cultures. Nature. 219: 509-510.

From: *Transgenic Plants: Current Innovations and Future Trends*
Edited by: C. Neal Stewart, Jr.

Chapter 4

In Planta Transformation

Georges Pelletier and Nicole Bechtold

Abstract

Several *in planta* methods of transformation have been described during the past thirty years. Most of them were not reproducible and the apparent positive results obtained were generally the result of artifacts or the use of methods of investigation leading to ambiguous interpretations Successful *Agrobacterium*-mediated gene transfer to *Arabidopsis thaliana* (and some related species) achieved by infiltration (vacuum or surfactant) of adult plants, led to many attempts to replicate the results in other species, which, unfortunately, have remained unsuccesssful. It was recently discovered that this particular transformation process targets the female gametophyte. In the future, a better understanding of the interaction between *Agrobacterium* and the female germ line of some Brassicaceaes could open possibilities for generalizing applications of this method.

Introduction

The very first attempts at plant transformation made use of "*in planta*" methods. These occurred before plant biologists developed reliable techniques for *in vitro* cell and tissue culture which allow the regeneration of entire plants from single cells. These first experiments were based on the assumptions that, similar to what was observed with some bacteria (Avery *et al.*, 1944), the incubation in defined conditions of certain plant organs or cells with extracted DNA could be sufficient to allow this DNA to enter the cell nucleus, to integrate into the chromosomes (Ledoux and Huart, 1968), and to lead to heritable modifications.

Seed imbibition-germination or pollination-fertilization were the two preferred processes during which purified DNA was applied. Marker genes and selection strategies existing at that time were often employed. During approximately 20 years, several publications appeared on different plant species but none of them led to confirmed results. In order to avoid the problems inherent with tissue culture, these attempts at *in planta* transformation continued even after *in vitro* culture methods were established. There remained difficulty in regeneration of entire plants, somaclonal variation, and the host specificity of *Agrobacterium*. In the mid-eighties, *Agrobacterium*-mediated transformation of *Arabidopsis thaliana* seeds was at first assessed to be similar to previous *in planta* methods; as a new, but probably not reproducible method! Subsequently "seed transformation" was replaced by a significantly more efficient and reliable method based on "vacuum infiltration", which allowed Agrobacteria to enter adult plant tissues. A final improvement came in the "dipping" method which uses a surfactant molecule instead of vacuum. Several independent approaches have proven that the female gametophyte and its genome are the target of the T-DNA in these *in planta* approaches during the gametogenesis-fertilization process. Only two other species, *Arabidopsis lasiocarpa* and Pakchoi (*Brassica rapa* L. ssp. *chinensis*), both belonging to the same family, Brassicaceae, have been proven to be similarly amenable to *in planta* transformation as *Arabidopsis thaliana*.

In Planta Direct Gene Transfer

The first transformation experiments were performed with DNA extracted from plants or bacteria presenting dominant or complementary characters and applied to seeds or pollen. Presumed transformants were visualized by screenings, or in some cases after a positive selection, based on the presence of a character present in the donor organism.

Germinating Seeds as Targets for DNA Transfer

In the first report of plant transformation (Hess, 1969), seedlings of a white flowering *Petunia hybrida* mutant were incubated with the DNA extracted from young leaves of red flowering petunias. A higher (27% versus 9%) percentage of the plants derived from the treated seeds showed red flowers (with variable intensity) as compared to control seeds treated with their own DNA. Some of the new anthocyanin synthesis loci in these plants were located through a genetic analysis of further generations on chromosomes carrying no known anthocyanin alleles, neither functional nor mutated (Hess, 1970). These results were never confirmed.

In the case of *Arabidopsis thaliana*, experiments suggested that exogenous DNA could be absorbed by germinating seedlings, translocated without denaturation in the plant towards floral organs, and recovered in the progeny (Ledoux *et al.*, 1971). Experiments based on genetic markers were performed in which seeds of thiamineless mutants of *Arabidopsis thaliana* corresponding to three different loci (*py*, *thi* and *tz*) were treated with DNA preparations from various bacterial species (Ledoux *et al.*, 1974). Corrections (plant growth without addition of thiamine) were obtained with a relatively high frequency (10^{-2} to 10^{-4}). Surprisingly, as in the previous experiments with petunia, the treated seeds gave rise directly to corrected plants, and the progeny obtained by selfing was entirely "transformed" without segregation.

Nawa *et al.* (1975) applied DNA from a homozygous cultivar of *Capsicum annuum*, bearing a defined allelic composition at four loci determining the shape and color of fruits, to germinating seeds of another cultivar bearing the opposite allelic composition. Change from recessive to homozygous or heterozygous dominant for one loci was observed at a low frequency (0.12 to 0.25%)in the progeny obtained by self pollination.

Treatment (24 h under low pressure followed by 3 day-germination in a solution of DNA + protamine) of wrinkled maize grains (su/su) with DNA originating from a variety with normal grains (Su/Su) was performed by Korohoda and Strzalka (1979). Resulting plants were self-pollinated and about 30% of them gave some grains with changes in color and shape. These modifications were transmitted to the progeny as homozygous or heterozygous traits.

Although the uptake and transient expression of DNA in seeds or embryos was subsequently confirmed in several instances (Töpfer *et al.*,1989; Senaratna *et al.*, 1991; Yoo and Jung, 1995), the reproducibility and the efficiency of all these methods is highly questionable.

Pollen as Vector for Plant Transformation

Another strategy was the treatment of pollen with DNA prior to or during pollination. It was supposed that this DNA could be incorporated into the sperm nucleus or transported by the pollen tube towards the egg cell. Hess *et al.* (1975) induced tumors on *Nicotiana glauca* seedlings obtained after pollination by *N. glauca* pollen treated with DNA isolated from *Nicotiana langsdorffii*. In the same way he tested the possibility of improving petunia growth on media containing lactose or galactose (Hess, 1978; 1979) or of obtaining red flowered petunias from a white-flowered pure line (Hess, 1980).

In a critical experiment (Sanford *et al.*,1985), pollen from multiple recessive lines was incubated with genomic DNA from multiple dominant lines and subsequently used to pollinate multiple recessive female plants. These experiments were conducted in corn and in tomato, plants which offered a large collection of multiple Mendelian marker genes showing a discrete variation. Contrary to previous reports, these authors were unable to confirm any transformation events.

Nevertheless, during the period following, a large series of experiments were performed on cereals, based on the same assumptions. Ohta (1986) obtained changes in the resulting endosperm by applying DNA from a multiple dominant maize genotype during self-pollination of the corresponding recessive genotype. Moreover, the transformed phenotype was sometimes observed in the next generation indicating the possibility of embryo cell transformation.

Using the pollen tube as a vector, plasmids containing the coding region of the *nptII* (neomycin-phosphotransferase II) gene were transferred into several varieties of wheat (Picard *et al.*,1988). From a total of 1961 mature grains harvested from 334 DNA-treated spikes, 19 gave kanamycin resistant plants. Some of them revealed expression of a NPT II activity and the kanamycin resistant trait was apparently transmitted to the progeny. Similarly, Luo and Wu (1989) deposited a DNA solution of a plasmid bearing the *nptII* gene onto the cut end of the style of rice plants before pollination and selected transformed plants on the basis of a root-tip DNA hybridization method (dot blot). Up to 20% of the seeds formed gave positive results.

Zeng *et al.* (1994) reported the successful introduction of the beta-glucuronidase (*uidA*) gene (with a rate of about 5%) into hexaploid wheat by the pollen-tube pathway method and confirmed the transformation by Southern hybridization and histochemical GUS staining.

A number of teams tried to repeat these experiments but, after a critical analysis of the results (Langridge *et al.*, 1992), were unable to provide definite proof of transformation and transmission to the progeny using this method.

Injection of DNA into Reproductive Organs

Soyfer (1980) induced normal starch production among the progeny of a waxy mutant of barley after microinjection of total cellular DNA of a normal barley genotype. Zhou *et al.* (1983) similarly succeeded in transferring different traits to species of cotton by microinjection of DNA of another species into the ovaries. A method based on delivery by microinjection of DNA carrying a selectable marker (*nptII*) into young floral tillers of rye was proposed by de la Pena *et al.* (1987). These authors postulated that the DNA was subsequently transported by the plant vascular system to the archesporial cells which were supposed to be in a competent stage about two weeks before meiosis. About one plant among 1000 harvested grains was resistant to kanamycin, showed NPT II activity and was positive in Southern hybridization. Croughan *et al.* (1988) reported a similar experiment in rice, but no confirmation was provided.

Other Methods

Methods based on *in vitro* particle bombardment were largely applied to structures such as embryos, embryogenic or meristematic cells having high regeneration capacities through tissue culture (Christou, 1995) or pollen grains before pollination. The *in planta* adaptation of particle bombardment methods to shoot meristems led to the reproducible transient expression of reporter genes in wheat (Sautter *et al.*, 1995) but not to transformants in the progeny even when reproductive organs were treated. The transformation of microspores, and their subsequent maturation *in vitro*, allowed the recovery of transgenic plants in the progeny after pollination in the case of tobacco (Touraev *et al.*, 1997). The *in vitro* maturation step remains the bottleneck of this *in vitro-in planta* hybrid method.

Using *in planta* electroporation of DNA into intact nodal meristems of leguminous plants, Chowrira *et al.* (1995) described transient expression in tissues of treated individuals and an apparent integration in their offspring.

Assessment of *In Planta* Direct Gene Transfer

During the past 30 years a number of teams have turned their attention to *in planta* direct DNA transfer in different species because of its apparent simplicity. In spite of very diverse methods, the general conclusion which can be drawn is negative : none of these methods can be considered as reproducible. Many of these results, for example, the first attemps at seed or pollen transformation, were later proven (Kleinhofs *et al.*, 1975; Redei *et al.*, 1977; Sanford *et al.*, 1985) to be simple errors or contaminations, or due to inappropriate selection markers. Nevertheless, despite persistent controversy and a general failure of these methods, research is still active in this area and is regularly revisited with new constructs and approaches. It was not totally useless, from a historical point of view, to describe these approaches, since, as we will discuss, the successes obtained in a very limited number of species through *in planta Agrobacterium*-mediated transformation partially followed, consciously or not, parallel ideas.

In Planta Agrobacterium-Mediated Transformation

The true beginning of transgenic plant technology took place in the early 1980s thanks to *Agrobacterium tumefaciens*-mediated gene transfer and reliable *in vitro* regeneration techniques. Initial successes were limited to tobacco, and for a certain time, to dicotyledonous plants. Reliable protocols permitting the transfer of DNA by *Agrobacterium*, without tissue culture and regeneration, emerged at the end of the 1980s, and after significant improvements, are now routinely used in *Arabidopsis* and a few related species (Bent, 2000).

Arabidopsis thaliana
Germinating Seeds
In planta transformation was first described in *Arabidopsis thaliana* by Feldmann and Marks (1987) and, as explained by these authors, followed up the results published by Ledoux *et al*. (1971) on the uptake of exogenous DNA in imbibing seeds. In this pioneer work, seeds were imbibed in a suspension of *Agrobacterium tumefaciens* bearing the *nptII* gene on T-DNA in Murashige and Skoog salts (MS) (Murashige and Skoog, 1962) with 4% sucrose and different vitamins, during 24 h at 28 °C. The imbibed seeds were sown in natural conditions and the plants that developed were then allowed to produce seeds by self-pollination. Among these seeds, some gave rise to entirely transformed plantlets which could be easily selected by germination on a kanamycin selective medium. The transformation frequency varied in the different treatments with an average of 1 transformant in the progeny of 100 plants derived from treated seeds (Feldmann, 1991). Kanamycin segregation and Southern blot analysis in the progeny of transformants confirmed the transformation. As the result of this low frequency of transformants, this first method was difficult to reproduce (Bouchez *et al*., 1993) but even so more than 17,000 T-DNA lines have been produced by the Feldmann group. An improvement of this protocol, by the sonication of the seeds before the co-culture, was described recently (Tomilov

et al., 1999). The microdamaging of the seeds apparently increases the transformation frequency.

Wounded Inflorescences
Chang *et al.* (1994) described an experiment in which *Arabidopsis* primary and secondary inflorescence shoots were cut off and the wounds inoculated with an overnight culture of *Agrobacterium*. Three inoculations were performed after the appearance of new shoots. Transformants were selected in the progeny. The method was reproduced by Bouchez *et al.* (1993) and Katavic *et al.* (1994). About 6.6% of the treated plants produce 1 to 9 transformants in their progeny. Transformation was confirmed by genetic and molecular analyses.

Flowering Plants
Bechtold *et al.* (1993) described the inoculation of flowering plants by vacuum infiltration of an *Agrobacterium* suspension in a MS medium supplemented with benzyl aminopurine (BA) (0.01 mg/l) and sucrose (50 mg/l). *Arabidopsis* plants at the early stage of flowering were uprooted, entirely immersed into the suspension and placed into a chamber where vacuum was applied for 20 min. The infiltrated plants were replanted and grown to produce seeds. Transformants were selected in the progeny. This method was based on the assumption that the stage at which the T-DNA transfer takes place is late in the development of the plant, either at the end of gametogenesis or at the zygote stage (Feldmann, 1991). It was assumed that the reduction of the duration and the physical distance between the inoculation and the transformation events could increase the transformation frequency. Thus, the transformation frequency obtained often exceeded 1% of the progeny of infiltrated plants but still varied among experiments. Genetic and Southern blot analyses confirmed unambigously the genetic transformation. This method was the first to be widely reproduced by the *Arabidopsis* scientific community before further improvements were developed.

A first variation consisted of infiltrating only the floral stems without uprooting the plants and in simplifying the infiltration medium (Bent *et al.*, 1994). The most important subsequent modification was the replacement of the vacuum by the addition of Silwet L-77 in the infiltration medium (Bent and Clough, 1998). This compound was first proven to promote T-DNA delivery into the cells of *in vitro* inoculated immature wheat embryos (Cheng *et al.*, 1997). Silwet L-77 is a non-toxic methylated silicone surfactant used in formulating pesticides which, like vacuum, allows the penetration of the Agrobacteria into the intercellular spaces. The "floral dip" transformation method (Clough and Bent, 1998) consists of simply dipping the flowers in a sucrose solution (50 mg/l) containing *Agrobacterium* and 0.05% Silwet L-77 for 2 min. This very simplified protocol allowed an increase in the reliability of the transformation frequency. It is now routinely possible to reach a transformation frequency of more than 1% of the progeny. Detailed descriptions of the protocols for vacuum and Silwet infiltration can be found in Bechtold and Pelletier (1998) and Bent and Clough (1998) respectively.

A similar inoculation technique, involves spraying an *Agrobacterium* suspension containing Silwet L-77 onto flowering plants (Chung *et al.*, 2000). This method could be of interest for plant species which are too large for dipping or vaccum infiltration, but not for *Arabidopsis* transformation. The same rates of transformation are obtained but it is more difficult to avoid *Agrobacterium* scattering.

Other Species

Similar approaches making use of *Agrobacterium* interaction with floral tissues have been described, both before and after the development of infiltration methods in *Arabidopsis thaliana*.

Petunia

Co-cultures of *Petunia* pollen and a wild strain of *Agrobacterium tumefaciens* were used by Hess and Dressler (1989). After pollination, seeds were harvested, germinated on a nutrient medium and calli were induced. These calli were subcultured on an hormone-free medium and two of them showed habituation and some molecular evidence of T-DNA transfer.

Wheat

The same group (Hess *et al.*, 1990) reported transformation experiments in wheat by pipetting *Agrobacterium* suspensions conferring kanamycin resistance into the spikelets at a stage when the anthers in the middle of the ear are just before dehiscence. Among harvested grains, 1% frequency of plants surviving on kanamycin-containing medium were recovered and positive in Southern hybridization was obtained. The authors interpreted there results by the induction of virulence by pollen exudates (flavonol glycosides) and the transformation of the male gametes. These results were not totally convincing and were suggested to be an artifact of the procedure

(Langridge *et al.*, 1992). Subsequently, the genetic transformation of wheat, mediated by *Agrobacterium tumefaciens,* was achieved by a tissue culture approach (Cheng *et al.*, 1997).

Medicago truncatula

Positive results were recently published using the vacuum infiltration method in *Medicago truncatula* (Trieu *et al.*, 2000). The results obtained differed significantly from those obtained with *Arabidopsis*: the most favourable stage corresponded to younger plants, siblings were frequently present among transformants, and a very high frequency was sometimes obtained. These results suggest that factors different from those acting in the case of *Arabidopsis* are playing an important role. These results have yet to be reproduced by other groups and require confirmation.

Pakchoï

Liu *et al.* (1996) and Cao *et al.* (2000) obtained transgenic *Brassica rapa* ssp. *chinensis* plants by vacuum infiltration with an *Agrobacterium tumefaciens* strain carrying a phosphinotricin resistance gene. Genetic and molecular analyses showed Mendelian transmission and the integration and expression of the transgene. Moreover, the dipping (Silwet L-77) method as an alternative to vacuum reproducibly gave the same transformation frequency in this species (0.05% of the seeds harvested from infiltrated plants, M. Cao, personal communication).

Other Arabidopsis Species

Tague (2001) examined the suitability of *in planta* transformation in different *Arabidopsis* species and in *Capsella bursa pastoris*. Among these species, *Arabidopsis lasiocarpa* behaved very similarly to *A. thaliana* and gave large numbers of transformed seeds through floral dipping and even more through vacuum infiltration.

These results demonstrate that the method originally devised for *Arabidopsis thaliana* can be applied to other Brassicaceaes, thus opening the possibility of adapting this method to a wider range of species and particularly those which are recalcitrant to standard *in vitro* methods.

Figure 1. *Arabidopsis* insertional mutagenesis developed at the INRA Versailles thanks to the infiltration method: multiplication in a greenhouse of 500 individual T-DNA insertion lines, out of the 60,000 produced to date.

When and Where Does *In Planta* Transformation Occur?

In infiltration methods, although transformation or transient expression may affect several somatic tissues (Rossi *et al.,* 1993; Bechtold *et al.,* 1993; Vaucheret, 1994; English *et al.,* 1997; Kapila *et al.,* 1997) the transformants obtained after self pollination of treated plants are systematically hemizygous for T-DNA insertions (Chang *et al.,* 1994; Feldmann, 1991; Bechtold *et al.,* 1993; Clough and Bent, 1998). It was therefore considered that "efficient transformation events" occur late in the development of the plant and affect either the germinal lines (male or female), the egg cells before their first division, or the young embryo cells (Feldmann *et al.,* 1994).

When seeds or wounded inflorescences are inoculated, it is hypothesized that the Agrobacteria spread to exposed surfaces and enter interstitial spaces, and are carried along in the plant during the development until flowering, at which point transformation could be induced.

Three groups (Ye et *al.,* 1999; Bechtold *et al.,* 2000; Desfeux *et al.,* 2000) tried to identify the cellular target of transformation and reached the conclusion that the female reproductive organs are this primary target but with the following differences. Ye *et al.* (1999) and Desfeux *et al.* (2000) used male-sterile and fertile counterparts in crosses after infiltration of one parent or the other. They observed no transformants in progeny after inoculation of the pollen donor and about 0.5% transformants after inoculation of the female donor. These experiments indicated that efficient transformation events occurred in female floral structures, but does not formally exclude transformation of the male gametophyte germinating in the female pistil or the male gamete during fertilization.

Bechtold *et al.* (2000) made use of a male gametophytic deficient mutant previously obtained by T-DNA insertional mutagenesis (Bonhomme *et al.,* 1998). This mutant is able to transmit this T-DNA (bearing a *nptII* gene) to the progeny only through its female gametes. After transformation by infiltration of this mutant with a second T-DNA vector bearing a *hpt* (hygromycin resistance) gene, the observation among the double transformants of a genetic linkage in coupling, but not in repulsion (with one exception), between the two markers demonstrated that the maternal chromosome set was the target of the T-DNA.

To identify the cellular target, the three groups used different promoter GUS constructs as described below.When the FMV promoter (constitutively expressed in pollen and ovules, Sanger *et al.,* 1990) fused to the *uidA*-intron reporter gene was used (Ye *et al.,* 1999), GUS staining after vacuum infiltration was observed in ovaries of unopened flowers, in up to 1% of pollen 3-5 days after infiltration, in up to 6% of the ovules 5-11 days after infiltration, and in seeds. Because the seed transformation frequency correlated very well with the frequency of GUS staining in ovules and with the results of the infiltration of the female parent in crosses, the authors suggested that the ovule is the target for vacuum infiltration transformation. No precise cellular target could be identified because GUS staining was observed in young ovaries, in ovules, in embryos, and entire seeds. Chimeric embryos were also observed.

Desfeux *et al.* (2000) observed no GUS staining in pollen after floral dipping with *Agrobacterium* containing LAT52::*uid A* (specifically expressed in pollen and young embryos, Twell *et al.* 1990) or ACT11::*uidA*-intron (expressed in pollen, ovules with a stronger expression after fertilization, and in embryos, Huang *et al.* 1997) gene fusions. On the other hand, GUS staining in entire ovules or localized within the ovules to the site of the embryo sacs were observed in unpollinated or pollinated flowers from 5-6 days after inoculation with the ACT11 promoter construct.

The SKP1 promoter is specifically expressed in both gametophytes (Drouaud *et al.*, 2000). Bechtold *et al.* (2000) observed no GUS staining in pollen in the case of a SKP1::*uidA* construct, after vaccum infiltration or floral dipping. In experiments in which several hundred transformants were selected from the same infiltrated inflorescences, no staining was observed in several thousands embryo sacs before fertilization. These observations suggest that the majority of efficient transformation events occur late in female gametophyte development.

Conclusions

In planta Agrobacterium-mediated transformation is the only reproducible *in planta* method but it is applicable only to a very limited number of species belonging to the Brassicaceae family. This method has been largely used for insertional mutagenesis (Choe and Feldman, 1998) and one can estimate that several hundred thousand independent T-DNA insertion lines have been generated by the *Arabidopsis* scientific community (Figure 1). This method represents a major tool for reaching an exhaustive knowledge of the function of the 26,000 genes of this species.

The simplicity of this method led many other laboratories to attempt its application to different and unrelated species but they have remained unsuccessful. Only recently some investigations have shed light on the targets of *Agrobacteria* and T-DNA. More insights and perhaps some key factors should result from a better understanding of the interaction of *Agrobacterium* with the fertilization processes of the *Arabidopsis* female gametophyte. In this perspective at least two factors have to be considered. Firstly to allow the access of *Agrobacteria* to ovules. It could be facilitated by the small size of *Arabidopsis* organs, but this does not appear to be a limiting factor since *B. rapa* (Pakchoï) is equally transformable although it has much larger flowers. (The fact that this species has a much less waxy epidermis than other Brassicas could be an advantage). In attempts to more widely apply the method, one must be able to check after treatment the presence of *Agrobacteria* in the vicinity of ovules. Secondly, the fact that successes seem to be reserved to the Brassicaceae family is in favour of the hypothesis that some substances they have in common could be very powerful inducers of virulence. These inducers, if they exist, remain to be determined.

A more general question is the possible relevance, from an evolutionary point of view, of this phenomenon in which *Agrobacterium* appears able to be able to transfer its T-DNA directly to the sexual progeny of a plant, at least in the case of *Arabidopsis*. In the seed transformation method (Feldmann and Marks, 1987), or just by spraying or immersing flowering plants (Bechtold *et al.*, 2000), i.e., conditions we can imagine to occur naturally, some transformants are obtained. May this sort of transfer exist in nature? Is it possible to reveal, in the genus *Arabidopsis*, as in the case of the genus *Nicotiana* (Furner *et al.*, 1986) some horizontal interkingdom genetic exchanges?

References

Avery, O.T., MacLeod, C. M., and McCarty, M. 1944. Studies on the chemical nature of the substance inducing transformation of pneumococcal types. J. Exp. Med. 79: 137-158.

Bechtold, N., Ellis, J., and Pelletier, G. 1993. *In planta Agrobacterium* mediated gene transfer by infiltration of adult *Arabidopsis thaliana* plants. C. R. Acad. Sci. Paris, Life Sciences. 316: 1194-1199.

Bechtold, N., and Pelletier, G. 1998. *In planta Agrobacterium* mediated transformation of adult *Arabidopsis thaliana* plants by infiltration. In: Arabidopsis protocols. J.M. Martinez-Zapater and J. Salinas, eds. Humana Press Inc. p. 256-266.

Bechtold, N., Jaudeau, B., Jolivet, S., Maba, B., Vezon, D., Voisin, R., and Pelletier G. 2000. The maternal chromosome set is the target of the T-DNA in the in planta transformation of *Arabidopsis thaliana*. Genetics. 155: 1875-1887.

Bent, A.F., Kunkel, B.N., Dahlbeck, D., Brown, K.L., Schmidt, R., Giraudat, J., Leung, J., and Staskawicz, B.J. 1994. RPS2 of *Arabidopsis thaliana*: a leucine-rich repeat class of plant disease resistance genes. Science. 265: 1856-1860.

Bent, A.F., and Clough, S.J. 1998. *Agrobacterium* germ-line transformation: transformation of *Arabidopsis* without tissue culture. Plant Molecular Biology Manual. B7:1-14.

Bent, A.F. 2000. *Arabidopsis in planta* transformation. Uses, mechanismes, and prospects for transformation of other species. Plant Physiol. 124: 1540-1547.

Bonhomme, S., Horlow, C., Vezon, D., De Laissardiere, S., Guyon, A., Ferault, M., Marchand, M., Bechtold, N., and Pelletier, G. 1998. T-DNA disruption of essential gametophytic genes is unexpectedly rare and cannot be inferred from segregation distortion criteria only in *Arabidopsis*. Mol. Gen. Genet. 260: 444-452.

Bouchez, D., Camilleri, C., and Caboche, M. 1993. A binary vector based on Basta resistance for in planta transformation of *Arabidopsis thaliana*. C. R. Acad. Sci. Paris, Life Sciences. 316: 1188-1193.

Cao, M. Q., Liu, F., Yao, L., Bouchez, D., Tourneur, C., Li, Y., and Robaglia, C. 2000. Transformation of Pakchoi (*Brassica rapa* L. *chinensis)* by *Agrobacterium* infiltration. Molecular Breeding. 6: 67-72.

Chang, S.S., Park, S.K., Kim, B.C., Kand B.J., Kim, D.U., and Nam, H.G. 1994. Stable genetic transformation of *Arabidopsis thaliana* by *Agrobacterium* inoculation *in planta*. Plant J. 5: 551-558.

Cheng, M., Fry, J.E., Pang, S., Zhou, H., Hironaka, C.M., Duncan, D.R., Conner, T.W., and Wan, Y. 1997. Genetic transformation of wheat mediated by *Agrobacterium tumefaciens*. Plant Physiol. 115: 971-980.

Choe, S., and Feldmann, K.A. 1998. T-DNA mediated gene tagging. In: Transgenic Plant Research. K. Lindsey, ed. Harwood Academic Publishers, Amsterdam. p. 57-73.

Chowrira, G.M., Akella, V., and Lurquin, P.F. 1995. Electroporation-mediated gene transfer into intact nodal meristems *in planta*. Generating transgenic plants without *in vitro* tissue culture. Mol. Biotechnol. 3: 17-23.

Christou, P. 1995. Strategies for variety-independent genetic transformation of important cereals, legumes and woody species utilizing particle bombardment. Euphytica. 85: 13-27.

Chung, M.H., Chen, M.K., and Pan, S.M. 2000. Floral spray transformation can efficiently generate *Arabidopsis* transgenic plants. Transgenic Research. 9: 471-476.

Clough, S.J., and Bent A. 1998. Floral dip: a simplified method for *Agrobacterium*-mediated transformation of *Arabidopsis thaliana*. Plant J. 16: 735-743.

Croughan, T.P., Destefano-Beltran, L ., Chu, O.R., and Jaynes J.M. 1988. Successful transformation of rice by direct DNA transfer. Plant Physiol. Supp. Abstract 821d. p.137.

de la Pena, A., Lörz, H., and Schell J. 1987. Transgenic rye plants obtained by injecting DNA into young floral tillers. Nature. 325: 274-276.

Desfeux, C., Clough, S.J., and Bent, A.F. 2000. Female reproductive tissues are the primary target of *Agrobacterium*-mediated transformation of *Arabidopsis thaliana*. Plant J. 16: 735-743.

Drouaud, J., Marrocco, K., Ridel, C., Pelletier, G., and Guerche, P. 2000. A *Brassica napus skp1*-like gene promoter drives GUS expression in *Arabidopsis thaliana* male and female gametophytes. Sex. Plant Reprod. 13: 29-35.

English, J., Davenport, G., Elmayan, T., Vaucheret, H., and Baulcombe, D. 1997. Requirement of transcription for homology-dependent virus resistance and trans-inactivation. Plant J. 12: 597-603.

Feldmann, K.A., and Marks, M.D. 1987. *Agrobacterium*-mediated transformation of germinating seeds of *Arabidopsis thaliana*: a non-tissue culture approach. Mol.Gen. Genet. 208: 1-9.

Feldmann, K.A. 1991. T-DNA insertion mutagenesis in *Arabidopsis*: mutational spectrum. Plant J. 1: 71-82.

Feldmann, K.A., Malmberg, R.L., and Dean, C. 1994. Mutagenesis in *Arabidopsis*. In: *Arabidopsis*. E. M. Meyerowitz and C.R. Sommerville, eds. Cold Spring Harbor Laboratory Press, Cold Spring Harbor, New York. p. 137-172.

Furner, I.J., Huffman, G.A., Amasino, R.M., Garfinkel, D.J., Gordon, M.P. and Nester, E.W. 1986. An *Agrobacterium* transformation in the evolution of the genus *Nicotiana*. Nature. 319: 422-427.

Hess, D. 1969. Versuche zur transformation an höheren pflanzen: induktion und konstante weitergabe der anthocyansynthese bei *Petunia hybrida*. Z. Pflanzenphysiol. 60: 348-358.

Hess, D. 1970. Versuche zur transformation an höheren pflanzen: genetische charakterisierung einiger mutmasslich transformierter pflanzen. Z. Pflanzenphysiol. 63: 31-43.

Hess, D. 1978. Genetic effects in *Petunia hybrida* induced by pollination with pollen treated with Lac transducing phages. Z. Pflanzenphysiol. 90: 119-132.

Hess, D. 1979. Genetic effects in *Petunia hybrida* induced by pollination with pollen treated with Gal transducing phages. Z. Pflanzenphysiol. 93: 429-436.

Hess, D. 1980. Investigations on the intra and interspecific transfer of anthocyanin genes using pollen as vectors. Z. Pflanzenphysiol. 98: 321-337.

Hess, D., and Dressler, K. 1989. Tumor transformation of *Petunia hybrida* via pollen cocultured with *Agrobacterium tumefaciens*. 1989. Botanica Acta. 102: 202-207.

Hess, D., Dressler, K., and Nimmrichter, R. 1990. Transformation experiments by pipetting *Agrobacterium* into the spikelets of wheat (*Triticum aestivum* L.). Plant Science. 72: 233-244.

Hess, D., Schneider, G., Lörz H., and Blaich, G. 1975. Investigations on the tumor induction in *Nicotiana glauca* by pollen transfer of DNA isolated from *Nicotiana langsdorffii*. Z. Pflanzenphysiol. 77: 247-254.

Huang, S., An, Y.Q., McDowel, J.M., McKinney, E.C., and Meagher, R.B. 1997. The *Arabidopsis ACT11* actin gene is strongly expressed in tissues of the emerging inflorescence, pollen, and developing ovules. Plant Mol. Biol. 33: 125-139.

Kapila, J., De Rycke, R., Van Montagu, M., and Angenon, G. 1997. An *Agrobacterium*-mediated transcient gene expression system for intact leaves. Plant Sci. 122: 101-108.

Katavic, V., Haughn, G.W., Reed, D., Martin, M., and Kunst L. 1994. *In planta* transformation of *Arabidopsis thaliana*. Mol. Gen. Genet. 245: 363-370.

Kleinhofs, A., Eden, F.C., Chilton, M.D., and Bendick, A.J. 1975. On the question of the integration of exogenous bacterial DNA into plant DNA. Proc. Nat. Acad. Sci. USA. 72: 2748-2752.

Korohoda, J., and Strzalka, K. 1979. High efficiency genetic transformation in maize induced by exogenous DNA. Z. Pflanzenphysiol. 94: 95-99.

Landridge, P., Brettschneider, R., Lazzeri, P., and Lörz, H. 1992. transformation of cereals via *Agrobacterium* and the pollen pathway: a critical assessment. Plant J. 2: 631-638.

Liu, F., Cao, M.Q., Yao, L., Robaglia, C., and Tourneur, C. 1996. In planta transformation of pakchoi (*Brassica campestris* L. ssp. *chinensis*) by infiltration of adult plants with *Agrobacterium*. Acta Hort. 467: 187-192.

Ledoux, L., and Huart, R. 1968. Integration and replication of DNA of *M. lysodeikticus* in DNA of germinating barley. Nature. 218: 1256-1259.

Ledoux, L., Huart, R., and Jacobs, M. 1971. Fate of exogenous DNA in *Arabidopsis thaliana*: translocation and integration. Eur. J. Biochem. 23: 96-108

Ledoux, L., Huart, R., and Jacobs, M. 1974. DNA-mediated genetic correction of thiamine-less *Arabidopsis thaliana*. Nature. 249: 17-21.

Luo, Z., and Wu, R. 1989. A simple method for the transformation of rice via the pollen tube pathway. Plant Mol. Biol. Rep. 6: 165-174.

Murashige, T., and Skoog, F. 1962. A revised medium for rapid growth and bioassays with tobacco tissue cultures. Physiol. Plant. 15: 473-497.

Nawa, S., Yamada, M.A., and Ohta, Y. 1975. Hereditary changes in *Capsicum annuum* L. induced by DNA treatment. Japan. J. Genetics. 50: 341-344.

Ohta, Y. 1986. High-efficiency genetic transformation of maize by a mixture of pollen and exogenous DNA. Proc. Natl. Acad. Sci. USA. 83: 715-719.

Picard, E., Jacquemin, J.M., Granier, F., Bobin, M., and Forgeois, P. 1988. Genetic transformation of wheat (*Triticum aestivum*) by plasmid DNA uptake during pollen tube germination. Proceedings of the 7th International Wheat Genetics Symposium, Cambridge, UK. p. 779-781.

Redei, G.P., Acedo, G., Weingarten, H., and Kier, L.D. 1977. Has DNA corrected genetically thiamineless mutants of *Arabidopsis*? In: Cell Genetics In Higher Plants. Dudits, D. *et al.*, eds. Akademiai Kiado, Budapest. p. 91-94.

Rossi, L., Escudero, J., Hohn, B., and Tinland, B. 1993. Efficient and sensitive assay for T-DNA dependent transient gene expression. Plant Mol. Biol. Rep. 11: 220-229.

Sanford,J.C., Skubik, K.A., and Reisch, B.I. 1985. Attempted pollen-mediated plant transformation employing genomic donor DNA. Theor. Appl. Genet. 69: 571-574.

Sanger, M., Daubert, S., and Goodmann, R.M. 1990. Characteristics of a strong promoter from figwort mosaic virus: comparison with the analogous 35S promoter from cauliflower mosaic virus and the regulated mannopine synthase promoter. Plant Mol. Biol. 14: 433-443.

Sautter, C., Leduc, N., Bilang, R., Iglesias, V.A., Gisel, A., Wen, X., and Potrykus, I. 1995. Shoot apical meristems as a target for gene transfer by microballistics. Euphytica, 85 : 45-51.

Senaratna, T., McKersie, B.D., Kasha, K., and Procunier, J.D. 1991. Direct DNA uptake during the imbibition of dry cells. Plant Science. 79: 223-228.

Soyfer, V.N. 1980. Hereditary variability of plants under the action of exogenous DNA. Theor. Appl. Genet. 58: 225-235.

Tague, B.W. 2001. Germ-line transformation of *Arabidopsis lasiocarpa*. Transgenic Res. 10: 259-267.

Tomilov, A.A., Tomilova, NB., Ogarkova, O.A., and Tarasov, V.A. 1999. Insertional mutagenesis of *Arabidopsis thaliana*: increase of germinating seeds transformation efficiency after sonication. Genetika. 35: 1214-1222.

Töpfer, R., Gronenborn, B., Schell, J. and Steinbiss, H.H. 1989. Uptake and transient expression of chimeric genes in seed-derived embryos. Plant Cell. 1: 133-139.

Touraev,A., Stöger, E., Voronin, V., and Heberle-Bors, E. 1997. Plant male germ line transformation. Plant J. 12: 949-956.

Trieu, A.T., Burleigh, S.H., Kardailsky, I.V., Maldonado-Mendoza,I.E., Versaw, W.K., Blaylock, L.A., Shin, H., Chiou, T-J., Katagi, H., Dewbre,

G.R., Weigel, D., and Harrison, M.J. 2000. Transformation of *Medicago truncatula* via infiltration of seedlings or flowering plants with *Agrobacterium*. Plant J. 22: 531-541.

Twell, D., Yamaguchi, J., and McCormick, S. 1990. Pollen-specific gene expression in transgenic plants: coordinate regulation of two different tomato gene promoters during microsporogenesis. Development. 109: 705-713.

Vaucheret, H. 1994. Promoter-dependant trans-activation in transgenic tobacco: kinetic aspects of gene silencing and gene reactivation. C. R. Acad. Sci. Paris, Life Sciences. 317: 310-323.

Ye, G-N., Stone, D., Pang, S-Z., Creely, W., Gonzalez, K., and Hinchee, M. 1999. Arabidopsis ovule is the target for *Agrobacterium in planta* vacuum infiltration transformation. Plant J. 19: 249-257.

Yoo, J., and Jung, G. 1995. DNA uptake by imbibition and expression of a foreign gene in rice. Physiol. Plant. 94: 453-459.

Zeng, J-Z., Wang, D-J., Wu, Y-Q., Zhang, J., Zhou, W-J., Zhu, X-P., and Xu N-Z. 1994. Transgenic wheat plants obtained with pollen tube pathway method. Science in China (series B). 37: 319-325.

Zhou, G., Weng, J., Zeng, Y., Huang, J., Qian, S., and Liu, G. 1983. Introduction of exogenous DNA into cotton embryos. Meth. Enzymology. 101: 433-481.

From: *Transgenic Plants: Current Innovations and Future Trends*
Edited by: C. Neal Stewart, Jr.

Chapter 5

Engineering the Chloroplast Genome for Biotechnology Applications

Henry Daniell and Muhammad Sarwar Khan

Abstract

Chloroplasts are known to have evolved from free-living microorganisms that were enslaved by plant cells. Chloroplasts are double membrane bound intracellular structures containing their own protein synthetic machinery but function in association with nucleus. Chloroplast transformation in higher plants is an extremely attractive approach for the development of transgenic traits that may be difficult to achieve via nuclear transformation. The chloroplast genome of higher plants is an ideal target for genetic engineering, since foreign proteins in transgenic chloroplasts accumulate to high levels due to the polyploid nature of chloroplast genome. Multiple genes may be expressed as polycistronic units in transgenic chloroplasts. Lack of pollen transmission in most cultivated crops results in natural gene containment, thus minimizing out-cross of transgenes to related weeds or crops and reducing the potential toxicity of transgenic pollen to non-target insects. Chloroplast transformation utilizes two targeting sequences that flank the foreign genes and insert them through homologous recombination at a precise, predetermined location in the chloroplast genome. This results in uniform transgene expression among transgenic lines and eliminates the "position effect" often observed in nuclear transgenic plants. Gene silencing, frequently observed in nuclear transgenic plants, has not been observed in transgenic chloroplasts. The ability to express foreign proteins at high levels in chloroplasts and chromoplasts, and to engineer foreign genes without the use of antibiotic resistant genes make this compartment ideal for development of edible vaccines or oral delivery of biopharmaceuticals. Moreover, the ability of chloroplasts to form disulfide bonds and to fold human proteins has opened the door to high-level production of biopharmaceuticals in plants. Furthermore, several foreign proteins or molecules observed to be toxic in the cytosol are non-toxic when compartmentalized within transgenic chloroplasts. This review highlights these and other recent accomplishments.

Introduction

The concept of chloroplast genetic engineering was developed in the mid-1980s when it became possible to regenerate green plants from albino protoplasts that had acquired green chloroplasts (as reviewed in Daniell and McFadden, 1987). Therefore, early investigations on chloroplast transformation in vascular plants focused on the development of *in organello* systems capable of efficient, prolonged protein synthesis and expression of foreign genes (Daniell and McFadden, 1987). However, after the development of the gene gun as a transformation device by John Sanford (as reviewed in Daniell, 1993), it was possible to transform intact chloroplasts. Chloroplast genetic engineering was first accomplished using autonomously replicating chloroplast vectors (Daniell *et al.,* 1990) and transient expression in chloroplasts (Daniell *et al.,* 1991), followed by stable integration of selectable marker genes into the tobacco chloroplast genome using the gene gun (Svab *et al.,* 1990; Svab and Maliga, 1993). However, only recently have genes conferring agronomically valuable traits been introduced into higher plants via chloroplast genetic engineering. For example, plants resistant to *Bacillus thuringiensis (Bt)*-sensitive insects were generated by integrating the *cryIAc* gene into the tobacco chloroplast genome (McBride *et al.,* 1995). Plants able to withstand even insects highly resistant to *Bt* were obtained by hyper-expression of the *cry2A* gene from engineered chloroplasts (Kota *et al.,* 1999). As reviewed below, chloroplasts have also been engineered recently to

Table 1. Fully sequenced plastid genome of organisms

Organism	Genome size (bp)	References
Algae		
Euglena gracilis	143,172	Hallick *et al.*, 1993
Cyanophora paradoxa	135,599	Stirewalt *et al.*, 1995
Odontella sinensis	119,704	Kowallik *et al.*, 1995
Porphyra purpurea	191,028	Reith and Munholland 1995
Chlorella vulgaris	150,613	Wakasugi *et al.*, 1997
Mesostigma viride	118,360	Lemieux *et al.*, 2000.
Land plants		
Marchantia polymorpha	121,024	Ohyama *et al.*, 1986
Nicotiana tabacum	155,844	Shinozaki *et al.*, 1986
Oryza sativa	134,525	Hiratsuka *et al.*, 1989
Epifagis virginiana	70,028	Wolfe *et al.*, 1992
Pinus thunbergii	119,707	Wakasugi *et al.*, 1994
Zea mays	140,387	Maier *et al.*, 1995
Arabidopsis thaliana	154,478	Sato *et al.*, 1999
Triticum estivum	134,540	Ogihara *et al.*, 2000

generate plants tolerant to bacterial and fungal diseases (DeGray *et al.*, 2001), drought (Lee et al., 2002) or herbicides (Daniell *et al.*, 1998; Iamtham and Day, 2000; Ye *et al.*, 2001; Lutz *et al.*, 2001).

Exceptionally high accumulation of foreign proteins (up to 46% of total soluble protein) has been reported recently for chloroplast transgenes (DeCosa *et al.*, 2001). This feature should make the compartment ideal for low-cost production of biopharmaceuticals. Several recent advances in chloroplast genome engineering make transgenic chloroplasts even more attractive as biopharmaceutical reactors (Daniell *et a*l., 2002). Such notable applications are, the stable expression of a protein based polymer having varied medical applications (Guda *et al.*, 2000), human somatotropin (Staub *et al.*, 2000), pharmaceutical peptides (DeGray *et al.*, 2001), human serum albumin (Daniell and Dhingra, 2002) antigens composed of functional oligomeric proteins with stable disulfide bridges (Daniell *et al.*, 2001a) and monoclonals (Daniell *et al.*, 2001b).

Chloroplast Genome Organization

Chloroplast number varies from 10-20 proplastids in meristematic cells, to several hundred in a mature leaf cell. All these provide suitable targets for transformation. Despite the variation in number and types of plastids, all

Table 2. Genes identified in land plant chloroplast genomes

Genes for the genetic system	
RNA polymerase	*rpoA, B, C1, C2*
Ribosomal RNAs	*rrn (16S, 23S, 4.5S, 5S)*
Ribosomal proteins	*rpl2, 14, 16, 20, 21, 22, 23, 32, 33, 36* *rps2, 3, 4, 7, 8, 11, 12, 14, 15, 16, 18, 19*
tRNAs	*trn* (30 tRNA genes)
Protease	*clpP*
Genes for the photosynthetic system	
Photosystem I	*psaA, B, C, I, J, M*
Photosystem II	*psbA, B, C, D, E, F, H, I, J, K, L, M, N, T*
Cytochrome b/f complex	*petA, B, D, G, L*
ATP synthase	*atpA, B, E, F, H, I*
Ribulose-1, 5-bisphosphate carboxylase	*rbcL*
Protochlorophyllide reductase	*chlB, chlL, chlN*
NADH-dehydrogenase	*ndhA, B, C, D, E, F, G, H, I, J, K*
Acetyl CoA carboxylase	*accD*
Genes of unknown function	*ycf1, 2, 3, 4, 5, 6, 9, 10*

plastids carry the same genetic material, i.e., a double stranded circular DNA molecule of 120 to 160 kb divided into singular small and large single copy regions separated by two inverted repeats (IR; Sugiura, 1992). Certain rare exceptions, such as legumes, lack IR. Chloroplasts generally contain 50-150 copies of the circular DNA molecules depending on the developmental stage. The chloroplast genome has been sequenced from a number of organisms (Table 1). It encodes about 120 proteins, both for protein synthesis as well as photosynthetic systems, and contains a number of open reading frames called *ycf*s (hypothetical chloroplast open reading frames) that have unknown functions (Table 2). In addition, several polypeptides encoded by the nuclear genome are synthesized on cytoplasmic ribosomes and are imported into chloroplasts.

Gene Expression in Chloroplasts

Chloroplast DNA of higher plants contains genes for the protein synthetic and photosynthetic functions of the organelle. Chloroplast gene expression has many similarities to gene expression in prokaryotes, from which chloroplasts are believed to have originated (as reviewed in Gruissem and Tonkyn, 1993). Chloroplast genes are transcribed by two distinct RNA polymerases: the plastid-encoded chloroplast RNA polymerase (PEP) and the nuclear-encoded plastid RNA polymerase (NEP). The RNA polymerase activity was characterized in its soluble and DNA-bound forms; the expression of the *rpo* genes was confirmed by detection of the corresponding subunits in highly purified enzyme preparations from maize chloroplasts (Hu *et al.,* 1991) and by western blotting of crude extracts from spinach chloroplasts with antisera raised against individual subunits (Briat *et al.,* 1987). Evidence for the expression of the *rpo* genes in the form of the corresponding RNA (Hudson *et al.,* 1988; Ruf and Kössel, 1988), and of specific proteins in soluble chloroplast extracts (Ruf and Kössel, 1988; Purton and Gray, 1989), confirmed that the core subunits of a chloroplast RNA polymerase are encoded in the chloroplast genome. The genes of the multi-subunit PEP core are encoded by the chloroplast genome, and are homologous to the eubacterial (*Escherichia coli*-like) DNA-dependent RNA polymerase α, β, and β', subunits (as reviewed in Gruissem and Tonkyn, 1993). The δ^{70}–like factors required for promoter recognition (Tiller *et al.,* 1991) are encoded in the nucleus (Tanaka *et al.,* 1997). The PEP promoters are similar to eubacterial δ^{70}-type promoters; the core is comprised of two hexameric sequences corresponding to the eubacterial –35 (TTGACA) and –10 (TATAAT) promoter elements. The hexamers are spaced 17-19 nucleotides apart; transcription initiates 5-7 nucleotides downstream of the –10 box sequence (as reviewed in Gruissem and Tonkyn, 1993).

The catalytic subunit of NEP is related to the mitochondrial and phage-type T3/T7 RNA polymerases (Lerbs-Mache, 1993). The PEP is putatively derived from the RNA polymerase of an ancestral bacterium. It is assumed that the phage-type chloroplast RNA polymerase evolved by duplication of the nuclear gene encoding the mitochondrial enzyme, and re-targeting of the gene product to the chloroplast (Hedtke *et al.,* 1997). Several chloroplast promoters have been shown to direct the transcription of genes in prokaryotic cells (Thompson and Mosig, 1988) and the expression of chloroplast genes in *E. coli* cell-free extracts allowed the identification and localization of many of the genes for photosynthetic components (Willey *et al.,* 1984). However, there are some cases where transcription initiation sites are not preceded by typical consensus elements (Gruissem *et al.,* 1986), e.g., *trnS* and *trnR* genes in spinach and a light-inducible *psbD/psbC* promoter in barley and wheat (Sexton *et al.,* 1992; Vera and Sugiura, 1995; Wada *et al.,* 1994). Recently, another non-consensus type promoter in the tobacco *atpB/E* operon has been reported (Kapoor *et al.,* 1997). Transcript levels from these promoters were decreased by cyclohexamide, a cytoplasmic protein synthesis inhibitor, providing further evidence for a non-consensus-type chloroplast promoter (Kapoor *et al.,* 1997). More recently, it has been shown that many chloroplast operons and genes have at least one promoter each for *E. coli*-like RNA polymerase and nuclear-encoded plastid RNA polymerase (Hajdukiewics *et al.,* 1997).

Transcription of chloroplast genes by one or both RNA polymerases reflects their functions. PEP transcribes Photosystem I and II genes; therefore it plays an important role in chloroplast gene expression. In the absence of the PEP, non-photosynthetic proplastids are still maintained, suggesting that essential housekeeping genes are transcribed by the NEP. Indeed, most non-photosynthetic genes have promoters for both RNA polymerases. Only a few genes are known to be transcribed exclusively from a NEP promoter. For example, *accD* encodes a subunit of the acetyl-CoA carboxylase in dicots (Hajdukiewics *et al.,* 1997), while this gene is encoded by nuclear DNA in monocots. The *rpoB* operon includes *rpoB, rpoC1* and *rpoC2* genes. Transcription of the *rpoB* operon is highly regulated (upregulated a thousand-fold), controlling the availability of the PEP. Thus, the *rpoB* NEP promoter plays a central role in chloroplast development. It is assumed that the phage-type plastid RNA polymerase evolved from the mitochondrial enzyme (Hedtke *et al.,* 1997) and the transcription of PEP genes by the NEP was probably a critical step in the nucleus taking control of the transcription of chloroplast genes, thereby fully enslaving chloroplasts.

Advancement in chloroplast genetic engineering technology has facilitated the study of chloroplast expression elements, including promoters and UTRs, analyzed by fusions with reporter genes (Monde *et al.,* 2000; Daniell *et al.,* 2002). Heterologous translation-enhancing sequences, for example, bacteriophage T7 gene 10 leader (G10L) sequence known to promote high-level protein accumulation in bacteria (Studier *et al.,* 1990), have been fused with reporter genes and shown to accumulate between ~16-18% of total soluble proteins (Kuroda and Maliga, 2001a; Khan and Maliga, 1999). Moreover, accumulation of an herbicide-tolerant EPSPS gene product was increased 200-fold when it was engineered to include the G*10*L sequence (Ye *et al.,* 2001). Such translational enhancement is attributed to complementarity between the Shine-Dalgarno (SD) sequence, sequences upstream of translation initiation codon and the anti- Shine-Dalgarno sequence (ASD) at the 3'-end of the chloroplast 16S rRNA (Kuroda and Maliga 2001b). The gene *10* coding sequence is partially complementary to the chloroplast 16SrRNA; however further increase in the complementarity by including chloroplast N-terminal amino acids sequence reduced translation efficiency. This reduced accumulation of protein is the result of decreased mRNA levels (up to 28%, Kuroda and Maliga, 2001a). Additional studies have confirmed this conclusion in which a translational fusion of the 14 N-terminal amino acids of the chloroplast *rbcL* and *atpB* genes to a reporter sequence resulted in different levels of reporter protein accumulation, but not attributable to differences in transcript abundance (Kuroda and Maliga, 2001b). Furthermore, silent mutations in the fused N-terminal coding sequences were found to decrease reporter protein accumulation without influencing RNA level (Kuroda and Maliga, 2001a). In a separate study, fusion of 14 amino acids N-terminal sequence of GFP with a synthetic EPSPS gene increased EPSPS protein accumulation (33-fold) without influencing steady-state transcript levels (Ye *et al.,* 2001). It is therefore desirable that when transgenes are required to be expressed at high levels in chloroplasts, that in addition to optimizing 5' UTR elements, sequences downstream of the translation initiation codon must be optimized as well.

Marker Genes for Chloroplast Engineering

Selectable Marker Genes to Manipulate Chloroplast Genomes

Several selectable marker genes confer resistance to drugs or herbicides in chloroplasts by detoxification of chemicals that affect chloroplast protein synthesis. A number of drugs like hygromycin, spectinomycin, streptomycin, kanamycin and phosphinothricin (PPT) encoded by *hph, aadA, nptII* and *bar* genes, respectively, have been used to select cells or plants carrying these genes during nuclear genetic engineering. Of these, genes encoding spectinomycin, streptomycin, and kanamycin resistance have been used to select cells with transformed chloroplast genomes (Svab and Maliga, 1993; Carrer *et al.*, 1993). During normal selection procedure, marker gene recipient cells go through phases of embryogenesis and organogenesis before regenerating into plants or organs, e.g., green shoots. During the time of embryogenesis and organogenesis, wild-type and transformed chloroplasts and chloroplast genome copies gradually sort out. Extended period of genome and organelle sorting yields chimeric plants consisting of sectors with wild-type and transformed cells. In the chimeric tissue, antibiotic resistance conferred by marker gene(s) is not cell-autonomous: transgenic and wild type sectors are both green due to phenotypic masking by the transgenic tissue. Such chimarism necessitates a second cycle of plant regeneration on a selective medium, depending upon the marker gene used. Another possibility to developing homoplasmic lines is to express lethal markers for selection that will clearly differentiate the marker gene non-recipient and recipient cells by killing cells containing untransformed chloroplasts (Daniell *et al.,* 2001c). Using non-lethal based selection genes like *aadA* may require additional reporter genes (Khan and Maliga, 1999). In the absence of a visual marker this is an inefficient process and ends up in heteroplastomic tissues and plants in most transformation events (Khan and Maliga, 1999). To facilitate this system, a gene providing visual selection and screening is required. Availability of such genes is discussed below.

Development of Marker-Free Chloroplast Transgenic Plants

Despite several advantages, one disadvantage of current chloroplast genetic engineering technology is the use of an antibiotic resistance gene, *aadA* (Svab and Maliga, 1993), as a selectable marker. The *aadA* gene product inactivates its selective agents, spectinomycin and streptomycin, by transferring the adenyl moiety of ATP to the antibiotics. Since these antibiotics are commonly used to control bacterial infection in humans and animals, there is concern that their over-use in other applications may lead to development of resistant bacteria (See Chapter 6). Therefore, several recent studies have explored strategies for engineering chloroplasts that are free of antibiotic-resistance markers. Recently, Daniell *et al.* (2001c) developed the spinach betaine aldehyde dehydrogenase (BADH) gene as an antibiotic free plant derived selectable marker to transform chloroplast genomes. The selection process involves conversion of toxic betaine aldehyde (BA) by the chloroplast-localized BADH enzyme to nontoxic glycine betaine, which also serves as an osmoprotectant (Rathinasabapathy *et al.,* 1994). Since the BADH enzyme is present only in chloroplasts of a few plant species adapted to dry and saline environments (Rathinasabapathy *et al.,* 1994; Nuccio *et al.,* 1999), it is suitable as a selectable marker in a large number of plants. Under BA selection, transformation of tobacco chloroplasts with a BADH-containing vector was 25-fold more efficient than selection for the vector *aadA* gene on spectinomycin. In addition, regeneration time was extremely rapid, with transgenic shoots appearing within 12 days in 80% of the leaf discs on BA selection, compared to 45 days in 15% of the discs on spectinomycin selection. This is the first report of genetic engineering of the chloroplast genome without the use of antibiotic resistance genes. Use of genes that are naturally present in plants for selection, in addition to gene containment, should ease public concerns over genetically modified crops.

Another approach to develop marker-free transgenic plants is to eliminate the antibiotic resistance gene after transformation. Strategies have been developed for removal of genes using either an engineered recombinase (Cre) and its target sites (*lox* sites) flanking the selection markers, or using endogenous chloroplast recombinases that delete the marker genes via engineered direct repeats (See Chapter 7). The former approach has been described recently by Hajdukiewicz *et al.* (2001). A green-fluorescent protein (GFP) reporter gene was introduced into the tobacco chloroplast genome. The GFP sequence was interrupted by the *aadA* selectable-marker gene flanked by *lox* sites, such that excision of *aadA* via recombination at the *lox* sites would generate intact GFP. By transforming the transgenic plants with a nuclear Cre transgene, the authors could monitor plastid gene excision events by looking for GFP fluorescence. This reporter gene activation assay demonstrated that excision of *aadA* occurred very early in plant development in all plastid genome copies. Subsequent elimination of the nuclear Cre gene by segregation resulted in marker-free plants stably expressing GFP in the chloroplast.

An alternative approach takes advantage of the endogenous chloroplast homologous recombination machinery. This approach was based on observations that short fragments of chloroplast DNA, engineered in transformation vectors as expression signals for chloroplast transgenes, may recombine with their homologous endogenous sequences to generate genome rearrangements or recombine with themselves and result in genetic instability for transgenes (Svab and Maliga, 1993; Eibl *et al.*, 1999). An experiment with *Chlamydomonas reinhardtii* showed that it is possible to exploit these recombination events to eliminate introduced selectable marker genes. Homologous recombination between two direct repeats, engineered to flank a selectable marker, enabled marker removal under non-selective growth conditions (Fischer *et al.*, 1996). Recently, Iamtham and Day (2000) applied a similar approach to tobacco chloroplasts. Two 174 bp direct repeats flanking the transgenes, *aadA* and *uidA*, were sufficient to generate *aadA*-free T_0 transplastomic lines while leaving a third unflanked transgene, *bar*, in the genome to confer herbicide resistance.

Reporter Genes to Manipulate Chloroplast Genomes

In addition to selectable marker genes, vital reporter genes undoubtedly contribute to the development of transformation technology by serving as tools for visual monitoring of transgene expression in transformed cells and tissues. A number of genes for example, *uidA*, *cat*, *nptII*, *ocs* and *luc* have been used to study gene expression, in plants as well as in animals, as reporters (Table 3). From these genes, *uidA* encoding β-glucuronidase has been successfully expressed transiently and stably in chloroplasts (Daniell *et al.*, 1991; Seki *et al.*, 1995; Staub and Maliga, 1994). However, histochemical

Daniell and Khan

Table 3. Foreign genes expressed in chloroplasts of higher plants. Only the first reports are included here, unless a variant or synthetic gene was used in subsequent investigations

Gene	Gene product	References
Selectable Markers and Reporters		
aadA	Aminoglycoside-3'-adenylyltransferase	Svab and Maliga, 1993
nptII	Neomycin phosphotransferase	Carrer *et al.*, 1993
codA	Cytosine deaminase	Serino and Maliga, 1997
BADH	Betaine aldehyde dehydrogenase	Daniell *et al.*, 2001c
uidA	β–glucoronidase	Staub and Maliga, 1994
cat	Chloramphenicol acetyl transferase	Daniell and McFadden, 1987; Daniell *et al.*, 1990
gfp	Green fluorescent protein	Hibberd *et al.*, 1998; Sidorov *et al.*, 1999
aadA:gfp	Selectable/screenable genes fusion product (FLARE-S)	Khan and Maliga, 1999
Plant Traits		
Herbicide resistance		
aroA	5-enol-pyruvyl shikimate-3- phosphate synthase	Daniell *et al.*, 1998; Ye *et al.*, 2001
bar	Bialaphos resistance gene	Iamtham and Day, 2000; Lutz *et al.*, 2001
Insect resistance		
Cry1Ac	*Bacillus thuringiensis* (Bt) toxin	McBride *et al.*, 1995
Cry2A	*Bacillus thuringiensis* (Bt) toxin	Kota *et al.*, 1999
Cry2Aa2 Operon	*Bacillus thuringiensis* (Bt) toxin	DeCosa *et al.*, 2001
Pathogen resistance		
*msi*99	Antimicrobial peptide (AMP)	DeGray *et al.*, 2001
Drought/salt (abiotic stress) tolerance		
tps1	Trehalose phosphate synthase	Lee *et al.*, 2001
BADH	Betaine aldehyde dehydrogenase	Daniell *et al.*, 2001c
Amino acid biosynthesis		
aroA	5-enol-pyruvyl shikimate-3- phosphate synthase	Daniell *et al.*, 1998; Ye *et al.*, 2001

ASA2	Anthranilate synthase (AS) α subunit	Zhang et al., 2001

Phytoremediation

merA	Mercuric ion reductase	Ruiz, 2001
merB	Organomercurial lyase	Ruiz, 2001

Non-plant Traits

Biopharmaceuticals

hST	Somatotropin, Human therapeutic protein	Staub *et al.*, 2000
HSA	Human serum albumin	Daniell and Dhingra, 2002
*msi*99	Antimicrobial peptide (AMP)	DeGray *et al.*, 2001
proinsulin	Human insulin α,β chains	Olga-Carmona Sanchez,2001
IFN α	Human interferon α5	Torres, 2001

Monoclonals

Guy's 13	For prevention of dental caries	Daniell et al., 2001b

Biomedical polymer

gvgvp 120	Bioelastic protein-based polymers	Guda *et al.*, 2000

Edible vaccines

*ctx*B	Cholera toxin B-subunit	Daniell *et al.*, 2001a

detection of GUS in plant organelles requires prolonged incubation because the envelope membranes of the organelles act as a selective barrier to substrate penetration. Moreover, chlorophyll bleaching is required to make GUS staining more effective in plants by using either ethanol or chloral hydrate. Furthermore, chemicals and physical procedures used in the staining disrupt cell ultrastructure (Baulcombe *et al.,* 1995).

The use of a non-toxic marker to identify transgenic cells after transformation is an effective procedure for discerning transformed cells/ organs and removing untransformed cells/tissues. The green fluorescent protein (GFP) of the jellyfish, *Aequorea victoria*, has recently been used as a reporter gene in plants (Baulcombe *et al.,* 1995; Chiu, *et al.,* 1996; Haseloff *et al.,* 1997). The *gfp* gene provides an easily scored cell-autonomous genetic marker in plants and has major uses in monitoring gene expression, protein localization and screening of transformation events at high resolution (Stewart 2001). GFP has been successfully expressed in *E. coli* and chloroplasts of tobacco, rice (Khan and Maliga, 1999), in different chloroplast types (Hibberd *et al.,* 1998) and in potato (Siderov *et al.,* 1999) using chloroplast as well as bacterial-specific expression signals (Hibberd *et al.,* 1998). The development of a gene encoding bifunctional proteins can minimize the use of multiple sets of promoters and terminators that result in chloroplast DNA fragment deletion through homologous recombination (due to the homology with chloroplast DNA or the presence of repeat sequences). Moreover such fusion genes facilitate both selection and screening of recipient cells. Such a fusion gene has been developed through translational fusion of the *aadA* and *gfp* genes called FLARE-S (Fluorescent antibiotic resistance enzyme conferring resistance to spectinomycin and streptomycin). This bifunctional protein facilitated chloroplast transformation in rice, where chloroplast transformation is not associated with a readily identifiable phenotype (Khan and Maliga, 1999).

Techniques to Deliver Foreign DNA into Chloroplast Genomes

In addition to the availability of efficient tissue culture systems, a method to deliver foreign DNA through the double membrane of the chloroplasts is needed. Successful transformation must facilitate efficient integration of the heterologous DNA without interfering with the normal function of the chloroplast genome and selection and screening of transgenic shoots. Work on gene expression (transient and or stable) in chloroplasts in several laboratories has provided different means of introducing foreign DNA into the chloroplast genome. These include polyethylene glycol (PEG) treatment (Golds *et al.,* 1993; Koop *et al.,* 1996), biolistic DNA delivery (Daniell *et al.,* 1990; Daniell *et al.,* 1991) and microinjection (Knoblauch *et al.,* 1999). Of these methods, only PEG treatment (Golds *et al.,* 1993; Koop *et al.,* 1996) and biolistic DNA delivery (Svab *et al.,* 1990; Svab and Maliga, 1993; Daniell *et al.,* 1998) have yielded stable chloroplast transformants. Of these two, only biolistic DNA delivery is being used to extend chloroplast transformation to different plastid types (Hibberd *et al.,* 1998; Khan and Maliga, 1999; Sidorov *et al.,* 1999; Ruf *et al.,* 2001).

Chloroplast Engineering Applications

Chloroplast Genetic Engineering for Insect Resistance
The use of commercial, nuclear transgenic crops expressing Bt toxins has escalated in recent years because of their advantages over traditional chemical insecticides. However, in crops with several target pests, each with varying

degrees of susceptibility to Bt (e.g., cotton), there is concern regarding the sub-optimal production of toxin, resulting in reduced efficacy and increased risk of the development of resistance to Bt by target insects. Another recently expressed environmental concern is the toxicity of transgenic pollen to non-target insects, including monarch butterflies (Losey *et al.,* 1999) although this study has been criticized as being premature and incomplete (Hodgson, 1999). Evolving levels of Bt resistance in insects could be dramatically reduced through engineering of the chloroplast genome in transgenic plants. An example of chloroplast genetic engineering include the expression of *Bt* genes, such as Cry2A, to express at high dosages of endotoxin to eliminate the possibility of low toxicity effects on insects resulting in the development of resistance (Kota *et al.,* 1999). Cry2A, is toxic to many caterpillars, such as the European corn borer and tobacco budworm, is quite different in structure/function from the Cry1A proteins (resulting in less cross resistance). Moreover, Cry2Aa2 protoxin expressing plants were used in insect essays and 100% mortality of Bt susceptible, Cry1A-resistant (20, 000 to 40,000-fold) and Cry2Aa2-resistant (330 to 393-fold) tobacco budworm, cotton bollworm and beet armyworm was observed (Kota *et al.,* 1999). Because Cry2A proteins are about half the size of Cry1A proteins, they should be expressed at even higher levels.

Chloroplast Genetic Engineering for Novel Pathways

In plant cells, nuclear mRNAs are translated monocistronically. This phenomenon poses a serious drawback when engineering multiple genes in plants (Bogorad, 2000). For example, in order to express the polyhydroxybutyrate polymer or Guy's 13 antibody, single genes were first introduced into individual transgenic plants, which were then hybridized to reconstitute the entire pathway or the complete protein (Navrath *et al.,* 1994; Ma *et al.,* 1995). Similarly, in a seven year-long effort, Ye *et al* (2001) introduced a set of three genes for a biosynthetic pathway that resulted in β-carotene expression in rice. In contrast to plant nuclear genes, most chloroplast genes are co-transcribed as polycistronic RNAs, which are subsequently processed to form mature translatable transcripts (Bogorad, 2000). Therefore, introduction of multiple chloroplast transgenes arranged in an operon should provide a unique opportunity to express entire pathways in a single transformation event. Most of the characters sought by plant breeders are quantitative traits that are controlled polygenetically.

The Bt *cry2Aa2* operon was used as a model system to test the feasibility of multigene operon expression in engineered chloroplasts (DeCosa *et al.* 2001). *Cry2Aa2* was the distal gene of a three-gene operon. The *orf* immediately upstream of *cry2Aa2* encoded a putative chaperonin that facilitates the folding of Cry proteins into stable cuboidal crystals (Ge *et al.*, 1998). Operon-derived Cry2Aa2 protein accumulated in transgenic chloroplasts as cuboidal crystals, to a level of 45.3% of the total soluble protein in mature leaves and remained stable even in senescing leaves (46.1%). These data suggest that large-scale production of foreign proteins within chloroplasts may be assisted by chaperonin-mediated folding. The crystals serve to enhance protein stability and facilitate single step purification of the protein. This event is the highest level of foreign gene expression ever reported in transgenic plants and was effective in killing insects that are exceedingly difficult to control (e.g., 10-day old cotton bollworm and beet armyworm larvae). Importantly, this study also showed that there was no insecticidal protein present in pollen, thus eliminating potential harm to non-target insects. This first demonstration of bacterial operon expression in transgenic chloroplasts opens the door to engineer novel pathways in a single transformation event. For example, we have engineered the mer operon to express mercuric ion reductase and organomercurial lyase to achieve phytoremediation via the chloroplast genome (Ruiz, 2001).

Chloroplast Genetic Engineering for Pathogen Resistance

On average, fungal and bacterial phytopathogens are responsible for 12-15% reduction of global crop production each year. The most disastrous case was that of bacterial blight which greatly damaged rice crop and resulted in up to 60% loss in India in the 1980s, and reached an epidemic stage in Pakistan in the 1990s. Moreover, in the 1990s there was a 10.1% loss of the global barley crop due to bacterial pathogens, resulting in a $1.9 billion loss. In 1994 in the United States, there was an estimated 44,600 metric ton reduction of soybean crop yield because of bacterial pathogens. In addition to causing severe losses in yield, fungal pathogens produce some of the deadliest mycotoxins, such as fumorisin and aflatoxin, thereby compromising food and feed safety. Many efforts have been made to combat these devastating pathogens. Plant breeding was introduced as a means to fight such diseases. However, results were limited due to the ability of the microbes to adapt to host defense mechanisms. Agrochemicals have been used but their application is limited by their toxicity to humans, animals, and the environment (Mourgues *et al.*, 1998). With the emergence of molecular biology, researchers have been able to elucidate many of the pathways and products in the plant response to phytopathogens. Anti-fungal peptides produced by various organisms have been cloned and studied. While progress made to date is promising for anti-fungal activity (Cary *et al.*, 2000), bacteria still maintain the ability to adapt to plant defenses. Therefore it is necessary to engineer the plant genome with such molecules that could bind to the bacterial outer surface and target them for lysis. Given the fact that the outer membrane is an essential and highly conserved part of all bacterial cells, it would seem highly unlikely that bacteria would be able to adapt (as they have against antibiotics) to resist the lytic activity of these peptides.

One class of molecules, termed AMP (anti-microbial peptide), that can control pathogens is an amphipathic alpha-helix that has affinity for negatively charged phospholipids commonly found in the outer-membrane of bacteria and fungi. Upon contact with membranes, individual peptides aggregate to form pores in the membrane, resulting in microbial lysis. Because of the concentration-dependent action of the AMP, it was expressed via the chloroplast genome to accomplish high dose delivery at the point of infection (De Gray *et al.,* 2001). Contrary to previous predictions, growth and development of the transgenic plants accumulating large amounts of AMP within transgenic chloroplasts was unaffected. *In vitro* assays with T_0, T_1 and T_2 plants confirmed that the AMP was expressed at high levels (21.5 to 43% of the total soluble protein) and retained biological activity against *Pseudomonas syringae,* a major plant pathogen. In addition, leaf extracts from transgenic plants inhibited growth of pregerminated spores of *Aspergillus flavus*, *Fusarium moniliforme* and *Verticillium dahliae* by more than 95% compared to untransformed plant extracts. *In situ* assays resulted in intense areas of necrosis around the point of infection in control leaves, while transformed leaves showed no signs of necrosis (200-800 µg of AMP was released from lysed chloroplasts at the site of infection). *In planta* assays with *Colletotrichum destructivum* showed necrotic anthracnose lesion in control leaves but not on transgenic leaves. T_1 *in vitro* assays against *Pseudomonas aeruginosa* (a multi-drug resistant human pathogen) displayed a 96% inhibition of growth as compared with untransformed plants. These results give a new option in the battle against phytopathogenic and drug-resistant human pathogenic microbes.

Chloroplast Genetic Engineering for Herbicide Resistance

Glyphosate is a potent, broad-spectrum herbicide that works by competitive inhibition of an enzyme in the aromatic amino acid biosynthetic pathway, 5-enol-pyruvyl shikimate-3- phosphate synthase (EPSPS). Unfortunately, like most commonly used herbicides, glyphosate does not distinguish crops from weeds, thereby restricting its usefulness. Engineering crop plants for resistance to the herbicide is a standard strategy to overcome this lack of herbicide selectivity. However, this approach raises the concern that escape of the engineered resistance gene via pollen may result in resistant weeds or may cause genetic pollution among other crops (Daniell, 1999; 2000; 2002). Engineering foreign genes through chloroplast genomes (maternally inherited in most of the cultivated crops) would provide a solution to this problem. Although pollen from plants with maternal chloroplast inheritance may contain metabolically active chloroplasts, the chloroplast DNA itself is lost during the process of pollen maturation and hence is not transmitted to the next generation (Daniell *et al.*, 2001; 2002). In addition, the target enzymes or proteins for many herbicides are compartmentalized within the chloroplast. The feasibility of engineering the chloroplast genome to confer herbicide resistance was first demonstrated by Daniell *et al.* (1998). By expressing a wild-type *EPSPS* nuclear gene from petunia within the tobacco chloroplast genome, plants resistant to glyphosate (10-fold higher than lethal concentration) were generated.

Recently, attempts have been made to further enhance the level of herbicide resistance by incorporating three different prokaryotic *EPSPS* genes (from *Achromobacter*, *Agrobacterium* and *Bacillus*) into the chloroplast genomes of tobacco (Ye *et al.*, 2001). By using different translational control sequences, authors succeeded in increasing expression for the *Agrobacterium EPSPS* (CP4) gene up to 250 fold over nuclear transgenic plants. Even though expression levels were enhanced more than 10,000 fold, tolerance to herbicide remained the same for native CP4 expressed in nuclear transgenic plants with expression level of 0.04% and chloroplast transgenic plants with expression level of 10% in total soluble protein. Similarly, increasing the amount of PAT (phosphinothricin acetyltransferase) by expressing bar genes (bacterial and synthetic) via the chloroplast genome did not improve resistance to PPT (phosphinothricin, the active ingredient in the herbicide Liberty) in transgenic plants (Lutz *et al.*, 2001). Moreover, high levels of expression did not help to improve direct selection of transgenic lines on medium containing PPT. The lack of correlation between expression levels and herbicide tolerance appears to be valid, irrespective of the compartment of expression, because EPSPS functions within chloroplasts whereas PAT functions in the cytosol. This observation is in contrast to other proteins where high level of protein expression proportionately enhances their function. For example, hyper-expression of Bt Cry proteins resulted in killing of even Bt-resistant insects (Kota et al. 1999) or concentration dependent killing of pathogenic bacteria and fungi by expressing anti-microbial peptides (DeGray *et al.*, 2001). One possible explanation may be that higher levels of expression of enzymes may not improve their activity proportionately.

Chloroplast Genetic Engineering to Overproduce Pharmaceutical Proteins

Although pharmaceutical proteins have been synthesized from plant nuclear transgenes, expression levels are generally low (Daniell *et al.*, 2001d). Expression of human proteins has been particularly disappointing: e.g. human interferon-α 0.000017% of fresh weight, HSA 0.02%, erythropoietin 0.0026%, human epidermal growth factor 0.001% of total soluble protein in nuclear transgenic plants (Daniell *et al.*, 2001d). Therefore, in order to exploit plant production of pharmacologically important proteins, it is important to increase expression levels of recombinant proteins. Chloroplasts, with their highly polyploid genomes offer an ideal compartment for overproduction of foreign proteins. An additional significant advantage in using chloroplasts is their potential to process eukaryotic proteins, including folding and formation of disulfide bridges. Such folding and assembly may minimize the need for highly expensive *in vitro* processing of pharmaceutical proteins after their extraction. For example, 60% of the total production cost of human insulin is associated with *in vitro* processing (Petrides *et al.,* 1995).

One example of pharmaceutical protein overproduction in chloroplasts is human somatotropin (hST), a therapeutic protein, is used in the treatment of hypopituitary dwarfism, Turner syndrome, chronic renal failure and HIV wasting syndrome. Staub *et al.* (2000) expressed hST in tobacco chloroplasts to levels between 0.2 and 7% of the total soluble protein in plants grown *in vitro*, depending upon the chloroplast translation signals used. Fusing ubiquitin to the hST sequence in place of its normal N-terminal signal peptide allowed *in vivo* processing of the protein to generate mature somatotropin. Chloroplast-expressed hST was shown to be correctly disulfide-bonded and biologically active, thereby demonstrating that chloroplasts have the necessary machinery to correctly fold heterologous eukaryotic proteins.

Human serum albumin (HSA) accounts for 60% of the total protein in blood serum and it is the most widely used intravenous protein in a number of human therapies. Pharmaceutically available HSA, currently produced from blood plasma, does not meet current needs. Therefore, other sources of HSA production are necessary. Regulation of HSA under the control of a Shine-Dalgarno sequence (SD), 5' *psbA* region or the *cry2Aa2* UTR resulted in different levels of expression in transgenic chloroplasts from seedlings: 0.8, 1.6, 5.9% of HSA in total soluble protein (tsp), respectively. On the other hand, a maximum of 0.02, 0.8 and 7.2% of HSA in tsp was observed in transgenic potted plants regulated by SD, *cry2Aa2* UTR or 5' *psbA* region respectively, demonstrating excessive proteolytic degradation, unless compensated by enhanced translation. The *psbA*-HSA expression was subject to developmental and light regulations, with the lowest expression observed in seedlings and maximal expression (11.1% tsp) under continuous illumination. HSA inclusion bodies observed in electron micrographs were so large that they increased the size of transgenic chloroplasts. Unfortunately, this led to gross underestimation of HSA because ELISA could be performed only in partially solubilized plant extracts. In spite of this underestimation, we report here the highest expression of a pharmaceutical protein and 500-fold higher than previous reports of HSA expression in nuclear transgenic plants. Formation of HSA inclusion bodies not only offered protection from proteolytic degradation but also provided a simple method of purification from other cellular proteins by centrifugation. HSA inclusion bodies could be readily solubilized to obtain monomeric form using appropriate reagents. This study also demonstrates that polycistrons are efficiently translated without an absolute requirement for transcript processing and that dicistrons are processed into monocistrons without the need for specialized intergenic sequences. The *cry2Aa2* UTR mediated expression in seedlings and chromoplasts, although as efficient as *psbA* 5' region, is independent of light regulation and should therefore facilitate expression of foreign genes in non-green tissues, thereby enabling oral delivery of pharmaceuticals. Regulatory regions used in this study should serve as a model system for enhancing expression of foreign proteins that are highly susceptible to proteolytic degradation and provide advantages in purification.

Chloroplast Genetic Engineering to Develop Edible Vaccines

Another major cause of the high cost of biopharmaceutical production is purification. For example, chromatography accounts for 30% of the production cost and 70% of the set-up cost for insulin production (Petrides *et al.,* 1995). Therefore, oral delivery of properly folded and fully functional biopharmaceuticals should cut down more than 90% of the production cost. Bioencapsulation of pharmaceutical proteins within plant cells offers protection against digestion in the stomach but allows successful delivery to the target tissues (Walmsley and Arntzen, 2000).

The B subunits of enterotoxigenic *Escherichia coli* (LTB) and cholera toxin of *Vibrio cholerae* (CTB) are candidate vaccine antigens. Recently, it has been shown that integration of an unmodified CTB coding sequence into the tobacco chloroplast genome resulted in accumulation of up to 4.1% of total soluble leaf protein (410-fold higher than the unmodified *LTB* gene expressed via the nuclear genome) as functional CTB oligomers (Daniell *et al.,* 2001a), Western blot analysis showed that chloroplast-synthesized CTB was properly assembled into oligomers and was antigenically identical to purified native CTB. In addition, binding assays confirmed that chloroplast-synthesized CTB binds to the intestinal membrane GM1-ganglioside receptor, indicating correct folding and disulfide bond formation of the plant-derived CTB pentamers. Chloroplast transgenic plants constitutively expressing CTB were morphologically indistinguishable from untransformed plants, in contrast to nuclear transgenic plants, which showed a stunted phenotype. Increased production of an efficient transmucosal carrier molecule and delivery system, like CTB, in transgenic chloroplasts makes plant-based oral pharmaceuticals commercially feasible. Furthermore, since the quaternary structure of many proteins is essential for their function, this investigation demonstrates the potential for other foreign multimeric proteins to be properly expressed and assembled in transgenic chloroplasts.

Chloroplast Genome Engineering of Edible Crop Plants
To exploit the chloroplast transformation technology to overproduce orally-delivered pharmaceuticals, it is important to extend this technology to edible crops. The recent development of chloroplast transformation for both potato (Sidorov *et al.*, 1999) and tomato (Ruf *et al.*, 2001) is particularly exciting. Both species are transformed using vectors originally developed for tobacco. This use is possible because of the chloroplast genome sequence conservation across many plant species. The expression levels of GFP in potato tissues was much higher (5% tsp) in leaves than in microtubers (0.05% tsp). Decreased expression in the edible part of the plant was possibly the result of lower copy numbers and overall attenuated transcription rates in amyloplasts compared to chloroplasts in leaves. The tomato plants with transformed chloroplasts yielded more encouraging results: high level expression of the chloroplast transgene (*aadA*) was observed in the chromoplasts of ripe red tomato fruits, which represented up to 50% of the expression levels in leaf chloroplasts (Ruf *et al.*, 2001). This study demonstrates for the first time the feasibility of expressing high-levels of foreign proteins in the plastids of edible plant organs.

Current Limitations of Chloroplast Genetic Engineering

One barrier to developing chloroplast transformation for cereals has been the regeneration of cereal plants from non-green embryonic cells rather than leaf mesophyll cells. Embryogenic cells contain only proplastids and not mature chloroplasts. It has been suggested that proplastids are smaller than the size of microprojectiles used for DNA delivery and therefore may pose problems in transformation experiments. However, Khan and Maliga (1999) demonstrated chloroplast transformation in rice using species-specific chloroplast transformation vectors equipped with the fluorescent selectable marker, FLARE-S. Shoots regenerated on streptomycin-containing media found positive for the transgene by PCR were inspected for fluorescence using confocal laser scanning microscopy. A mixture of transformed and wild-type chloroplasts was observed in leaf tissues. A major limitation of developing genetically stable homoplasmic lines is perhaps the low level of gene expression in non-green chloroplasts due to the low copy number of genomes and the low rate of protein synthesis. Identification of promoters and UTRs active in non-green tissues are necessary to overcome this limitation.

Another limitation to extend chloroplast transformation to other crop species is the lack of information on genome sequences necessary to locate intergenic sequences for the integration of transgenes. However, since the chloroplast genome is largely conserved across many plant species, the strategy of using targeting sequences from one species to transform the chloroplast genome of another was first developed (Daniell *et al.*, 1998). Both potato (Sidorov *et al.*, 1999) and tomato (Ruf *et al.*, 2001) were transformed using chloroplast vectors with tobacco targeting sequences and tobacco was transformed with petunia targeting sequences (DeGray *et al.*, 2001), although all these belong to the same family. However, the efficiency of transformation is less for heterologous chloroplast vectors than species-specific vectors (DeGray *et al.*, 2001).

Yet another challenge is delivering foreign DNA through the chloroplast envelope membranes. Of the two methods, gene gun mediated DNA delivery and PEG-treated protoplasts used to transform chloroplast genome of higher plants, only particle bombardment has proven to be very efficient. This may be due to the ability of metal particles to effectively penetrate and deliver foreign DNA across the double membrane of chloroplast envelope. PEG treatment of protoplasts has been used to stably transform tobacco chloroplast genomes (Koop *et al.*, 1996); although this is laborious, it is inexpensive when compared to particle bombardment and not as encumbered on the intellectual property. Yet another method of direct DNA delivery to plastids is microinjection. Although this method is effective for foreign gene expression in chloroplasts (Knoblauch *et al.*, 1999), application of this technology to stably transform higher plant chloroplast genomes awaits further investigation. Development of new methods of gene delivery systems into chloroplasts should facilitate further progress in this field.

References

Baulcombe, D.C., Chapman, S., and Cruz, S.S. 1995. Jellyfish green fluorescent protein as a reporter for virus infections. Plant J. 7: 1045-1053.

Bogorad, L. 2000. Engineering chloroplasts: an alternative site for foreign genes, proteins, reactions and products. Trends Biotechnol. 18: 257-263.

Briat, J-F., Bisanz-Seyer, C., Laulhere, J-P., Lerbs, S., Lescure, A-M., and Mache, R. 1987. The RNA polymerase from chloroplasts and its use for *in vitro* transcription of plastid genes. Plant Physiol. 25: 273.

Carmona-Sanchez, O. 2001. Expression of the human proinsulin gene in transgenic chloroplasts. M.S. thesis, University of Central Florida, Orlando, USA.

Carrer, H., Hockenberry, T.N., Svab, Z. and Maliga, P. 1993. Kanamycin resistance as a selectable marker for plastid transformation in tobacco. Mol. Gen. Genet. 241: 49-56.

Cary, J., Rajasekaran, K., Jaynes, J. and Cleveland, T. 2000. Transgenic expression of a gene coding a synthetic antimicrobial peptide results in inhibition of fungal growth *in vitro* and *in planta*. Plant Sci. 154: 171-181.

Chiu, W-L., Niwa, Y., Zeng, W., Hirano, T., Kobayashi, H., and Sheen, J. 1996. Engineered gfp as a vital reporter in plants. Curr. Biol. 6: 325-330.

Daniell, H. 1993. Foreign gene expression in chloroplasts of higher plants mediated by tungsten particle bombardment. Methods in Enzymol. 217: 536-556.

Daniell, H. 1997. Transformation and foreign gene expression in plants mediated by microprojectile bombardment. Meth. Mol. Biol. 62: 453-488.

Daniell, H. 1999. New tools for chloroplast genetic engineering. Nature Biotechnol. 17: 855-856.

Daniell, H. 2000. Genetically modified food crops: current concerns and solutions for next generation crops. Biotechnology and Genetic Engineering Reviews 17: 327-352.

Daniell, H. 2002. Molecular strategies for gene containment in GM crops. Nature Biotechnol. 20: 581-586.

Daniell, H. and McFadden, B.A. 1987. Uptake and expression of bacterial and cyanobacterial genes by isolated cucumber etioplasts. Proc. Natl. Acad. Sci. USA. 84: 6349-6353

Daniell, H., and Dhingra, A. 2002. Multigene engineering: dawn of an exciting new era in biotechnology. Current Opin. Biotechnol. 13: 136-141.

Daniell, H., Datta, R., Verma, S., Gray, S., and Lee, S.B. 1998. Containment of herbicide resistance through genetic engineering of the chloroplast genome. Nat. Biotechnol. 16: 345-348.

Daniell, H., Vivikananda, J., Nielsen, B.L., Ye, G.N., Tewari, K.K., and Sanford, J.C. 1990. Transient foreign gene expression in chloroplasts of cultured tobacco cells after biolistic delivery of chloroplast vectors. Proc. Natl. Acad. Sci. USA. 87: 88-92.

Daniell, H., Krishna, M., and McFadden, B.F. 1991. Transient expression of β-glucuronidase in different cellular compartments following biolistic delivery of foreign DNA into wheat leaves and calli. Plant Cell Rep. 9: 615-619.

Daniell, H., Lee, S.B., Panchal, T. and Wiebe, P.O. 2001a. Expression of cholera toxin B subunit oligomers in transgenic tobacco chloroplasts. Journal of Molecular Biology. 311: 1001-1009.

Daniell, H., Dhingra, A. and Fernandex-San Millan, A. 2001b. Chloroplast transgenic approach for the production of antibodies, biopharmaceuticals and edible vaccines. Proc. Int. Cong. Photosynth. Brisbane, S40-04, Australia

Daniell, H., Muthukumar, B. and Lee, S.B. 2001c. Marker free transgenic plants: engineering the chloroplast genome without the use of antibiotic selection. Curr Genet. 39:109-116

Daniell, H., Streatfield, S., and Wycoff, K. 2001d. Medical molecular farming: production of antibodies, biopharmaceuticals and edible vaccines in plants. Trends Plant Sci. 6: 219-226.

Daniell, H., Khan, M.S., and Allison, L. 2002. Milestones in chloroplast genetic engineering: an environmentally friendly era in biotechnology. Trends Plant Sci. 7: 84-91.

DeCosa, B., Moar, W., Lee, S.B., Miller, M. and Daniell, H. 2001. Hyper-expression of the Bt *Cry2Aa2* operon in chloroplasts leads to formation of insecticidal crystals. Nat. Biotechnol. 19: 71-74.

DeGray, G., Smith, F., Sanford, J. and Daniell, H. 2001. Hyper-expression of an antimicrobial peptide via the chloroplast genome to confer resistance against phytopathogenic bacteria. Plant Physiology. 127: 852-862.

Eibl, C., Zou, Z., Beck, A., Kim, M., Mullet, J. and Koop, H-U. 1999. *In vivo* analysis of plastid psbA, rbcL and rpl32 UTR elements by chloroplast transformation: tobacco plastid gene expression is controlled by modulation of transcript levels and translation efficiency. Plant J. 19: 333-345.

Fischer, N., Stampacchia, O., Redding, K. and Rachaix, J.D. 1996. Selectable marker recycling in the chloroplast. Mol. Gen. Genet. 251: 373-380.

Ge, B., Bideshi, D., Moar, W. and Federici, B. 1998. Differential effects of helper proteins encoded by the *cry*2A and *cry*11A operons on the formation of Cry2A inclusions in *Bacillus thuringiensis*. FEMS Microbiol. Lett. 165: 35-41.

Golds, T., Maliga, P., and Koop, H-U. 1993. Stable plastid transformation in PEG-treated protoplasts of *Nicotiana tabacum*. Bio/Technology 11: 95-97.

Gruissem, W., and Tonkyn, J.C. 1993. Control mechanisms of plastid gene expression. Crit. Rev. Plant Sci. 12: 19-55.

Gruissem, W., Elsner-Menzel, C., Latshaw, S., Narita, J.O., Schaffer, M. A., and Zurawski, G. 1986. A subpopulation of spinach chloroplast tRNA genes does not require upstream promoter elements for transcription. Nucl. Acids Res. 14: 7541-7556.

Guda, C., Lee, S.B. and Daniell, H. 2000. Stable expression of biodegradable protein based polymer in tobacco chloroplasts. Plant Cell Rep. 19, 257-262

Hajdukiewics, P., Allison, L.A., and Maliga, P. 1997. The two plastid RNA polymerases encoded by the nuclear and plastid compartments transcribe distinct groups of genes in tobacco plastids. EMBO J. 16: 4041-4048.

Hallick, R. B., Hong, L., Drager, R.G., Favreau, M. R., Monfort, A., Orsat, B., Spielmann, A., and Stutz, E. 1993. Complete sequence of *Euglena gracilis* chloroplast DNA. Nucl. Acids Res., 21: 3537-3544.

Haseloff, J., Siemering, K.R., Prasher, D.C., and Hodge, S. 1997. Removal of a cryptic intron and subcellular localization of green fluorescent protein are required to mark transgenic *Arabidopsis* plants brightly. Proc. Natl. Acad. Sci. USA. 94: 2122-2127.

Hedtke, B., Borner, T., and Weihe, A. 1997. Mitochondrial and chloroplast phage-type RNA polymerases in Arabidopsis. Science. 277: 809-811.

Hibberd, J.M., Linley, P.J., Khan, M.S., and Gray, J.C. 1998. Transient expression of green fluorescent protein in various plastid types following microprojectile bombardment. Plant J. 16: 627-632.

Hiratsuka, J., Shimada, H. Wittier, R. Ishibashi, T. Sakamoto, M. Mori, M. Kondo, C. Honji, Y. Sun, C. Meng, B. Li, Y. Kanno, A. Nishizawa, Y. Hirai, A. Shinozaki, K., and Sugiura, M. 1989. The complete sequence of rice (*Oryza sativa*) chloroplast genome: Intermolecular recombination between distinct tRNA genes accounts for a major plastid DNA inversion during the evolution of the cereals. Mol. Gen. Genet. 217: 185-194.

Hodgson, J. 1999. Monarch Bt-corn paper questioned. Nature Biotechnol. 17: 627.

Hu, J. Troxler, R. F., and Bogorad, L. 1991. Maize chloroplast RNA polymerase: the 78 kilodalton polypeptide is encoded by the plastid *rpoC1* gene. Nucl. Acids Res. 19: 3431-3434.

Hudson, G. S., Holton, T. A., Whitfeld, P. R., and Bottomley, W. 1988. Spinach chloroplast *rpoB/C* genes encode three subunits of the chloroplast RNA polymcrase. J. Mol. Biol. 200: 639-654.

Iamtham, S. and Day, A. 2000. Removal of antibiotic resistance genes from transgenic tobacco plastids. Nat. Biotechnol. 18: 1172-1176.

Kapoor, S., Suzuki, J.Y. and Sugiura, M. 1997. Identification and functional significance of a new class of non-consensus-type plastid promoters. Plant J. 11: 327-337.

Khan, M.S., and Maliga, P. 1999. Fluorescent antibiotic resistance marker to track plastid transformation in higher plants. Nature Biotechnol. 16: 627-632.

Knoblauch, M., Hibberd, M., Gray, J.C., and Bel, A.J.E.V. 1999. A galinstan expansion femtosyringe for microinjection of eukaryotic organelles and prokaryotes. Nature Biotechnol. 17: 906-909.

Koop, H-U., Steinmuller, K., Wagner, H., Robler, C., and Sacher, L. 1996. Integration of foreign sequences into the tobacco plastome via polyethylene glycol-mediated protoplast transformation. Planta 199: 193-201.

Kota, M., Daniell, H., Varma, S., Garczynski, F., Gould, F and Moar, W.J. 1999 Overexpression of the *Bacillus thuringiensis* Cry2A protein in chloroplasts confers resistance to plants against susceptible and Bt-resistant insects. Proc. Natl. Acad. Sci. USA. 96: 1840-1845

Kowallik, K.V., Stoebe, B. Schaffran, I. Kroth, P., and Freier U. 1995. The chloroplast genome of a chlorophyll A+C-containing *Odontella sinensis*. Plant Mol. Biol. Rep. 13: 336-342.

Kuroda, H. and Maliga, P. 2001a. Complementarity of the 16S rRNA penultimate stem with sequences downstream of the AUG destabilizes the plastid mRNAs. Nucleic Acids Res. 29: 970-975.

Kuroda, H. and Maliga, P. 2001b. Sequences downstream of the translation initiation codon are important determinants of the translation efficiency in chloroplasts. Plant Physiol. 125: 430-436

Lee, S.B., Kwon, H-B., Kwon, S-J., Park, S-C., Jeong, M-J.,Han, S-E., Daniell, H. and Byun, M-O. 2002. Drought tolerance conferred by the yeast trehalose-6 phosphate synthase gene engineered via the chloroplast genome. Transgenic Research. In press.

Lemieux, C. Otis, C. and Turmel, M. 2000. Ancestral chloroplast genome in *Mesostigma viride* reveals an early branch of green plant evolution. Nature 403: 649-652.

Lerbs-Mache, S. 1993. The 110-kDa polypeptide of spinach plastid DNA-dependent RNA polymerase: Single-subunit enzyme or catalytic core of multimeric enzyme complexes? Proc. Natl. Acad. Sci. USA. 90: 5509-5513.

Losey, J.E., Rayor, L.S., Carter, M.C. 1999. Transgenic pollen harms monarch larvae. Nature 399: 214.

Lutz, K.A., Knapp, J.E. and Maliga, P. 2001. Expression of bar in the plastid genome confers herbicide resistance. Plant Physiol. 125: 1585-1590.

Ma, J., Hiatt, A., Hein, M., Vine, N., Wang, F., Stabila, P., Van Dolleweerd, C., Mostov, K. and Lehner, T. 1995. Generation and assembly of secretory antibodies in plants. Science. 268: 716-719.

Maier, R.M., Neckermann, K. Igloi, G.L., and Kössel H. 1995. Complete sequence of the maize chloroplast genome - gene content, hotspots of divergence and fine turning of genetic information by transcript editing. J. Mol. Biol. 251: 614-628.

McBride, K. E., Svab, Z., Schaaf, D.J., Hoga, P.S., Stalker, D.M. and Maliga, P. 1995. Amplification of a chimeric *Bacillus* gene in chloroplasts leads to an extraordinary level of an insecticide protein in tobacco. Biotechnology 13: 362-365.

Monde, R.A., Greene, J.C., and Stern, D.B. 2000. The sequence and secondary structure of the 3'-UTR affect 3'-end maturation, RNA accumulation, and translation in tobacco chloroplasts. Plant Mol. Biol. 44: 529-542.

Mourgues, F.; Brisset, M.N.; Cheveau, E. 1998. Strategies to improve plant resistance to bacterial diseases through genetic engineering. Trends Biotechnol. 16: 203-210.

Navrath, C., Poirier, Y. and Somerville, C. 1994. Targeting of the polyhydroxybutyrate biosynthetic pathway to the plastids of *Arabidopsis thaliana* results in high levels of polymer accumulation. Proc. Natl. Acad. Sci. USA. 91: 12760-12764.

Nuccio, M.L., Rhodes, D., McNeil, S.D. and Hanson, A.D. 1999. Metabolic engineering of plants for osmotic stress tolerance. Curr. Opinion Plant Biol. 2:128-134.

Ogihara, Y., Isono, K., Kojima, T., Endo, A., Hanaoka, M., Shina, T., Terachi, T., Utsugi, S., Murata, M., Mori, N., Takumi, S., Ikeo, K., Gojobori, T., Murai, R., Murai, K., Matsuoka, Y., Ohnishi, Y., Tajiri, H. and Tsunewaki, K. 2000. Chinese spring wheat (*Triticum estivum* L.) chloroplast genome: complete sequence and contig clones. Plant Mol. Biol. Rep. 18: 243-253.

Ohyama, K., Fukuzawa, H. Kohchi, T. Shirai, H. Sano, T. Sano, S. Umesono, K. Shiki, Y. Takeuchi, M. Chang, Z. Aota, S. Inokuchi, H., and Ozeki, H. 1986. Chloroplast gene organization deduced from complete nucleotide sequence of liverwort *Marchantia polymorpha* chloroplast DNA. Nature. 322: 572-574.

Petrides, D., Sapidou, E., Calandranis, J.. 1995. Computer-Aided process analysis and economic evaluation for biosynthetic human insulin production-A case Study. Biotechnology and Bioengineering. 48: 529-541.

Purton, S., and Gray, J.C. 1989. The plastid rpoA gene encoding a protein homologous to the bacterial RNA polymerase alpha subunit is expressed in pea chloroplasts. Mol. Gen. Genet. 217: 77-84.

Rathinasabapathy, B., McCue, K.F., Gage, D.A. and Hanson, A.D. 1994. Metabolic engineering of glycine betaine sythesis: Plant betaine aldehyde dehydrogenases lacking typical transit peptides are targeted to tobacco chloroplasts where they confer aldehyde resistance. Planta. 193: 155-162

Rieth, M., and Munholland, J. 1995. Complete nucleotide sequence of the *Porphyra purpurea* chloroplast genome. Plant Mol. Biol. Rep. 13: 333-335.

Ruf, M., and Kössel, H. 1988. Structure and expression of the gene coding for the a-subunit of DNA-dependent RNA polymerase from the chloroplast genome of *Zea mays*. Nucl. Acids Res. 16: 5741-5754.

Ruf, S., Hermann, M., Berger, I.J., Carrer, H. and Bock, R. 2001. Stable genetic transformation of tomato plastids: high-level foreign protein expression in fruits. Nature Biotechnol. 19: 870-875.

Ruiz, O. 2001 Phytoremediation of mercury and organomercurials via chloroplast genetic engineering, M.S. thesis, University of Central Florida, Orlando,USA.

Sato, S. Nakamura, Y. Kaneko, T. Asamizu, E. and Tabata, S. 1999. Complete structure of the chloroplast genome of *Arabidopsis thaliana*. DNA Res. 6: 283-290.

Seki, M., Shigemoto, N., Sugita, M., Sugiura, M., Koop, H-U., Irifune, K., and Morikawa, H. 1995. Transient expression of β-glucuronidase in plastids of various plant cells and tissues delivered by a pneumatic particle gun. J. Plant Res. 108: 235-240.

Serino, G. and Maliga, P. 1997 A negative selection scheme based on the expression of cytosine deaminase in plastids. Plant J. 12, 697-701

Sexton, T.B., Christopher, D.A., and Mullet, J.E. 1992. Light induced switch in barley *psbD-psbC* promoter utilization: A novel mechanism regulating chloroplast gene expression. EMBO J. 9: 4485-4494.

Shinozaki, K., M. Ohme, M. Tanaka, T. Wakasugi, T. Hayashida, N. Zaita, J. Chunwongse, J. Obokata, K. Yamaguchi-Shinozaki, C. Ohto, K. Torazawa, B.Y. Meng, M. Sugita, H. Deno, T. Kamogashira, K. Yamada, J. Kusuda, F. Takaiwa, A. Kato, N. Tohdoh, H. Shimada, and M. Sugiura. 1986. The complete nucleotide sequence of the tobacco chloroplast genome; its gene organization and expression. EMBO J., 5: 2043-2049.

Sidorov, V.A., Kasten, D., Pang, S.Z., Hajdukiewics, P.T.J., Staub, J.M. and Nehra, N.S. 1999. Stable chloroplast transformation in potato: use of green fluorescent protein as a plastid marker. Plant J. 19: 209-216.

Staub, J.M. and Maliga, P. 1994. Translation of psbA mRNA is regulated by light via the 5'-untranslated region in tobacco plastids. Plant J. 6: 547-553.

Staub, J.M., Gracia, B., Graves, J., Hajdukiewics, P.T.J., Hunter, P., Paradkar, V., Schlitter, M., Carrol, J.A., Spatola, L., Ward, D., Ye, G. and Russell, D.A. 2000. High-yield production of a human therapeutic protein in tobacco chloroplasts. Nat. Biotechnol. 18: 333-338.

Stewart, C.N. Jr. 2001. The utility of green fluorescent protein in transgenic plants. Plant Cell Rep. 20: 376-382.

Stirewalt, V.L., Michalowski, C.B. Löffelhardt, W. Bohnert, H.J. and Bryant, D.A. 1995. Nucleotide sequence of the cyanelle genome from *Cyanophora paradoxa*. Plant Mol. Biol. Rep. 13: 327-332.

Studier, F.W., Rosenberg, A.H., Dunn, J.J. and Dubendorff, J.W. 1990. Use of T7 RNA polymerase to direct expression of cloned genes. Methods Enzymol. 185: 60-89.

Sugiura, M. 1992. The chloroplast genome. Plant Mol. Biol. 19: 149-168.

Svab, Z., and Maliga, P. 1993. High frequency plastid transformation in tobacco by selection for a chimeric aadA gene. Proc. Natl. Acad. Sci. USA. 90: 913-917.

Svab, Z., Hajduckiewicz, P., and Maliga, P. 1990. Stable transformation of plastids in higher plants. Proc. Natl. Acad. Sci. USA. 87: 8526-8530.

Tanaka, K., Tozawa, Y., Mochizuki, N., Shinozaki, K., Nagatani, A., Wakasa, K., and Takahashi, H. 1997. Characterization of three cDNA species encoding plastid RNA polymerase sigma factor heterogeneity in higher plant plastids. FEBS Lett. 413: 309-313.

Thompson, R.J., and Mosig, G. 1988. Integration host factor (IHF) represses a Chlamydomonas chloroplast promoter in *E. coli*. Nucl. Acids Res. 16: 3313-3326.

Tiller, K., Eisermann, A., and Link, G. 1991. The chloroplast transcription apparatus from mustard. Evidence for three different transcription factors which resemble bacterial sigma factors. Eur. J. Biochem. 198: 93-99.

Torres, M. 2001. Expression of interferon α 5 in transgenic tobacco chloroplasts. M.S. thesis, University of Central Florida, Orlando, USA.

Vera, A., and Sugiura, M. 1995. Chloroplast rRNA transcription from structurally different tandem promoters: an additional novel type promoter. Curr. Genet. 27: 280-284.

Wada, T., Tunoyama, Y., Shiina, T., and Toyoshima, Y. 1994. *In vitro* analysis of light-induced transcription in the wheat *psbD/C* gene cluster using plastid extracts from dark grown and short-term-illuminated seedlings. Plant Physiol. 104: 1259-1267.

Wakasugi, T., Tsudzuki, J. Ito, S. Nakashima, K. Tsudzuki, T., and Sugiura, M. 1994. Loss of all *ndh* genes as revealed by sequencing the entire chloroplast genome of black pine, *Pinus thumbergii.* Proc. Natl. Acad. Sci. USA. 91: 9794-9798.

Wakasugi, T., Nagai, T., Kapoor, M., Sugita, M., Ito, M., Ito S., Tsudsuki, J., Nakashima, K., Tsudsuki, T., Suzuki, Y., Hamada, A., Onta, T., Inamura, A., Yoshinaga, K., and Sugiura, M. 1997. Complete nucleotide sequence of the chloroplast genome from the green alga *Chlorella vulgaris*: the existence of genes possibly involved in chloroplast division. Proc. Natl. Acad. Sci. USA. 94. 5967-5972.

Walmsley, A.M. and Arntzen, C.J. 2000. Plants for delivery of edible vaccines. Curr. Opin. Biotechnol. 11: 126-129.

Willey, D.L., Howe, C.J., Auffert, A.D., Bowman, C.M., Dyer, T.A., and Gray, J.C. 1984. Location and nucleotide sequence of the gene for cytochrome f in wheat chloroplast DNA. Mol. Gen. Genet. 194: 416-422.

Wolfe, K.H., Morden, C.W., and Palmer, J.D. 1992. Function and evolution of a minimal plastid genome from a nonphotosynthetic parasitic plant. Proc. Natl. Acad. Sci. USA. 89: 10648-10652.

Ye, G-N., Hajdukiewics, P.T.J., Broyles, D., Rodriguez, D., Xu, C.E., Nehra, N.S. and Staub, J.M. 2001. Plastid-expressed 5-enolpyruvylshikimate-3-phosphate synthase genes provide high level glyphosate tolerance in tobacco. Plant J. 25: 261-270.

Ye, X., Al-Babili, S., Kloti, A., Zhang, J., Lucca, P., Beyer, P. and Potrykus, I. 2000. Engineering the provitamin A (β-carotene) biosynthetic pathway into (carotenoid-free) rice endosperm. Science. 287: 303-305.

Zhang, X.H., Brotherton, J.E., Widhom, J.M., Portis, A.R. 2001. Targeting a nuclear anthranilate synthase α subunit gene to the tobacco plastid genome results in enhanced tryptophan biosynthesis. Return of a gene to its pre-endosymbiotic origin. Plant Physiol. 127: 131-141

From: *Transgenic Plants: Current Innovations and Future Trends*
Edited by: C. Neal Stewart, Jr.

Chapter 6

Antibiotic Resistance Genes in Transgenic Plants: Their Origins, Undesirability and Technologies for Their Elimination From Genetically Modified Crops

Anil Day

Abstract

Plant transformation technologies use antibiotic resistance genes as markers to identify the small fraction of transgenic cells that have taken up trait genes. In addition to plant selectable marker genes, vector-localised genes such as the ampicillin resistance *bla*(TEM1) gene, can also integrate into the chromosomes of transgenic plants. Integration of vector sequences is particularly problematic when whole plasmids integrate into plant nuclear DNA following their transfer into cells by artificial DNA delivery methods such as particle bombardment. Microbial resistance to antibiotics threatens the success of infectious disease treatment and prevention in the 21st century. While the risk of horizontal transfer of antibiotic resistance genes is minuscule, their elimination from genetically manipulated crops provides a simple solution for ending the continuing debate over the likelihood of pathogen acquisition of plant-derived antibiotic resistance genes. To avoid the presence of antibiotic resistance genes in transgenic crops, they can be removed once they have served their purpose or they can be replaced with alternative marker genes. These two approaches are not mutually exclusive and can be combined where needed to avoid safety evaluations on each new marker gene.

This chapter reviews technologies for removing antibiotic resistance genes from transgenic plants and describes an expanding list of alternative marker genes that do not require antibiotic selection. Plastid engineering illustrates the ease with which both antibiotic resistance genes and vector sequences can be removed from plants using homologous recombination. Efficient marker gene excision technologies and alternative marker genes combine for a better toolkit for the next generation of transgenic crops. This toolkit will facilitate multiple rounds of transformation with the best marker for a particular crop and allow the removal of all excess foreign DNA from a crop. As a consequence the focus of attention will shift from the marker genes to the all important trait genes that are responsible for the added value of genetically manipulated crops.

Genetically Manipulated Crops: Potential and Current Limitations

The success of genetically modified (GM) plants is due to the selection of foreign genes that confer novel and desirable traits on crops. However, it is clear that the widespread use of GM technology in agriculture will not only be determined by the engineered traits themselves but also to a large extent by the overall precision of methods used to transfer foreign genes to plants. The present generation of gene transfer technologies often result in excess foreign DNA sequences in GM crops. This superfluous DNA is not required for the functionality of a trait gene and can even be highly undesirable in the final crop. In particular, the unnecessary presence of antibiotic resistance genes in GM crops has been criticised. Within the European Economic Community, directive 2001/18/EC dictates the gradual elimination of antibiotic markers (that pose a risk to human and animal health) in genetically manipulated organisms by the end of 2004 for commercial releases and the end of 2008 for research purposes. This chapter will discuss the origin(s) of antibiotic resistance genes in transgenic plants, the perceived risks associated with their widespread dissemination into the environment, and review the range of current and emerging methodologies that can be used to eliminate these controversial genes from GM crops.

Gene Transfer Methodologies that Introduce Antibiotic Resistance Genes into GM Crops

The reason(s) behind the presence of antibiotic resistance genes in GM crops requires an understanding of gene transfer technologies. Genetic engineering of plants can be broken down into three basic steps: i) cloning trait genes in *Escherichia coli* plasmids, ii) delivery of trait genes into plant cells, iii) integration of trait genes into plant DNA and identification of transformed cells.

Cloning Trait Genes in *Escherichia coli* Plasmids

The first step involves propagating and amplifying the trait gene in the bacterium *E. coli*. To achieve this the trait gene is inserted into a cloning vector, which is usually a plasmid. Plasmids are small circular self-replicating DNA molecules that are easily purified from bacteria. Two vector DNA components are necessary for plasmid maintenance in *E. coli*: first, DNA sequences necessary for plasmid replication and partition at cell division, and second, a bacterial marker gene, which is expressed in *E. coli*. Bacterial marker genes are needed because commonly-used plasmid cloning vectors are unstable and would normally be lost. When bacterial marker genes are combined with selective growth regimes, the only bacterial cells capable of growth are plasmid-containing cells that can express marker genes. Prokaryotic regulatory elements, such as promoters and transcriptional terminators, are used to express bacterial marker genes in *E. coli*. Such genes are unlikely to be expressed well in plant nuclei where eukaryotic regulatory elements are required. Invariably, antibiotic resistance genes are used to maintain plasmids in *E. coli* (Table 1). Common bacterial plasmid cloning vectors, such as the pUC and pBluescript derivatives of pBR322, contain the *bla*(TEM1) gene as the selectable marker (Yanisch-Perron *et al.*, 1985; Short *et al.*, 1988). The *bla*(TEM1) gene product, β-lactamase, breaks down the following β-lactam antibiotics: the narrow-spectrum cephalosporins, cefamandole, and cefoperazone and all the anti-gram-negative-bacterium penicillins except temocillin. These antibiotics inhibit the synthesis of bacterial cell walls. Typically, plasmids are maintained by growing laboratory-adapted *E.coli* strains in nutrient broth containing 25 to 200 micrograms of

Table 1. Examples of vector-localised antibiotic resistance genes found in *E. coli* cloning plasmids. Binary Ti vectors for *Agrobacterium*-based transformation (Bin-Ti) and suppliers of commercially available plasmids are indicated. [1]Plasmids with more than one vector resistance gene.

Marker Gene	Gene Product	Selective Antibiotic	Cloning vectors	References
bla (TEM1)	β-lactamase	ampicillin	pUC series, pBlueScript, pBR322[1], pBR325[1], pUN121[1], pACYC177[1], pAT153[1], pPCV001 (Bin-Ti), pC22 (Bin-Ti)[1], pET (Stratagene), LITMUS (New England Biolabs), pRSET (Invitrogen)	Yanisch-Perron *et al.*, 1985; Short *et al.*, 1988; Bolivar and Backman, 1979; Twigg and Sherratt, 1980; Hellens *et al.*, 2000
tetA	tetracycline efflux protein	tetracycline	pBR322[1], pBR325[1], pACYC184[1], pAT153[1], pUN121[1], pJJ1881 (Bin-Ti)	Bolivar and Backman, 1979; Twigg and Sherratt, 1980; Nilsson *et al.*, 1983; Hellens *et al.*, 2000
cat	chloramphenicol acetyl transferase	chloramphenicol	pBR325[1], pACYC184[1] pPZP111 (Bin-Ti)	Bolivar and Backman, 1979; Hellens *et al.*, 2000
nptI [*aph(3')-I*]	neomycin phosphotransferase	kanamycin	pACYC177[1] pGreen 0029 (Bin-Ti)	Bolivar and Backman, 1979 Hellens *et al.*, 2000
nptIII [*aph(3')-III*]	neomycin phosphotransferase	kanamycin	pBIN 19 (Bin-Ti)	Bevan, 1984
aadA [*ant 3''*-1a]	aminoglycoside 3" adenyl transferase	streptomycin or spectinomycin	pOMEGA pC22 (Bin-Ti)[1]	Prentki and Krisch, 1984; Hellens *et al.*, 2000

ampicillin per milliliter of broth. It is not uncommon to find two or three antibiotic resistance genes in cloning vectors. Inserting foreign DNA into one of these genes knocks it out, rendering bacteria sensitive to the corresponding antibiotic, thereby allowing rapid identification of plasmids that have taken up foreign DNA. For example, pBR322 contains *bla*(TEM1) plus a tetracycline resistance gene whilst pBR325 contains genes conferring resistance to ampicillin *bla*(TEM1), chloramphenicol and tetracycline (Bolivar and Backman, 1979). Commonly used vector-localised markers for *Agrobacterium* based plant transformation (see section below) encode resistance to kanamycin, gentamycin, spectinomycin plus streptomycin, and tetracycline (Table 1, plasmids marked Bin-Ti; Hellens *et al.*, 2000). Laboratory-adapted *E. coli* strains contain deleterious mutations to minimize the possibility of their survival in the environment following accidental release into the environment.

Delivery of Trait Genes into Plant Cells

There are two general approaches for delivering a trait gene, propagated on *E. coli* plasmids, into plant cells. In the first approach, small plasmids, usually containing two to three foreign genes, are purified away from the 100 to 500 fold larger *E. coli* genome, which encodes about four thousand genes (Blattner *et al.*, 1997). Purified plasmid DNA is then transformed into plant cells using a variety of methods. Electroporation, chemicals (often calcium or polyethylene glycol) and, less frequently, microinjection are used to introduce DNA into protoplasts (Potrykus, 1991). Protoplasts are plant cells whose tough walls have been removed with cell-wall degrading enzymes such as cellulase. Bombardment with plasmid-coated microprojectiles (particle gun method), usually 0.6 to 1 micron gold or tungsten particles, is a particularly effective method for delivering DNA into a variety of responsive organs from a range of plant species, including groups such as cereals, which are difficult to transform stably by other means (Barcelo *et al.*, 1994). These gene delivery systems can be grouped on the basis that they are artificial and utilise purified plasmid DNA. They can be distinguished from the second gene transfer approach involving natural gene delivery systems, such as *Agrobacterium*-based transformation, that have been manipulated by genetic engineers (discussed below). Plant viruses are not covered in this review since infection, which is transient, does not require the use of antibiotic resistance genes. Artificial gene delivery systems introduce the entire plasmid into plant cells. This means bacterial vector DNA will accompany the trait gene into plant cells.

A trait gene can be purified away from vector sequences using restriction enzymes and agarose gel electrophoresis before plant transformation

(Sambrook *et al.*, 1989). The method is simple and effective but can only be used if convenient restriction sites release the trait gene from the vector such that it appears as a distinct band, well separated from vector bands, after size fractionation on agarose gels. Unfortunately, purification of trait genes by gel electrophoresis is time consuming and most procedures can only handle relatively small amounts of plasmid DNA. When high concentrations of the trait gene are required for a transformation protocol it may be impractical to replace whole plasmids, composed of vector plus trait gene, (which can be prepared in large amounts) with purified trait genes. The efficiency of vector removal by gel purification is variable and is rarely greater than 90%. Several cycles of gel purification can increase the purity of an insert. More problematic is the fact that gel-purified preparations of genes are often less effective than whole plasmids in producing transformed plants and are unlikely to be used in recalcitrant species where transformation is inefficient.

Natural delivery systems exploit the ability of the soil bacterium *Agrobacterium tumefaciens* or less commonly its relative *Agrobacterium rhizogenes* to transfer DNA into the nucleus of plant cells (Hellens *et al.*, 2000). The transfer of genes from *Agrobacterium* to plant cells is a natural process that has been exploited by genetic engineers. First, the plasmid containing the trait gene is shuttled into *Agrobacterium* from *E. coli* by bacterial conjugation, a natural process by which bacteria exchange plasmids, or by electroporation. Engineered *Agrobacterium* strains can then deliver the trait gene into plant chromosomes if it is flanked by specific DNA sequences, called T-DNA borders, which are recognized by the gene transfer machinery present in *Agrobacterium*. In essence the T-DNA borders delimit the boundaries of a DNA cassette (called Transfer DNA or T-DNA), which is transferred into chromosomes housed in the nucleus of plant cells. Since vector sequences such as bacterial antibiotic resistance genes are located outside this cassette only the trait gene, lying within the T-DNA border, should be transferred to plant cells. The presence of bacterial vector sequences in the nuclear chromosomes of transgenic plants is therefore less of a problem with *Agrobacterium*-mediated gene transfer than artificial gene delivery systems, such as particle bombardment, which utilize purified plasmids. However, the *Agrobacterium* gene transfer machinery cannot be assumed to be 100% effective in excluding vector sequences. Regions outside the T-DNA borders have been found in 20-75% of transgenic plants (Martineau *et al.*, 1994; Kononov *et al.*, 1997; Smith *et al.*, 2001). These findings mean that the presumed absence of vector sequences in plants generated using *Agrobacterium*, particularly the antibiotic resistance genes such as *bla*(TEM1) used to maintain plasmids in bacteria (Table 1), will need to be checked by molecular techniques involving PCR or blot hybridization analysis.

Integration of Foreign Trait Genes into Plant DNA and Identification of Transformed Cells

Plant cells contain three genomes composed of double-stranded DNA, which are located in distinct sub-cellular compartments. The largest is the nuclear genome, is composed of around 20,000 to 40,000 genes, and organised as multiple linear chromosomes (Walbot, 2000). A circular genome typically containing around 110 to 130 genes is found in crop plastids (Hiratsuka *et al.*, 1989). The chloroplast, which is responsible for photosynthesis, is the most familiar member of the plastid family of organelles. Leaf mesophyll cells typically contain around 100 chloroplasts with 100 plastid genomes each making a total of 10,000 plastid genomes per cell. Plastids are descended from ancient cyanobacteria and their genomes and expression apparatus share many similarities with those found in bacteria. This is demonstrated by the fact that a number of plastid promoters function in *E.coli*. The compatibility of the expression apparatus of plastids and bacteria has important consequences with respect to risk assessments on plastid-localised antibiotic resistance genes. The third genome, which is also circular, codes for around 60 genes and is found in mitochondria (Unseld *et al.*, 1997). Mitochondria are probably descended from an ancestor of the *Rickettsia* group of α-proteobacteria, which are obligate intracellular parasites. The vast majority of plant genetic engineering involves the integration of trait genes into nuclear chromosomes. More recently it has also been possible to integrate trait genes into the plastid genome of a handful of flowering plants. Mitochondrial transformation is restricted to yeast (Johnston *et al.*, 1988) and algae (Randolph-Anderson *et al.*, 1993). There are no current methodologies for transforming plant mitochondria.

The combined processes of delivery of foreign genes into plant cells and their subsequent integration into either nuclear or plastid DNA are extremely inefficient (see Chapter 5). The result is that the vast majority of plant cells remain untransformed and will regenerate into plants alongside the few that have been transformed. Even using plants such as tobacco, which are readily transformed with foreign DNA, it is not practical to search through the hundreds and thousands of untransformed plants to find the handful that are transgenic. Selection techniques combined with marker genes are needed to identify the small fraction of cells that take up and express foreign genes after their stable integration into plant DNA. Some success has been achieved with screenable markers such as the green fluorescent protein (Molinier *et al.*, 2000), which rely on visual screens for a reporter gene product rather than selection based on marker genes. The incidence of chimeric plants composed of mixtures of wild-type and transgenic cells might be expected to be higher when screenable markers are used.

Table 2. Examples of plant selectable markers. [1]Also confers resistance to the antibiotic bialaphos

		Marker Gene	Gene Product	Selective Agent	References
Nuclear transformation	Resistance to antibiotics and other antimicrobics	*npt*II (*aph(3')*II)	neomycin phosphotransferase	kanamycin, geneticin (G418)	Bevan *et al.*, 1983 Herrera-Estrella *et al.* 1983a
		aadA	aminoglycoside 3" adenylyl transferase	spectinomycin, streptomycin	Svab *et al.*, 1990a
		hph	hygromycin phosphotransferase	hygromycin B	Waldron *et al.*, 1985 Komari *et al.* 1996
		cat	chloramphenicol acetyl transferase	chloramphenicol	Herrera-Estrella *et al.* 1983b
		aacC3, aacC4	gentamycin-3-N-acetyl transferase	gentamycin	Hayford *et al.*, 1988
		Ble	bleomycin sequestration protein	bleomycin	Hille *et al.*, 1986
		DHFR	dihydrofolate reductase	methotrexate	Herrera-Estrella *et al.* 1983a
		Sul	dihydropteroate synthase	sulfonamides	Guerineau *et al.* 1990
	Herbicide Resistance	[1]*Bar*	phosphinothricin acetyl transferase	glufosinate-ammonium (phosphinothricin)	DeBlock *et al.*, 1987 Barcelo *et al.* 1994
		EPSPS	5-enolpyruvylshikimate-3-phosphate synthase	glyphosate	Shah *et al.*, 1986 Clemente *et al.*, 2000
		Als	acetolactate synthase	chlorsulfuron	Haughn *et al.*, 1988 Li *et al.* 1992
		Bxn	bromoxynil specific nitrilase	bromoxynil	Stalker *et al.*, 1988
Plastid transformation	Antibiotic Resistance	*aadA*	aminoglycoside 3" adenyl transferase	spectinomycin, streptomycin	Svab and Maliga,1993
		*npt*II	neomycin phosphotransferase	kanamycin,	Carrer *et al.*, 1993
		16S rDNA	mutant 16S plastid ribosomal DNA	spectinomycin	Svab *et al.*, 1990b

The most common marker genes used to transform plants confer resistance to either antibiotics or herbicides. Table 2 lists examples of plant marker genes and the agents used to select cells expressing them. Antibiotics are natural compounds made by microbes that are toxic to other organisms. Like antibiotics, synthetic antimicrobial compounds such as the sulfonamides or the dihydrofolate reductase inhibitors (e.g. trimethoprim, methotrexate) have been used to control infections but in most situations they have been replaced by safer antibiotics, which are less toxic. Methotrexate is principally an anti-cancer drug. Resistance markers to antibiotics and synthetic antimicrobics are listed in Table 2 since both classes of drugs are used to treat microbial infections. Because of the overlapping role of these compounds, resistance markers to antimicrobics might be a more appropriate term but is not commonly used in transgenic research. Not all antibiotics are used as anti-microbial agents. The highly toxic bleomycin breaks DNA and is mainly used as an anticancer drug.

By far the most frequently used marker for plant nuclear transformation is the bacterial neomycin phosphotransferase II gene (*npt*II) or *aphA2* gene encoding the enzyme aph(3')II, which confers resistance to the aminoglycosides kanamycin, neomycin, paromomycin, ribostamycin, butirosin and geneticin (G418) but not the newer aminoglycosides amikacin or netilmycin (Nap *et al.*, 1992). These aminoglycosides are inhibitors of protein synthesis in bacteria and organelles (Nap *et al.*, 1992). Typically, kanamycin or G418 are used to select plant cells that have taken up the *npt*II gene. A wider range of marker genes is available for nuclear transformation as compared to plastid transformation (Table 2). The aminoglycoside 3" adenylyl transferase (*aadA*) gene, also referred to as the *ant*(3")-1a gene, is an efficient marker for plastid transformation and confers resistance to spectinomycin and streptomycin, which inhibit plastid protein synthesis (Svab and Maliga, 1993). Mutant plastid ribosomal RNA genes that confer resistance to spectinomycin can also be used to select plastid transformants (Svab *et al.*, 1990b). However, they are less efficient markers than *aadA* and are not widely used. Marker genes integrated into nuclear chromosomes are endowed with short eukaryotic regulatory elements typical of plant nuclear genes to form an expression cassette. Commonly used 5' and 3' regulatory elements are the 35S promoter from Cauliflower Mosaic Virus (CaMV) and the terminator region of the nopaline synthase gene from *Agrobacterium* T-DNA (Hellens *et al.*, 2000). Non-viral promoters are also used widely and their use overcomes the potential destabilization of trangenes linked to viral promoters following field infection of GM crops by the viruses providing these promoters Rfu (Al-Kaff *et al.*, 2000). Examples are the nopaline synthase promoter and the maize ubiquitin promoter used in cereal transformation (Christensen and Quail, 1996). Eukaryotic regulatory elements would not be expected to function in bacteria such as *E. coli*. However, some eukaryotic promoters such as the CaMV 35S promoter are expressed in bacteria (Assaad and Signer, 1990; Vancanneyt *et al.*, 1990). Introns typical of plant nuclear genes can be inserted into expression cassettes to increase the expression of marker genes in plant nuclear chromosomes (Luehrsen and Walbot, 1991; Christensen and Quail, 1996). These nuclear introns will also prevent expression of nuclear marker genes in bacteria (Vancanneyt *et al.*, 1990). A common 5' plastid regulatory element is composed of the plastid 16S ribosomal RNA promoter, to drive transcription, linked to a bacterial ribosome-binding site, for effective translation. These prokaryotic-like plastid regulatory elements also function efficiently in bacteria but would not be expected to function well in the eukaryotic nuclear-cytosolic environment. However, accidental delivery of large numbers of plastid expression cassettes, containing the plastid *psbA* promoter and *npt*II gene, into the nucleus can lead to kanamycin resistance (Ye *et al.*, 1996).

The Problem of Antibiotic Resistance Genes in Transgenic Plants
The increase in incidence of antibiotic resistant strains of pathogenic bacteria is a major challenge for modern medicine. Widespread, and in some cases, indiscriminate uses of antibiotics in medicine and agriculture have acted as the selective agent responsible for the build up of large pools of resistant bacteria and resistance genes (Davies, 1994). Microbial resistance threatens the success of infectious disease treatment and prevention in the 21st century. The World Health Organization has developed a global strategy for containment of anti-microbial resistance addressed at policy-makers and managers (WHO, 2001). Against this worrying increase in resistant "superbugs" any risk, however small, that could promote microbial acquisition of antibiotic resistance is viewed as unacceptable.

As discussed in the section on gene transfer methodologies the antibiotic resistance genes found in GM crops can either be derived from the plasmid vector or plant marker gene. Nuclear markers, vector genes and plastid markers have unique features that require each of these classes to be considered separately with respect to biosafety. The *npt*II nuclear marker is frequently used to select kanamycin-resistant plant cells (Nap *et al.*, 1992), while a typical vector gene is the *bla*(TEM1) gene conferring ampicillin resistance (Yanisch-Perron *et al.*, 1985). The preferred marker gene for plastid transformation is the *aadA* gene, which confers resistance to spectinomycin and streptomycin (Svab and Maliga, 1993). In this review the *npt*II, *bla*(TEM1) and *aadA* genes will be used as examples to illustrate the different risks associated with commonly used nuclear, vector and plastid markers. Since, the risk associated with a particular gene is a combination of the resistance determinant itself as well as its designated role within a construct, the examples chosen only provide a general guide to the differences between nuclear-, vector- and plastid-localized antibiotic resistance genes.

Both trait and marker gene products are subject to the normal regulatory process to ensure they do not constitute a hazard in food (see Chapter 9). This includes testing for allergens and toxins (Kuiper *et al.*, 2001; see Chapter 9). There are two further concerns raised by antibiotic resistance genes. The first is whether antibiotic resistance genes in GM crops can be transmitted to pathogenic bacteria (discussed below). The second is the inactivation of an oral dose of antibiotic by high levels of an antibiotic resistance enzyme present in a GM product. Strong expression of nuclear or plastid marker genes can lead to high levels of antibiotic resistance enzymes in the cytosol or plastid. For example, the *aadA* gene product can accumulate to 0.5 to 5% of the total soluble protein in plastid transformants (Maliga, 2001). Vector genes, such as *bla*(TEM1*)* or plastid marker genes such as *aadA*, integrated into nuclear DNA are unlikely to be expressed well in most plants. However, this needs to be checked for individual plants to rule out the small risk of *bla*(TEM1) expression mediated by plant regulatory sequences flanking the chromosomal integration site and the possibility of copy number enhanced expression (Ye *et al.*, 1996). Both NPTII and the *aadA* gene product require ATP as a co-factor. Low levels of ATP in the gut will reduce the activities of these enzymes in the digestive tract (Nap *et al.*, 1992).

It can be argued that plant-derived proteins would be rapidly hydrolyzed in the digestive tract and as such pose little risk of degrading an antibiotic. However, it is far better to remove this risk altogether by, for example, using harsh treatments (high temperatures, chemical extraction) that will destroy enzymes in GM-plant derived foods. In some cases, it may be difficult to control the appearance of foreign gene products in particular foods. For example, foreign gene products expressed in pollen could end up in honey where they can be stored for at least six weeks (Eady *et al.*, 1995). A more effective strategy is to eliminate antibiotic resistance enzymes in GM products by removing their source genes from GM crops.

The Risk of Antibiotic Resistance Transfer from GM Crops to Bacteria
The scenario of events leading to transfer of these genes to pathogenic bacteria can be broken down into five steps: i) release of DNA from GM crops or their products by processes such as natural or facilitated (such as silage making) decomposition mediated by soil microbes and fungi or digestion in the animal gut, including invertebrates such as earthworms (Thimm *et al.*, 2001), or human gut, ii) uptake of intact antibiotic resistance genes by competent bacteria in the environment, gut or oral cavity, iii) replication and

expression of resistance genes in bacteria, iv) selection and multiplication of bacteria containing plant-derived antibiotic resistance genes, v) bacteria-to bacteria spread of resistance genes.

Release of DNA from GM Crops

The presence of antibiotic resistance genes in food products from GM crops has been a particular matter for concern. Molecular analyses, using the sensitive polymerase chain reaction, have allowed the detection of foreign genes in whole meal from GM crops such as maize and soybean (Vollenhofer *et al.*, 1999). To degrade DNA in food requires harsh conditions such as high temperatures and pressurized steam. Procedures such as the chemical expulsion and extrusion of oilseeds degrade DNA completely (Chiter *et al.*, 2000). DNA in food is degraded by the acidic environment of the stomach together with the soup of digestive enzymes in the gastrointestinal tract derived from the host and resident microbes. Intact genes are unlikely to persist for long periods in this environment but their persistence will be affected by diet and feeding/fasting regimes. When mice were fed purified plasmids, fragments of up to 976 base pairs were found in blood two to eight hours after feeding (Schubbert *et al.*, 1997). This study showed that DNA could cross the intestinal wall. In another study on farm animals only short DNA fragments (< 200 base pairs) derived from plant plastids (non-transgenic plastids) were detected in the blood lymphocytes of cows (Einspanier *et al.*, 2001). Plant DNAs were not found in the muscle, liver, spleen and kidneys of cattle. Maize plastid DNA fragments were detected in these organs from chicken but none were found in eggs (Einspanier *et al.*, 2001). The fragmented DNA found in the blood and organs of animals represent food-derived nutrients that are further metabolized. There is no evidence to suggest that this DNA integrates into the genome of animal cells. While intact plasmids have been shown to survive in sheep saliva for 24 hours they are degraded after a minute in fluid from a sheep's rumen (Duggan *et al.*, 2000). Experiments with an artificial rumen (Kivaisi *et al.*, 1990) and other artificial gut compartments may help to track the fate of antibiotic resistance genes under a range of conditions where diet and microbes can be altered relatively easily.

The persistence of DNA in the environment may vary depending on the locality. Silage effluent has been reported to degrade DNA rapidly (Duggan *et al.*, 2000) whereas clay has been reported to stabilize DNA (Demaneche *et al.*, 2001). The adsorption of DNA by clay not only protects it but also prevent its uptake by bacteria (Demaneche *et al.*, 2001). The number of antibiotic resistance genes present in a GM crop will also determine the frequency of release and subsequent persistence of genes in the external environment. Particle bombardment can introduce many copies, often ten or more, into the nucleus of a GM plant. Plastid transformants can contain as many as 50,000 copies of the *aadA* gene per cell. The possible spread of resistance genes through industrial wastes is more applicable to genetically manipulated industrial bacteria than to GM crops (Doblhoff-Dier *et al.*, 2000).

DNA Uptake By Competent Bacteria

Uptake of DNA by bacteria is dependent on their competence. Some bacteria, such as the *Acinetobacter calcoaceticus* (Lorenz *et al.*, 1992) and *Bacillus subtilis* (Dubnau, 1991), can become competent by natural processes such as starvation. Other bacteria such as *E. coli* do not take up plasmid DNA as readily and require highly artificial conditions before they can be transformed with DNA. Even using the best laboratory protocols the fraction of competent *E. coli* cells that are competent to take up DNA is rarely above a few percent (Hanahan, 1983; Dower *et al.*, 1988).

Replication and Expression of Resistance Genes in Bacteria

Once an antibiotic resistance gene enters a bacterial cell from a GM crop it is stabilized by integration into a DNA molecule capable of replication, namely, the main bacterial genome or a plasmid (Nielsen *et al.*, 1998). Homologous recombination, which is promoted by shared DNA sequences between donor transforming DNA (from the plant) and resident target DNA (in the bacterium), will promote the integration event. Homologous recombination in *E. coli* requires minimum identical stretches of 27 base pairs between donor and target DNAs (Shen and Huang, 1986). A drop of 100% to 90% sequence identity between donor and target results in a 40-fold decrease in homologous recombination frequency (Shen and Huang, 1986). Recombination between divergent DNA is enhanced in bacterial strains and species with less stringent mismatch repair systems (Rayssiguier *et al.*, 1989). An antibiotic resistance gene may be transferred along with bacterial vector sequences that contain an origin of DNA replication. Such vector sequences are often found integrated alongside antibiotic resistance genes in the chromosomes of nuclear transgenic plants generated by artificial transformation methods such as particle bombardment (see above and Figure 1B). An origin of replication would allow autonomous replication of an attached antibiotic resistance gene following circularization of DNA.

In relatively recent experimental studies (Gebhard and Smalla, 1998; de Vries and Wackernagel, 1998), transfer of *npt*II gene sequences from DNA extracted from transgenic plants to *Acinetobacter* sp was detected at frequencies of 4×10^{-8} and 5×10^{-9} per treated bacterial cell. These experiments used sensitive assays where plant derived *npt*II sequences entered *Acinetobacter*, repaired deleted *npt*II genes and allowed expression of kanamycin resistance. Many other experiments failed to detect *npt*II gene transfer because the bacteria lacked homologous DNA, in this case deleted *npt*II genes, to capture incoming DNA (Nielsen *et al.*, 1998). These results suggest that any DNA sequences that are common to bacteria and transgenic plants, such as vector elements, will promote stable gene transfer of plant markers to bacteria (Nielsen *et al.*, 1998). Vector elements might therefore be crucial for the stabilization of plant-derived antibiotic resistance genes in bacteria.

An additional factor that might reduce the uptake of plant DNA is the presence of restriction endonucleases in bacteria, which degrade foreign DNA. DNA methylation typical of plant nuclear genes might prevent their digestion by restriction endonucleases. However, this methylation pattern might make some plant genes susceptible to enzymes that cleave methylated DNA such as the products of the *E. coli mcrA, mcrB* and *mrr* genes (Raleigh *et al.*, 1988). Plastid DNA, which is relatively unmethylated, appears to be susceptible to most purified restriction endonucleases, which are commercially available.

The uptake and replication of marker genes in bacteria will not in itself lead to antibiotic resistance. For resistance a plant-derived gene has to be expressed in bacteria. Vector components such as the *bla*(TEM1) gene contain prokaryotic regulatory elements and will be expressed in *E. coli* and a range of related bacteria. Similarly, the plastid localized *aadA* gene with prokaryotic-like regulatory elements will also be expressed in bacteria such as *E. coli*. It should be noted that *E. coli* regulatory elements are unlikely to function in all eubacteria. This includes obligate anaerobes such as members of the genus *Bacteroides*, which predominate in the human gut and outnumber bacteria such as *E. coli* by about 1000 to one. Some eukaryotic regulatory elements, such as the 35S promoter, will function in *E. coli* (Assaad and Signer, 1990). The presence of nuclear introns in a number of plant marker genes reduces the likelihood of expression in bacteria. Since vector and plastid marker genes are expressed well in *E. coli*, the theoretical risk of their transfer from transgenic plants to bacteria is greater than that for most nuclear markers. In the case of the plastid *aadA* gene adding RNA editing sites (Corneille *et al.*, 2000) can reduce this risk. Such genes would only be expressed in plastids since bacteria lack the post-transcriptional enzymatic machinery required for RNA editing. Inclusion of codons that are rarely used in *E. coli* will also reduce expression of the *aadA* gene in this particular microbe (Lutz *et al.*, 2001). However, any variation in codon usage between different bacterial species will undermine this containment strategy.

Gene transfer from plants to bacteria is so rare that most experiments with GM plants have failed to detect it (Nielsen *et al.*, 1998). The uptake of plant transgenes by bacteria has not been observed in field experiments involving transgenic plants (Nielsen *et al.*, 1998). Success has only been achieved in laboratory experiments using DNA extracts and by increasing the sensitivity of detection using specially engineered strains of bacteria with homology to plant transgenes to capture transforming DNA (see above). The estimate of gene transfer from this artificial situation is probably several orders of magnitude higher than the actual rate for most nuclear localized marker genes. There are as yet no estimated rates of transfer of plastid-localized transgenes to bacteria. The high copy number of plastid genes might be expected to promote this process. One of the current problems of examining horizontal gene transfer from nuclear transgenic plants to bacteria is its very low frequency making it extremely difficult to study. If experiments on *aadA* transfer from plastids to bacteria are successful and lead to detectable frequencies of gene transfer they will help to define the key steps needed to establish a population of resistant bacteria in a range of environments, including gut microflora. This information can be used to reduce the likelihood of horizontal gene transfer from GM crops to bacteria.

Selection and Multiplication of Bacteria Containing Plant-Derived Antibiotic Resistance Genes

The combined processes of uptake, stabilization and expression of plant-derived antibiotic resistance genes by bacteria is an extremely rare event. Moreover, the small numbers of antibiotic-resistant cells generated by this process are unlikely to pose a hazard unless their fitness is increased relative to sensitive cells. In the absence of antibiotic selection, an antibiotic resistance marker provides no fitness advantage, and the persistence of small numbers of resistant cells within a much larger population of sensitive cells is likely to be highly unstable (Thomson, 2001). Fluctuations in the rates of cell growth and division within a bacterial population would tend to eliminate small numbers of antibiotic resistant cells.

Bacteria-to- Bacteria Spread of Resistance Genes

Widespread and indiscriminate use of antibiotics in medicine and as prophylactics and growth promoters in poultry, swine and cattle (Corpet, 2000) has resulted in reservoirs of resistant bacteria in humans and animals (WHO, 2001). The antibiotic resistance genes in transgenic plants are derived from resistant bacteria that are already prevalent in the environment and it is these resistant bacteria, that have and will continue to act as the most likely source of new outbreaks of disease that do not respond to standard antibiotics. These bacteria might already be pathogens such as a clone of *Salmonella typhimurium* DT104, associated with diarrhea, which is resistant to ampicillin, chloramphenicol, streptomycin, and tetracycline (WHO, 2001). Alternatively, benign resistant bacteria can spread and contact pathogens through poor hygiene, insufficient cooking, and the handling of live animals and carcasses. This spread increases the opportunities for benign bacteria to transfer their resistance genes to pathogenic bacteria.

The small hypothetical risk of antibiotic resistance gene transfer from GM crop to pathogenic bacteria has to be placed in context with the very real risk of the same antibiotic resistance gene being transmitted to pathogens via bacteria-to-bacteria spread from existing reservoirs of resistant bacteria in the environment (JETACAR, 1999). Intercellular transmission of resistant markers amongst bacteria, involving conjugation or viruses, is very efficient. Self-transmissible and mobilizable DNA elements, such as plasmids and transposons, are particularly efficient at disseminating antibiotic resistance genes (Davies, 1994; Shoemaker *et al.*, 2001).

The Case for Removal of Antibiotic Resistance Marker Genes from GM Crops
Ending the Debate over the Magnitude of Risk Associated with Antibiotic Resistance Genes in Transgenic Crops
The possible hazards associated with the *npt*II gene in transgenic research has been scrutinized in detail, given its early and widespread use as a nuclear marker in plant genetic engineering (Nap *et al.*, 1992; Fuchs *et al.*, 1993). Existing bacterial resistance has limited the use of kanamycin in medicine (Flamm *et al.*, 1993; JETACAR, 1999). Even if a plant derived *npt*II gene were transferred to an enteric bacterium, the resulting strain would comprise part of an already existing kanamycin-resistant population strain that do not have a fitness advantage over the normal gut microflora (JETACAR, 1999). The therapeutic use of other aminoglycoside antibiotics, such as neomycin, has also decreased largely due to the availability of less toxic broad range β-lactam antibiotics (Davies, 1994). Expression of NPTII is not the sole determinant of kanamycin resistance. Other aminoglycoside resistance enzymes, such as NPTI, NPTIII, NPTIV, NPTV and NPTVI (Nap *et al.*, 1992), will also inactivate kanamycin. Reduced antibiotic uptake, enhanced antibiotic efflux and absence of antibiotic target sites provide alternative resistance mechanisms (Davies, 1994). Only three of 184 kanamycin resistant colonies isolated from a coastal plain scheme contained the *npt*II gene (Leff

et al., 1993). In a separate study, the *npt*II gene was predominantly found in kanamycin-resistant bacteria isolated from sewage samples rather than river-water or soil (Smalla *et al.*, 1993).

The product of the plastid *aadA* marker gene inactivates spectinomycin and streptomycin. Streptomycin is occasionally used to treat resistant Mycobacteria responsible for tuberculosis (Coninx *et al.*, 1999). Spectinomycin is used in the treatment of gonorrhea if penicillin cannot be used due to allergy or bacterial resistance (Ison, 1996). Kanamycin, spectinomycin and streptomycin are poorly absorbed by the gut and in normal medical use these antibiotics are administered by injection. This minimizes the exposure of gut microbes to these antibiotics in humans. Unfortunately, antibiotics are used widely as growth promoters and prophylactic agents in agriculture thereby providing selection for the emergence of resistant microbes in the animal gut (Khachtourians, 1998). For example, over a 14 month period the subtherapeutic addition of penicillin, sulfamethazine, chlorotetracycline, oxytetracycline and neomycin to animal feed (0.25 to 1 kg per ton of feed) and drinking water, and therapeutic use of sulfamethazine for diarrhea, was associated with the emergence of antibiotic-resistant *E. coli* O157:H7 in dairy cattle (Shere *et al.*, 1998). When kanamycin is administered orally in animal feed, the majority (97%) passes through the gut and into the environment (Nap *et al.*, 1992). Soil inactivation of antibiotics in animal manure would prevent their long-term persistence in their environment (Nap *et al.*, 1992).

The presence of the vector derived *bla*(TEM1) gene in GM crops is more controversial. The anti-gram-negative-bacterium penicillins, including ampicillin, are dispensed widely in the community, often taken orally and used to kill a range of bacteria responsible for common infections. Ampicillin resistance in more than 50% of *E. coli* clinical isolates is due to the *bla*(TEM1) gene (Livermore, 1995; Frere, 1995). These examples illustrate the differences in identified risks associated with each antibiotic resistance gene. Risk is increased when a specific antibiotic is used widely and when its use is essential to control the spread of a particularly harmful disease. Antibiotic resistance genes in GM crops only add to this risk if these genes can enter, replicate and be expressed in bacteria and the resulting bacteria are selected by antibiotics to create reservoirs of resistant microbes. Based on current knowledge the resulting combined risk is likely to be small, even for highly amplified plastid-localised *aadA* genes.

The plant-to-plant spread of antibiotic resistance genes leading to wider dispersal of antibiotic resistance genes into the environment is a further issue (Gressel, 2000). The likelihood of crop-to-weed spread might be enhanced by the presence of linked trait genes, such as herbicide and insect resistance that could confer a selective advantage in particular environments. Antibiotic resistance genes would then hitchhike with advantageous trait genes to weeds. A recent study showed that the use of multiple herbicides within a locality where three types of herbicide resistant oil-seed rape were grown led to rapid pollen-mediated crop-to-crop spread of resistance genes resulting in triple herbicide gene stacking (Hall *et al.*, 2000). Transgenic plants containing the *npt*II gene alongside herbicide or insect resistance genes are not uncommon. For example, transgenic corn line Aphis # 96-317-01p contains the *npt*II gene as well as the *cryIA* and *EPSPS* genes, for resistance to European corn borer and glyphosate, respectively (Crops no longer regulated by USDA, http://www.isb.vt.edu/CFDOCS/biopetitions1.cfm). Special conditions, such as ongoing selection for a GM trait, might be required for the establishment and continued presence of GM crop traits in weeds. The degree of escape of a GM trait to weeds will depend on the crop species. Growing outbreeding crops in regions containing sexually compatible weedy relatives will enhance transgene escape. For example, herbicide resistant transgenic oilseed rape outcrossed with field mustard weeds to produce resistant hybrids (Mikkelsen *et al.*, 1996). Conditions that favor crop-to-weed dispersal of antibiotic resistance genes are undesirable since they allow plant marker genes to gain access to an increased diversity of plant-microbial niches.

Whilst there is little scientific support to indicate that transgenic crops with antibiotic markers will exacerbate the existing problem of resistance in pathogenic bacteria, this possibility cannot be ruled out with absolute certainty. Any lingering doubts, however small, are compounded by our limited ecological knowledge on routes of inter-specific bacteria-to-bacteria spread of resistance genes from resistant pockets of bacteria and the fact that the calculated risks of uptake of plant DNA by bacteria are limited to those that can be cultured in laboratories. In one study, less than 0.2% of the bacteria present in an environmental sample could be cultivated (Leff *et al.*, 1993). Notwithstanding the fact that large quantities of products from GM crops containing antibiotic resistance genes have been consumed without any obvious deleterious effects the arguments in favour of antibiotic marker gene removal have triumphed. This has resulted in new European Union legislation (Directive 2001/18/EC) to ensure their absence from transgenic crops. Removal of antibiotic resistance markers from GM crops eliminates, with one stroke, all perceived hazards associated with these genes, without in any way reducing the efficacy of trait genes. This simple solution is highly desirable since it moves the focus of scientific and public attention away from the excess foreign DNA in GM crops towards trait genes. After all, it is the properties conferred by trait genes that are the main purpose of genetic engineering.

Removing Marker Genes Allows Repeat Cycles of Plant Transformation, Reduces the Possibility of the Duplication of Expression Elements and May Benefit Plant Fitness

Removal of antibiotic resistance genes from GM crops, once they have served their purpose as plant transformation markers, has a number of additional advantages. It facilitates the repeated use of a single resistance marker to allow the step-wise introduction of multiple trait genes into a transgenic crop. This marker recycling strategy will become increasingly important in the age of multi-gene engineering. The alternative of using different marker genes to introduce multiple trait genes into crops is not always practical. For some crops, only a few effective marker genes are available, thereby limiting the number of transformation cycles for introducing trait genes. Also, in some instances, alternative marker genes, such as herbicide resistance genes, may be undesirable in the final GM crop. Inclusion of herbicide resistance genes, when not required, quite unnecessarily adds another foreign gene to be considered by the regulatory process. Under favorable conditions an herbicide resistance gene might escape to other crops or weeds taking the trait gene with it (see the section above). Whilst the consequences of foreign gene escape to other plants may be limited, it might be prudent to simplify risk assessments by removing any foreign genes not required in the final GM crop.

Combining multiple trait genes with several marker genes can lead to duplication of regulatory elements. Multiple copies of a regulatory element in plant nuclei can lead to gene silencing of associated trait genes (Matzke *et al.*, 2000). When regulatory elements are limited, reducing the number of marker genes will ease the possibility of their duplication. If a marker gene is expressed to very high levels, for instance, the *aadA* gene product can comprise as much as 5% of the total soluble protein in leaf chloroplasts, its removal will avoid the energetic costs associated with its production and any deleterious concentration-dependent affects of the *aadA* product on plant fitness (Purrington and Bergelson, 1997). Such considerations relate particularly to cases where maximising overall yield of a transgenic protein is important such as the proposed use of plants as sustainable factories for making pharmaceutical proteins (Giddings *et al.*, 2000).

Strategies for Avoiding Antibiotic Resistance Genes in GM Crops

There are two approaches for avoiding the presence of antibiotic resistance genes in transgenic plants. In the first method, antibiotic resistance genes are still used to transform plant cells, but once transgenic plants are produced they are removed, leaving behind trait gene(s). In the second approach, antibiotic resistance genes are left out of the transformation process altogether by substituting them with alternative marker genes. Elimination of antibiotic resistance genes from transgenic plants is not the sole reason for developing alternative marker genes. Importantly, new marker genes such as the *E. coli* mannose phosphate isomerase (phosphomannose isomerase) gene can improve transformation efficiencies in some species (Joersbo *et al.*, 1998). The two approaches of antibiotic resistance gene removal by excision or marker substitution are not mutually exclusive and can be combined. Combining methodologies will allow a particularly efficient marker to be used repeatedly for multi-gene engineering, and can circumvent unforeseen hazards and the consequent need for safety tests on new markers in GM crops.

Removal of Antibiotic Resistance Genes from Transgenic Plants Subsequent to Their Use as Selectable Markers

A large amount of effort has gone into developing methodologies for removing plant selectable markers, such as the *npt*II gene, from transgenic plants. Schemes for the excision of vector sequences from transgenic plants have not been reported widely despite the fact they can contain antibiotic resistance genes and bacterial origins of replication (Figure 1B). The problem of bacterial vector sequences in transgenic crops is less of a problem with *Agrobacterium*-based plant transformation systems (see above). Within the emerging technologies to excise antibiotic resistance genes from transgenic plants, schemes exploiting efficient homologous recombination pathways mediated by native plant enzymes deserve special attention because they can remove both vector and plant selectable markers. Further advantages of using host recombinases include simplicity and the elimination of any unintended detrimental effects of foreign site-specific enzymes (see below) on non-target genes. Moreover, the use of native recombinases eliminates the requirement to remove foreign site-specific recombinases subsequent to their use.

Figure 1. Schemes available for removing marker genes from transgenic plants. Foreign DNA can integrate into the plastid genome (A), or nuclear genome (B). Methods for excising plant marker genes from the plastid (A) and nucleus (C) are shown. Cre, R, Gin and FLP refer to genes encoding site-specific recombinases. These genes can either be introduced by transformation in the T_0 generation or by sexual crosses in the T_1 generation. Vector sequences are absent in plastid transformants but are present in the majority of nuclear transgenic plants transformed by artificial delivery systems and a variable fraction (normally 20-75%) of plants transformed with *Agrobacterium*. Vector DNA includes a vector marker gene and a bacterial origin of replication (ori). For simplicity integrated vector sequences are not shown in C.

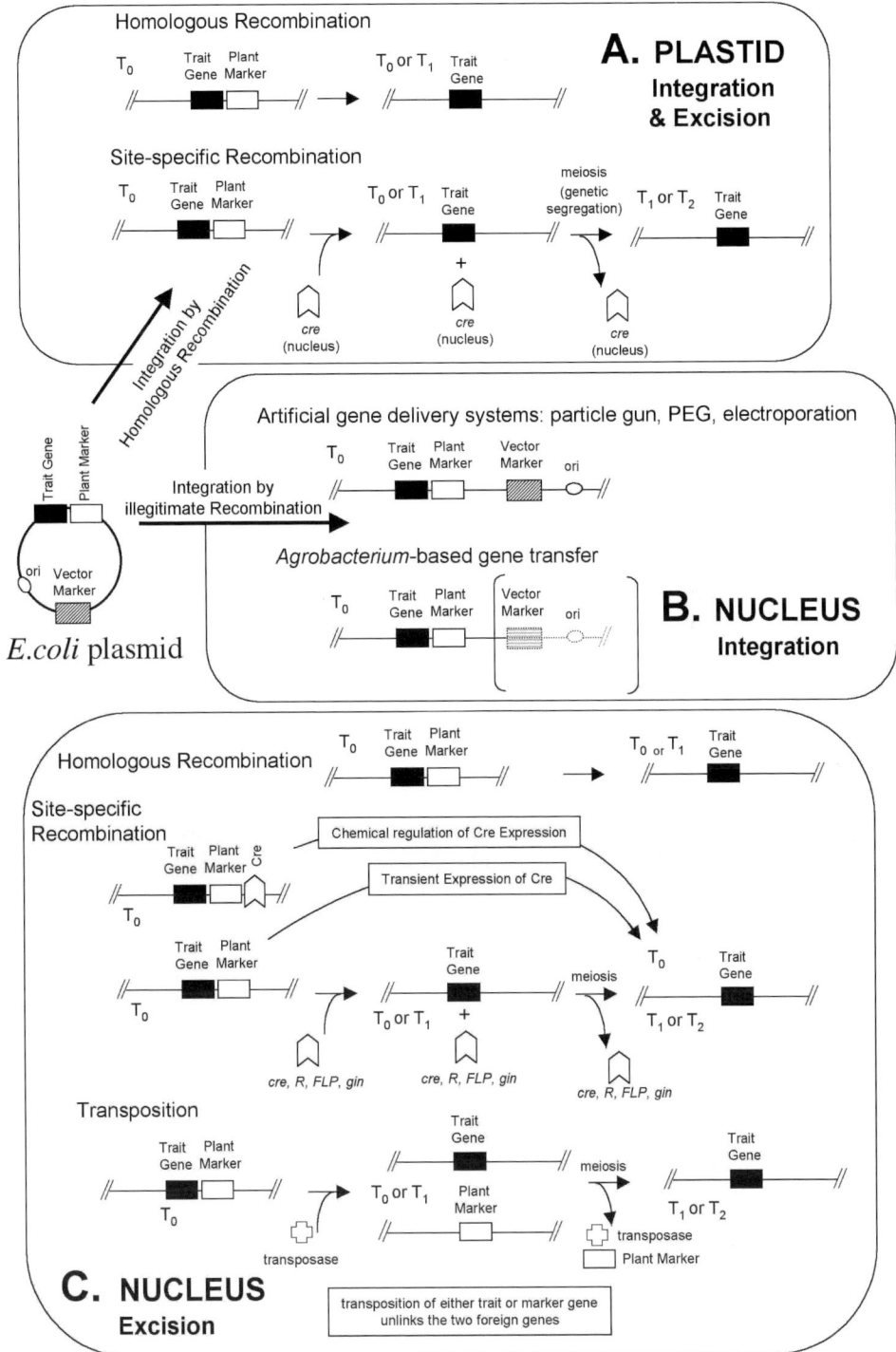

Native Plant Enzymes Responsible for Homologous DNA Recombination Provide a Natural Resource for Removing Both Vector Sequences and Antibiotic Marker Genes from Plants

Clean gene transformation technologies allow precise integration of trait gene(s) without vector sequences or marker genes. This ideal has been achieved recently (Iamtham and Day, 2000) by exploiting the efficient homologous DNA recombination pathway in plastids (Figure 1A). Homologous recombination allows targeting and precise integration of trait and marker gene(s) mediated by shared DNA sequences (homologous DNA) between the target site and the regions flanking the foreign gene(s). Trait and marker genes embedded in homologous DNA integrate into plastid DNA by a double recombination event in regions flanking the foreign genes. This results in exclusion of non-homologous vector DNA and its eventual loss due to its inability to replicate and segregate in plastids.

Excision of the *aadA* gene was achieved by placing two copies of a short identical DNA sequence to either side of it, which are orientated in the same direction. These direct sequences provide the homology necessary for homologous recombination. When recombinases act on direct repeats, the intervening DNA, in this case the *aadA* gene plus one copy of the repeat, are removed. Since trait genes are located outside direct repeats they are not excised. This natural gene excision process, which takes place throughout the transformation process, is controlled by selection. Loss of *aadA* is prevented by antibiotic selection until recombinant plastid genomes, containing *aadA* plus trait genes, replace wild-type plastid genomes. Selection for *aadA* is then relaxed to allow the accumulation of *aadA*-free plastid genomes. Cytoplasmic sorting, involving segregation of *aadA*-free plastid genomes from *aadA* plastid genomes during plant growth gives rise to cells, organs and eventually shoots that are *aadA*-free. Although, *aadA*-free, herbicide resistant plastid transformants have been isolated in the T_0 generation by vegetative propagation *in vitro*, a reduction in plastid copy number in eggs or zygotes, the so called organelle bottleneck (Bergstrom and Pritchard, 1998), facilitates the isolation of *aadA*-free plants in the T_1 and subsequent generations. The percentage of *aadA*-free seedlings can be high. When three 418 base pair repeats were involved in excising *aadA* from the plastid 24% of the progeny were *aadA*-free (Iamtham and Day, 2000). The frequency of co-transformation of plasmids into the plastids and nucleus of the same cell by particle bombardment appears to be low (our unpublished results). However, it is prudent to undertake several rounds of crosses involving fertilization of transgenic plants with pollen from wild-type plants to retain recombinant plastid genomes (plastids are maternally inherited) while substituting-in nuclear chromosomes from non-transgenic plants.

Homology-based gene excision would theoretically also work in plant mitochondria. Although foreign DNA has not yet been integrated into higher plant mitochondria, homologous recombination is responsible for integration of foreign DNA within the yeast mitochondrial genome (Johnston *et al.*, 1988). Moreover, homology-based mitochondrial DNA recombination has been observed in plant mitochondria (Palmer and Shields, 1984; Kao *et al.*, 1992). Non-homologous DNA recombination, also called illegitimate DNA recombination, is the predominant pathway in plant nuclei, ruling out efficient homology-based gene targeting and facile marker gene excision in the nucleus. Among less complex eukaryotes, including fungi and unicellular protozoans, such as the ciliates, it is not uncommon to find high rates of homologous recombination within the nucleus. Homologous recombination also predominates in the nucleus of the moss *Physcomitrella patens* (Schaefer, 2001). Long-term experiments to down-regulate pathways of non-homologous recombination and up-regulate homologous recombination within plant nuclei are underway in the model plant *Arabidopis thaliana* (Personal communication, Dr. Chris West, University of Manchester). Similar approaches are underway in cell-based mammalian systems (Vasquez *et al.*, 2001). Even if an efficient homologous DNA recombination pathway can be engineered within the plant nucleus, it will be important to ensure that this process does not have any unintended deleterious consequences on plant growth.

There is one report on the use of spontaneous homologous recombination acting on direct repeats to excise the *npt*II gene from the nucleus of tobacco plants (Zubko *et al.*, 2000). In this case the direct repeats were composed of the 352 base pair attachment (att) P region of the lambda bacteriophage of *E. coli*. Homologous recombination between attP sites results in *npt*II excision and trait gene retention. Low rates of recombination between the attP repeats meant that when excision of *npt*II was detected (in two out of eleven kanamycin resistant calli), it accounted for about 10 to 20% of recombination events. In the majority of cases, *npt*II was excised together with the trait gene by illegitimate recombination events. Homologous recombination allows marker-free plants to be isolated in the T_0 generation (Figure 1C), which is particularly suited to vegetatively propagated species (Zubko *et al.*, 2000). Whilst this example shows that homology-based recombination can be used to excise nuclear marker genes from plants the frequency is too low (Risseeuw *et al.*, 1995; Kumar and Fladung, 2001) for efficient homology-based integration (gene targeting), which would also result in exclusion of non-homologous vector sequences.

Site-Specific Recombinases Acting on Target Sites Flanking Marker Genes (See Chapter 7)

Two-component systems comprised of site-specific recombinases and their short target DNA sequences have been used to excise antibiotic resistance genes from plants (Figure 1C). The Cre recombinase from the *E. coli* bacteriophage P1 together with its 34 bp *loxP* target site have been used extensively in plants (Dale and Ow, 1991; Russell *et al.*, 1992). Additional well-characterised site-specific recombinases are available for meditating gene excision in plants and these include the FLP recombinase and *FRT* target sites from the yeast 2 micron plasmid (Kilby *et al.*, 1995), the R recombinase and *Rs* target sites from the pSR1 plasmid of *Zygosaccharomyces rouxii* (Onouchi *et al.*, 1995; Sugita *et al.*, 2000), the Gin recombinase and target sites from bacteriophage Mu (Maeser and Kahmann, 1991), and the recombinase and target sites from *Streptomyces* bacteriophage phiC31 (Thorpe and Smith, 1998). Site-specific recombinases are particularly useful when the rates of homologous recombination are low as in the nucleus of higher plants. Their use provides a high degree of control over the timing of gene excision. Unlike gene excision systems based on native homologous recombination enzymes, foreign site-specific recombinases need to be removed from plants once they have been used. Use of site-specific recombinases also requires the evaluation of potential non-target recombination events resulting in unpredictable and permanent genome rearrangements.

To make use of site-specific recombinases, their target sites must flank antibiotic resistance genes as direct repeats. These direct repeats define the borders of a gene elimination cassette. In the absence of site-specific recombinases, antibiotic marker genes are stable and can be used as selectable markers. Once transgenic plants have been isolated using antibiotic selection, expression of the appropriate site-specific recombinase results in excision of its cognate elimination cassette. Trait genes that are located outside elimination cassettes are retained. Foreign recombinases were first demonstrated to be a viable system for removing marker genes from transgenic plants by using the Cre/*loxP* site-specific excision system (Dale and Ow, 1991). To induce excision of *loxP* elimination cassettes, transformed plants were crossed with transgenic plants expressing Cre recombinase or re-transformed with plasmid constructs expressing Cre. Once antibiotic resistance genes have been removed by the action of Cre recombinase, the *cre* gene is no longer required and can be removed from the trait gene by genetic segregation. As an alternative, transient expression of the *cre* gene can also lead to excision of the elimination cassette (Gleave *et al.*, 1999).

This avoids the need to remove a stable copy of the *cre* gene after it has served its purpose (Figure 1C). Fusion of Cre to the *Agrobacterium* infection protein VirF results in its transport into plant cells early in the *Agrobacterium* infection process (Vergunst *et al.*, 2000). This provides a novel route for introducing the Cre protein, but not its gene, into plant cells (Hohn *et al.*, 2001). The inclusion of the counterselectable *cod*A gene with a marker gene in an elimination cassette facilitates isolation of marker-free plants, which can be selected because they are 5-fluorocytosine-resistant (Gleave *et al.*, 1999). Plants containing *cod*A gene are sensitive to 5-fluorocytosine because the *cod*A gene product converts it to toxic 5-fluorouracil.

In the above approaches, the Cre recombinase gene and *loxP* elimination cassettes were introduced into plants as separate events. This was necessary to prevent premature excision of *loxP* elimination cassettes, which would be catalysed by any Cre recombinase present. In a recent improvement to the method tight regulation of Cre recombinase gene expression (Zuo *et al.*, 2001) has allowed the *lox P* elimination cassette, *cre* and trait genes to be introduced into plants as a single contiguous DNA sequence. The method allows marker-free plants to be isolated in the T_0 generation (Figure 1C). Tight regulation was achieved by linking the *cre* gene to *lexA* operator sites fused to a mini CaMV35S promoter. The XVE transcription activator protein (Zuo *et al.*, 2000), comprised of the DNA-binding domain of the bacterial LexA repressor, an acidic transactivating domain, and the regulatory region of the human estrogen receptor, is constitutively transcribed in plants. Expression of Cre is tightly regulated by the transactivating activity of the XVE protein, which is switched on by binding of XVE protein to estradiol. By placing *XVE*, *nptII* and *cre* within the elimination cassette all three foreign genes were excised when Cre recombinase was expressed.

The Cre/*lox* system has also been used to excise marker genes from transgenic plastids (Figure 1A). This has been achieved by first transforming plastids with marker genes located in *loxP* elimination cassettes. Excision of the elimination cassette was then achieved by introducing a nuclear-coded plastid-targeted Cre-recombinase into these plants by genetic crosses or transformation (Hajdukiewicz *et al.*, 2001; Corneille *et al.*, 2001). Excision of elimination cassettes from the majority of plastid genomes led to rapid generation of marker-free plastid transformants despite the multicopy nature of plastid DNA. Unpredictably, Cre-induced recombination between *loxP* sites in plastids also led to the stimulation of undesirable plastid DNA rearrangements (Hajdukiewicz *et al.*, 2001; Corneille *et al.*, 2001). Once marker-free plastid transformants are produced the nuclear *cre* gene can be removed by genetic segregation.

Inserting Foreign Genes Between the Termini of Transposable Elements
Plant nuclear genomes are littered with discrete DNA sequences called
transposable elements that are capable of replicating themselves and moving
to new chromosomal locations. Class II transposable elements transpose via
DNA rather than RNA intermediates and encode site-specific DNA
recombinases called transposases. Class II elements include the Activator
family of transposable elements (McClintock, 1984), originally isolated from
maize, that are capable of transposition in a range of plants including dicots.
Excision of Activator is followed by its re-insertion elsewhere in the genome.
This 'cut and paste' mechanism used by conservative, as opposed to
replicative, class II transposable elements is useful for removing antibiotic
resistance markers from plants. In effect, excision of transposable elements
is based on two components, namely a transposase enzyme encoded by the
element and its target DNA sequences located at the element's termini.

Exploitation of class II transposable elements to remove antibiotic
resistance genes from transgenic plants has largely focussed on the well-
characterised maize Activator (Ac) element. In order to re-locate a foreign
gene in the plant nucleus it is flanked by the terminal DNA sequences of
Activator. This combination of foreign gene flanked by Ac-termini is called
a re-location cassette. When Ac-transposase is expressed in the plant the
foreign gene is excised and re-integrates elsewhere in the nuclear genome.
This is normally achieved in two steps by first transforming plants with the
relocation cassette and subsequently introducing the gene expressing Ac-
transposase by sexual crosses or a second round of plant transformation
(Figure 1C).

Either the trait gene or the antibiotic resistance gene can be placed in a
re-location cassette. About 50% of transposition events move the relocation
cassette well away from its original integration site (Yoder and Goldsbrough,
1994). This movement unlinks trait genes from marker genes in transgenic
plants (Goldsbrough *et al.*, 1993; Yoder and Goldsbrough, 1994). Genetic
segregation ensures that isolation of marker-free plants containing the trait
gene amongst the progeny of these plants. The proportion of marker-free
plants will depend on the genetic distance between trait and marker genes,
which will be variable since transposition cannot be controlled and Activator
often transposes to a linked site. Fortuitous abortive transposition events
(approx 10% of excision events) involving excision without subsequent
integration of a re-location cassette containing an antibiotic marker gene
can also give rise to marker-free plants (Yoder and Goldsbrough, 1994).

A possible drawback to this approach is that undesirable rearrangements
at the excision or integration sites might accompany transposition of the
relocation cassette. More commonly, transposition of a trait gene can result

in a cleaner insertion site than the original integration event. Since transposable elements are important mediators of genome instability the expression of transposase could activate native transposable elements and lead to unforeseen genome rearrangements.

Genetic Segregation of Trait and Marker Genes After Their Independent Transfer into Individual Plant Cells by Co-transformation

Genetic segregation of antibiotic resistance genes from a foreign trait gene is possible when they are located at unlinked chromosomal loci in the nucleus (Figure 2). Standard genetic crosses will then produce a fraction of progeny that only contain the trait gene without antibiotic resistance genes due to independent assortment of genes at meiosis (Mendel's Second Law). When trait and selectable marker genes are cloned into separate plasmids and subsequently co-transformed into plants a proportion of plants will contain both genes. In some of these co-transformed plants the trait and selectable marker gene will be integrated at well-separated chromosomal loci. Although simple in theory, this approach is made difficult by the fact that linkage rather than non-linkage of marker and trait gene is common. This is particularly true of artificial delivery systems where independently introduced plasmids are often found joined together in complex patterns at a single chromosomal integration site.

Use of *Agrobacterium* mediated transformation, which usually results in 'cleaner' insertions of foreign DNA, is more amenable to integrating trait and marker genes at distant sites of chromosomes (Depicker *et al.*, 1985; McKnight *et al.*, 1987; Deblock and Debrouwer, 1991). When mixtures of nopaline C58 *A. tumefaciens* strains were used to infect plants, 37% of transgenic plants received genes from both strains. Unfortunately, in about 80% of these co-transformants the genes were integrated at linked loci (DeBlock and Debrouwer, 1991). Use of binary plasmids containing two T-DNAs in an octopine LBA4404 *A. tumefaciens* strain resulted in a high frequency (47%) of co-transformation of marker and trait genes (Komari *et al.*, 1996). Trait and marker genes were readily segregated in half of these co-transformants. One of the problems with this co-transformation approach to isolate marker-free plants is that it quadruples the number of primary transformants required. Normally, only a fraction of nuclear transgenic plants contain a trait gene expressed at the correct levels, which is also stably inherited. This is due to integration of foreign genes at many chromosomal sites including unfavourable chromosomal locations. The additional requirement of non-linkage of trait and marker genes in this fraction of desirable transgenic plants might prove to be difficult particularly in crops where transformation efficiencies are low (Ow, 2001).

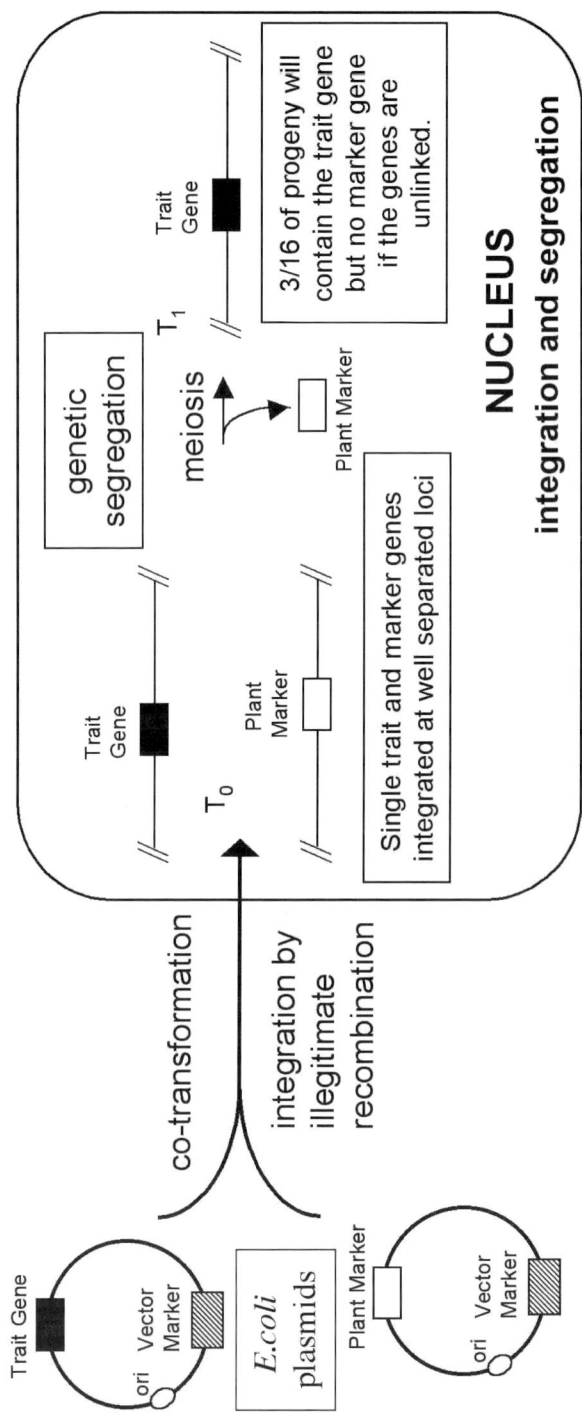

Figure 2. Marker-free nuclear transgenic plants produced by co-transformation and genetic segregation. Trait and plant marker genes are on separate plasmids. The scheme illustrates an ideal case where one copy of each gene is integrated and the two genes are unlinked. Selfing of this heterozygous plant allows the isolation of marker-free plants in the T$_1$ generation. For *Agrobacterium*-based gene transfer independent delivery of trait gene and plant marker gene can also be achieved by localizing them to separate T-DNAs within a single plasmid (Komari *et al.*, 1996). Integrated vector sequences (see Fig. 1B) are not shown.

Substitution of Antibiotic Resistance Genes with Alternative Marker Genes

Replacing Vector-Localised Antibiotic Resistance Genes with Other Less Controversial Marker Genes

Vector-localised antibiotic resistance marker genes, such as *bla(TEM1)*, that are used for the gene cloning steps in *E. coli* can be replaced with a range of alternative marker genes. *E. coli* has a rich history of genetics and a range of auxotrophic mutations have been isolated. Strains with recessive mutant genes that render them unable to grow on media lacking amino acids, such as leucine, histidine or tryptophan, can be rescued with wild-type copies of these genes located on plasmids. Selection for prototrophic growth ensures the maintenance of plasmids in auxotrophic *E. coli* strains. The *leuB* (Razin and Carbon, 1977), *hisB* (Struhl *et al.*, 1976) and *trpE* (Kahn *et al.*, 1979) mutants provide examples of strains in which plasmids containing the wildtype versions of these genes can be propagated. In these cases, selection is based on complementation of mutant genes in the *E. coli* genome with plasmid-located functional copies. It is also possible to restore functionality to a mutant gene by suppressing the mutation using a plasmid-localised suppressor gene (Seed, 1983). Suppressor genes such as *supF*, which encode an altered transfer RNA, have the advantage of being very small. The use of complementation or suppression to maintain plasmids in *E. coli* relies on rescuing mutant auxotrophic strains that are unable to grow on simple salts media lacking specific amino acids. This is a disadvantage compared to antibiotic selection, which is not strain dependent, and allows rapid growth of bacteria to high densities in rich media.

Most plasmid cloning has been done in *E. coli* K12 strains, which are unable to grow on media containing ribitol as the sole carbon source. The presence of three ribitol (*rtl*) utilization genes from *E. coli* strain C, co-ordinately expressed as an operon, allows *E. coli* K12 to utilise ribitol for growth. Plasmids containing the *rtl* operon provide a new selection scheme for maintaining plasmids in *E. coli* K12 strains before they are transferred into plants (LaFayette and Parrott, 2001). Like contemporary antibiotic selection schemes ribitol selection will allow plasmids to be introduced into a wide variety of *E. coli* K12 strains. This should facilitate widespread use of plasmids utilizing ribitol selection by the research community. Once integrated into the plant nuclear genome, vector genes such as the *E. coli HisB* gene and the *rtl* operon are unlikely to be expressed. In the remote event that these genes were taken up by wild-type bacteria in the environment or gut they are unlikely to confer a selective advantage on the transformed host and would be lost.

Plant Selectable Markers Not Based on Antibiotic Selection

A variety of selection schemes based on non-antibiotic marker genes are available for transforming plants. They can be divided into negative and positive selection schemes based on the mode of action of the selective agent (Table 3). Negative selection schemes utilise agents such as antibiotics and herbicides that kill untransformed cells thereby revealing resistant transgenic cells, which proliferate. Positive selection schemes provide a growth advantage to transformed cells by introducing genes that allow them to utilise a component of the plant growth medium, such as a carbon source, which cannot be metabolised by non-transformed cells (Joersbo, 2001), or the ability to grow and regenerate in the absence of hormones (Ebinuma *et al.* 1997). The development of new selection systems based on novel marker genes provides a valuable resource for transgenic research beyond the elimination of antibiotic resistance markers. An increased range of marker genes in the genetic engineer's tool-kit will facilitate development of efficient transformation methods for any given crop. New selectable markers such as the phosphomannose isomerase gene have been reported to be more efficient than conventional antibiotic resistance markers. An added advantage is the agents used for positive selection are not hazardous. When substituting antibiotic resistance markers with alternative markers it will be important to rule out any adverse effects on plant metabolism. In some cases it may be advisable to excise non-antibiotic marker genes and other selectable markers (described below) as well as screenable markers using the procedures described above.

As an alternative to negative and positive selection schemes based on introducing foreign genes into plants it is worth considering transformation schemes based on complementation of recessive mutant genes with their wild type alleles. Such approaches are widely used to select transformed cells in haploid organisms including bacteria, and eukaryotes such as algae and fungi. For example, the *ARG7* (Debuchy *et al.*, 1989) and *NIT1* (Blankenship and Kindle, 1992) genes rescue arginine-requiring and nitrate reductase mutants in the green alga *Chlamydomonas reinhardtii* allowing their use as selectable markers for nuclear transformation. While these schemes are attractive since they involve native plant genes they are difficult to apply to higher plants where gene redundancy combined with polyploidy make it difficult to isolate auxotrophic mutants. The further task of crossing these mutant genes into crops before they can be transformed makes this complementation approach too unwieldy for generating transgenic crops. Nonetheless, plant transformation schemes based on complementation of mutants might be feasible in a limited number of plant species given the isolation of nitrate reductase (Vaucheret *et al.*, 1990), isoleucine-requiring (Sidorov *et al.*, 1981) and tryptophan-requiring mutants (Last *et al.*, 1991).

Table 3. Plant selectable markers not based on antibiotics or herbicides

	Marker Gene	Gene Product	Selective Agent	References
Negative selection	2-DOG-6-phophatase	2-deoxyglucose-6-phophatase	2-deoxyglucose	Kunze et al., 2001
	HemL	glutamate-1-semialdehyde aminotransferase	Gabaculine	Gough et al., 2001
	TDC	tryptophan decarboxylase	4-methyl tryptophan	Goddijn et al., 1993
	DHPS	dihydrodipicolinate synthase	S-aminoethy L-cysteine	Perl et al. 1993
	AK	lysine/threonine desensitised, aspartate kinase	high levels of lysine and threonine	Perl et al. 1993
	BADH (cDNA) plastid transformation	betaine aldehyde dehydrogenase	betaine aldehyde	Daniell et al., 2001
Positive selection	PMI	phosphomannose isomerase	Mannose	Joersbo et al., 1998
	xylA	xylose isomerase	Xylose	Haldrup et al., 1998
	ipt	isopentenyl transferase	hormone-independent shoot proliferation	Ebinuma et al., 1997
	rol	Ri plasmid oncogenes root loci	hormone-independent root proliferation	Cui et al., 2000 Ebinuma and Komamine, 2001
	uidA	β–glucuronidase	glucuronide benzyladenine	Joersbo and Okkels, 1996

Non-Antibiotic Markers Requiring Negative Selection

Herbicide resistance genes work by negative selection and provide alternatives to antibiotic resistance genes. Marker gene and herbicide combinations such as the *bar* gene with glufosinate (Barcelo *et al.*, 1994), mutant acetolactate synthase genes with chlorsulforon (Li *et al.*, 1992), and *EPSPS* gene with glyphosate (Clemente *et al.*, 2000) have all been used to isolate nuclear transgenic plants (Table 2). Although the *bar* and *EPSPS* genes are expressed in plastids they are inefficient plastid markers and primary plastid transformants cannot be selected with glufosinate or glyphosate (Lutz *et al.*, 2001; Ye *et al.*, 2001). However, secondary herbicide selection can be introduced very early in plastid transformation schemes once primary transformants containing the *aadA* gene have been selected with spectinomycin and before plants have been regenerated. Use of secondary herbicide selection makes the *aadA* gene redundant and allows it removal by homologous recombination (Iamtham and Day, 2000).

Plant selectable markers that do not require antibiotics or herbicides are listed in Table 3. Recently, the gene encoding 2-deoxyglucose-6-phosphatase from bakers' yeast was used to select transgenic plant cells grown in the presence of 2-deoxyglucose (Kunze *et al.*, 2001). The endogenous plant enzyme (hexokinase) converts 2-deoxyglucose to 2-deoxyglucose-6-phosphate, which is toxic to plant cells. The introduced yeast phosphatase reverses this step allowing the survival of transgenic cells. In another recent report the mutant glutamate-1-semialdehyde aminotransferase encoded by the *hem*L gene of cyanobacterium *Synechococcus* PCC6301 strain GR6, linked to a plastid-targeting signal, was used as a selectable marker for nuclear transformation in tobacco. Transgenic tobacco plants containing the GR6 *hem*L gene are resistant to gabaculine, an inhibitor of glutamate-1-semialdehyde-aminotransferase, which prevents the synthesis of the heme precursor 5-aminolevulinic acid (Gough *et al.*, 2001). Other promising schemes have been described in the past but have not yet been widely used. For example, the tryptophan decarboxylase gene from *Catharanthus roseus* detoxifies the tryptophan analog 4-methyl tryptophan to non-toxic 4-methyl tryptamine. Transgenic plant cells are resistant to 4-methyl tryptophan (Goddijn *et al.*, 1993). The *E. coli* dihydrodipicolinate synthase gene confers resistance to S-aminoethy L-cysteine and can be used as a selectable marker in plants (Perl *et al.*, 1993). A bacterial desensitized aspartate kinase allowed selection of potato transformants based on resistance to lysine plus threonine in the regeneration and rooting media (Perl *et al.*, 1993).

The cDNA encoding spinach betaine aldehyde dehydrogenase (BADH) is a new plastid transformation marker that does not require antibiotic selection (Daniell *et al.*, 2001). Selection is achieved by placing plant material on media containing betaine aldehyde, which is toxic and kills untransformed cells. BADH converts betaine aldehyde to non-toxic glycine betaine. The selection scheme may also be helped by the fact that glycine betaine is an osmoprotectant and its presence in transgenic cells may have a positive influence on plant growth (Daniell *et al.*, 2001).

Non-Antibiotic Markers Using Positive Selection

A number of markers allow positive selection including the phosphomannose isomerase gene from *E. coli*, the xylose isomerase gene of *Thermoanaerobacterium thermosulforogenes*, and the isopentenyl transferase (*ipt*) gene from *Agrobacterium tumefaciens* (Table 3). When the phosphomannose isomerase marker gene is used with D-mannose as carbon source, untransformed cells take up mannose and convert it to mannose-6-phosphate, which accumulates to detrimental levels. The phosphomannose isomerase in transgenic cells converts mannose-6-phosphate to fructose-6-phosphate, which can be metabolised and confers a growth advantage on transformed cells (Joersbo *et al.*, 1998). The phosphomannose isomerase gene has been used to transform a number of major crops including sugar beet, potato, oilseed rape, maize and wheat (Joersbo, 2001). The widespread use of mannose selection will require evaluation of the finding that the phosphomannose isomerase gene plays a role in the virulence of mammalian fungal pathogens (Wills *et al.*, 2001) and the observation that D-mannose induces cellular changes resembling programmed cell death (Stein and Hansen, 1999). The xylose isomerase gene of *T. thermosulforogenes* allows positive selection when transformed cells are placed on medium containing D-xylose as the sole carbon source (Haldrup *et al.*, 1998). The xylose isomerase gene has been shown to be an effective marker for transformation of potato and tomato (Haldrup *et al.*, 1998).

The isopentenyl transferase (*ipt*) gene from *A. tumefaciens* leads to high cytokinin levels and can be selected on the basis of rapid shoot proliferation of transformed material on hormone-free media (Ebinuma *et al.*, 1997). Once transformed shoots have been selected the *ipt* gene needs to be removed due to undesirable effects of *ipt* gene expression on plant development. This was first achieved by placing the *ipt* gene within a modified Activator transposable element (Ebinuma *et al.*, 1997). Excision of the *ipt*-containing Activator element without subsequent integration led to the isolation of marker-free plants. In an improvement to the method excision of the *ipt* gene was mediated by the site-specific recombination system R/RS from *Zygosaccharomyces rouxii* (Sugita *et al.*, 1999). Chemical regulation of R recombinase activity was recently achieved by linking its gene to an inducible glutathione-S-transferase promoter from *Zea mays* (Sugita *et al.*, 2000). Alternatively, instead of excising the *ipt* gene, *ipt* expression can be switched off using a tightly regulated dexamethasone-inducible promoter once transformation has been achieved (Kunkel *et al.*, 1999). The *rol* genes from *Agrobacterium rhizogenes* increase auxin sensitivity resulting in root proliferation. Selection is based on transformed cells producing roots (Chriqui *et al.,* 1996). Subsequent excision of *rol* genes using the R/*RS* site-specific recombinase system allowed the generation of shoots and transgenic plants (Cui *et al.*, 2000; Ebinuma and Komamine, 2001). The *E. coli* β-glucuronidase (GUS) gene, widely used as a reporter of gene expression, can also be used to select transformed plants when grown on media in which the cytokinin benzyladenine is replaced by a glucuronide derivative (Joersbo and Okkels, 1996). GUS in transformed cells cleaves off the glucuronide moiety to produce active cytokinin that enables shoot proliferation. Untransformed cells do not divide as rapidly because they are unable to carry out the conversion of the cytokinin derivative to its active form. This selection system is not widely used due to the commercial unavailability of glucuronide benzyladenine (Joersbo and Okkels, 1996).

Future Directions

The implementation of technologies for the removal of antibiotic resistance genes from GM crops is a mandatory first step in re-focusing the debate on engineered traits and away from the excess foreign DNA, which is an unnecessary by product of earlier transformation technologies. This chapter has reviewed a range of methods for removing antibiotic resistance genes from crops. In some cases, these methods have been developed in model plants and the next step is to implement them in crops. The removal of antibiotic resistance genes from transgenic plants is only one aspect to removing excess DNA from transgenic plants. Any foreign DNA sequences that are not required for the functionality of desired traits are superfluous and their continued presence in a GM plant should be identified early in a transgenic research program. These sequences could include, for example, non-antibiotic plant marker genes, such as herbicide resistance genes, bacterial origins of replication or foreign genes encoding site-specific DNA recombinases. Whilst these sequences are not as controversial as antibiotic resistance markers, their early elimination, where practical, could pay dividends in terms of reduced safety evaluation tests and simplifying the regulatory process needed for the approval of a GM crop. The development of new markers and gene excision systems, coupled to intellectual property rights, should bring rewards to those investing in these technologies when the next generation of GM crops, lacking excess DNA, gains more rapid regulatory approval and enhances public confidence in the safety of GM crops.

References

Al-Kaff, N.S., Kreike, M.M., Covey, S.N., Pitcher, R., Page, A.M., and Dale, P.J. 2000. Plants rendered herbicide-susceptible by cauliflower mosaic virus-elicited suppression of a 35S promoter-regulated transgene. Nature Biotechnol. 18: 995-999.

Assaad, F.F., and Signer, E.R. 1990. Cauliflower mosaic-virus P35S promoter activity in *Escherichia coli*. Mol. Gen. Genet. 223: 517-520.

Barcelo, P., Hagel, C., Becker, D., Martin, A., and Lorz, H. 1994. Transgenic cereal (tritordeum) plants obtained at high efficiency by microprojectile bombardment of inflorescence tissue. Plant J. 5: 583-592.

Bergstrom, C.T., and Pritchard, J. 1998. Germline bottlenecks and the evolutionary maintenance of mitochondrial genomes. Genetics. 149: 2135-2146.

Bevan, M.W., Flavell, R.B., and Chilton, M.D. 1983. A chimaeric antibiotic resistance gene as a selectable marker for plant cell transformation. Nature 304: 184-187

Bevan, M.W. 1984. Binary *Agrobacterium* vectors for plant transformation. Nucleic Acids Res. 12: 8711-8721.

Blankenship, J.E., and Kindle, K.L. 1992. Expression of chimeric genes by the light regulated CabII-1 promoter in *Chlamydomonas reinhardtii*: a *cabII/nit1* gene functions as a dominant selectable marker in a nit1-nit2 strain. Mol. Cell. Biol. 12: 5268-5279.

Blattner, F.R., Plunkett, G., Bloch, C.A., Perna, N.T., Burland, V., Riley, M., ColladoVides, J., Glasner, J.D., Rode, C.K., Mayhew, G.F., Gregor, J., Davis, N.W., Kirkpatrick, H.A., Goeden, M.A., Rose, D.J., Mau, B., and Shao, Y. 1997. The complete genome sequence of *Escherichia coli* K-12. Science. 277: 1453-1474.

Bolivar, F., and Backman, K. 1979. Plasmids of *Escherichia coli* as cloning vectors. Meth. Enzymol. 68: 245-267.

Carrer, H., Hockenberry, T.N., Svab, Z., and Maliga, P. 1993. Kanamycin resistance as a selectable marker for plastid transformation in tobacco. Mol. Gen. Genet. 241: 49-56.

Chiter, A., Forbes, J.M., and Blair, G.E. 2000. DNA stability in plant tissues: implications for the possible transfer of genes from genetically modified food. FEBS Lett. 481: 164-168.

Chriqui, D., Guivarch, A., Dewitte, W., Prinsen, E., and van Onkelen, H. 1996. *Rol* genes and root initiation and development. Plant and Soil. 187: 47-55.

Christensen, A.H., and Quail, P.H. 1996. Ubiquitin promoter-based vectors for high-level expression of selectable and/or screenable marker genes in monocotyledonous plants. Transgenic Res. 5: 213-218.

Clemente, T.E., LaVallee, B.J., Howe, A.R., Conner-Ward, D., Rozman, R.J., Hunter, P.E., Broyles, D.L., Kasten, D.S., and Hinchee, M.A. 2000. Progeny analysis of glyphosate selected transgenic soybeans derived from *Agrobacterium*-mediated transformation. Crop Sci. 40: 797-803.

Coninx, R., Mathieu, C., Debacker, M., Mirzoev, F., Ismaelov, A., de Haller, R., and Meddings, D.R. 1999. First-line tuberculosis therapy and drug-resistant *Mycobacterium tuberculosis* in prisons. Lancet. 353: 969-973.

Corneille, S., Lutz, K., and Maliga, P. 2000. Conservation of RNA editing between rice and maize plastids: are most editing events dispensable? Mol. Gen. Genet. 264: 419-424.

Corneille, S., Lutz, K., Svab, Z., and Maliga, P. 2001. Efficient elimination of selectable marker genes from the plastid genome by the CRE-*lox* site-specific recombination system. Plant J. 27: 171-178.

Corpet, D.E. 2000. Mechanism of antimicrobial growth promoters used in animal feed. Revue De Medecine Veterinaire. 151: 99-104.

Cui, M.L., Takayanagi, K., Kamada, H., Nishimura, S., and Handa, T. 2000. Transformation of *Antirrhinum majus* L. by a rol-type multi- auto-transformation (MAT) vector system. Plant Sci. 159: 273-280.

Dale, E.C., and Ow, D.W. 1991. Gene-transfer with subsequent removal of the selection gene from the host genome. Proc. Natl. Acad. Sci. USA. 88: 10558-10562.

Daniell, H., Muthukumar, B., and Lee, S.B. 2001. Marker free transgenic plants: engineering the chloroplast genome without the use of antibiotic selection. Curr. Genet. 39: 109-116.

Davies, J. 1994. Inactivation of antibiotics and the dissemination of resistance genes. Science. 264: 375-382.

de Vries, J., and Wackernagel, W. 1998. Detection of *nptII* (kanamycin resistance) genes in genomes of transgenic plants by marker-rescue transformation. Mol. Gen. Genet. 257: 606-613.

Deblock, M., Botterman, J., Vandewiele, M., Dockx, J., Thoen, C., Gossele, V., Movva, N.R., Thompson, C., Vanmontagu, M., and Leemans, J. 1987. Engineering herbicide resistance by expression of a detoxifying enzyme. EMBO J. 6: 2513-2518.

Deblock, M., and Debrouwer, D. 1991. Two T-DNAs co-transformed into *Brassica napus* by a double *Agrobacterium tumefaciens* infection are mainly integrated at the same locus. Theor. Applied Genet. 82: 257-263.

Debuchy, R., Purton, S., and Rochaix, J.D. 1989. The argininosuccinate lyase gene of *Chlamydomonas reinhardtii*: an important tool for nuclear transformation and for correlating the genetic and molecular maps of the *ARG7* locus. EMBO J. 8: 2803-2809.

Demaneche, S., Jocteur-Monrozier, L., Quiquampoix, H., and Simonet, P. 2001. Evaluation of biological and physical protection against nuclease degradation of clay-bound plasmid DNA. Applied and Environmental Microbiol. 67: 293-299.

Depicker, A., Herman, L., Jacobs, A., Schell, J., and Vanmontagu, M. 1985. Frequencies of simultaneous transformation with different T- DNAs and their relevance to the *Agrobacterium* plant cell interaction. Mol. Gen. Genet. 201: 477-484.

Doblhoff-Dier, O., Bachmayer, H., Bennett, A., Brunius, G., Cantley, M., Collins, C., Collard, J.M., Crooy, P., Elmqvist, A., Frontali-Botti, C., Gassen, H.G., Havenaar, R., Haymerle, H., Lamy, D., Lex, M., Mahler, J.L., Martinez, L., Mosgaard, C., Olsen, L., Pazlarova, J., Rudan, F., Sarvas, M., Stepankova, H., Tzotzos, G., and Wagner, K. 2000. Safe biotechnology 10: DNA content of biotechnological process waste. Trends Biotechnol. 18: 141-146.

Dower, W.J., Miller, J.F., and Ragsdale, C.W. 1988. High-efficiency transformation of *Escherichia coli* by high voltage electroporation. Nucleic Acids Res. 16: 6127-6145.

Dubnau, D. 1991. Genetic competence in *Bacillus subtilis*. Microbiol. Rev. 55: 395-424.

Duggan, P.S., Chambers, P.A., Heritage, J., and Forbes, J.M. 2000. Survival of free DNA encoding antibiotic resistance from transgenic maize and the transformation activity of DNA in ovine saliva, ovine rumen fluid and silage effluent. FEMS Microbiol. Letts. 191: 71-77.

Eady, C., Twell, D., and Lindsey, K. 1995. Pollen viability and transgene expression following storage in honey. Transgenic Res. 4: 226-231.

Ebinuma, H., and Komamine, A. 2001. MAT (Multi-Auto-Transformation) Vector System. The oncogenes of *Agrobacterium* as positive markers for regeneration and selection of marker-free transgenic plants. *In Vitro* Cell. Develop. Biol. Plant. 37: 103-113.

Ebinuma, H., Sugita, K., Matsunaga, E., and Yamakado, M. 1997. Selection of marker-free transgenic plants using the isopentenyl transferase gene. Proc. Natl. Acad. Sci. USA. 94: 2117-2121.

Einspanier, R., Klotz, A., Kraft, J., Aulrich, K., Poser, R., Schwagele, F., Jahreis, G., and Flachowsky, G. 2001. The fate of forage plant DNA in farm animals: a collaborative case-study investigating cattle and chicken fed recombinant plant material. European Food Res. Technol. 212: 129-134.

Flamm, R.K., Phillips, K.L., Tenover, F.C., and Plorde, J.J. 1993. A survey of clinical isolates of Enterobacteriaceae using a series of DNA probes for aminoglycoside resistance genes. Mol. Cell. Probes. 7: 139-144.

Frere, J.M. 1995. Beta-lactamases and bacterial resistance to antibiotics. Mol. Microbiol. 16: 385-395.

Fuchs, R.L., Ream, J.E., Hammond, B.G., Naylor, M.W., Leimgruber, R.M., and Berberich, S.A. 1993. Safety assessment of the neomycin phosphotransferase-Ii (NptII) protein. Bio/Technology. 11: 1543-1547.

Gebhard, F., and Smalla, K. 1998. Transformation of *Acinetobacter sp.* strain BD413 by transgenic sugar beet DNA. Appl. Environ. Microbiol. 64: 1550-1554.

Giddings, G., Allison, G., Brooks, D., and Carter, A. 2000. Transgenic plants as factories for biopharmaceuticals. Nature Biotechnol. 18: 1151-1155.

Gleave, A.P., Mitra, D.S., Mudge, S.R., and Morris, B.A.M. 1999. Selectable marker-free transgenic plants without sexual crossing: transient expression of cre recombinase and use of a conditional lethal dominant gene. Plant Mol. Biol. 40: 223-235.

Goddijn, O.J.M., Schouten, P.M.V., Schilperoort, R.A., and Hoge, J.H.C. 1993. A chimeric tryptophan decarboxylase gene as a novel selectable marker in plant cells. Plant Mol. Biol. 22: 907-912.

Goldsbrough, A.P., Lastrella, C.N., and Yoder, J.I. 1993. Transposition mediated repositioning and subsequent elimination of marker genes from transgenic tomato. Bio/Technology. 11: 1286-1292.

Gough, K.C., Hawes, W.S., Kilpatrick, J., and Whitelam, G.C. 2001. Cyanobacterial GR6 glutamate-1-semialdehyde aminotransferase: a novel enzyme-based selectable marker for plant transformation. Plant Cell Rep. 20: 296-300.

Gressel, J. 2000. Molecular biology of weed control. Transgenic Res. 9: 355-382.

Guerineau, F., Brooks, L., Meadows, J., Lucy, A., Robinson, C., and Mullineaux, P. 1990. Sulfonamide resistance gene for plant transformation. Plant Mol Biol 15: 127-136.

Hajdukiewicz, P.T.J., Gilbertson, L., and Staub, J.M. 2001. Multiple pathways for Cre/lox-mediated recombination in plastids. Plant J. 27: 161-170.

Haldrup, A., Petersen, S.G., and Okkels, F.T. 1998. The xylose isomerase gene from *Thermoanaerobacterium thermosulfurogenes* allows effective selection of transgenic plant cells using D-xylose as the selection agent. Plant Mol. Biol. 37: 287-296.

Hall, L., Topinka, K., Huffman, J., Davis, L., and Good, A. 2000. Pollen flow between herbicide-resistant *Brassica napus* is the cause of multiple-resistant *B. napus* volunteers. Weed Sci. 48: 688-694.

Hanahan, D. 1983. Studies on transformation of *Escherichia coli* with plasmids. J. Mol. Biol. 166: 557-580.

Haughn, G.W., Smith, J., Mazur, B., and Somerville, C. 1988. Transformation with a mutant *Arabidopsis* acetolactate synthase gene renders tobacco resistant to sulfonylurea herbicides. Mol. Gen. Genet. 211: 266-271.

Hayford, M.B., Medford, J.I., Hoffman, N.L., Rogers, S.G., and Klee, H.J. 1988. Development of a plant transformation selection system based on expression of genes encoding gentamicin acetyltransferases. Plant Physiol. 86: 1216-1222.

Hellens, R., Mullineaux, P., and Klee, H. 2000. A guide to *Agrobacterium* binary Ti vectors. Trends Plant Sci. 5: 446-451.

Herrera-Estrella, L., Deblock, M., Messens, E., Hernalsteens, J.P., Vanmontagu, M., and Schell, J. 1983a. Chimeric genes as dominant selectable markers in plant cells. EMBO J. 2: 987-995.

Herrera-Estrella, L., Depicker, A., Van Montagu, M., and Schell, J. 1983b. Expression of chimaeric genes transferred into plant cells using a Ti-plasmid-derived vector. Nature. 303: 209-213.

Hille, J., Verheggen, F., Roelvink, P., Franssen, H., van Kammen, A., and Zabel, P. 1986. Bleomycin resistance: a new dominant marker for plant cell transformation. Plant Mol. Biol. 7: 171-176.

Hiratsuka, J., Shimada, H., Whittier, R., Ishibashi, T., Sakamoto, M., Mori, M., Kondo, C., Honji, Y., Sun, C.-R., Meng, B.-Y., Li, Y.-Q., Kanno, A., Nishizawa, Y., Hirai, A., Shinozaki, K., and Sugiura, M. 1989. The complete nucleotide sequence of the rice (*Oryza sativa*) chloroplast genome: Intermolecular recombination between distinct tRNA genes accounts for a major plastid DNA inversion during the evolution of cereals. Mol. Gen. Genet. 217: 185-194.

Hohn, B., Levy, A., and Puchta, H. 2001. Elimination of selection markers from transgenic plants. Cur. Opin. Biotechnol. 12: 139-143.

Iamtham, S., and Day, A. 2000. Removal of antibiotic resistance genes from transgenic tobacco plastids. Nature Biotech. 18: 1172-1176.

Ison, C.A. 1996. Antimicrobial agents and gonorrhoea: Therapeutic choice, resistance and susceptibility testing. Genitourinary Med. 72: 253-257.

JETACAR. 1999. The use of antibiotics in food-producing animals: antibiotic resistant bacteria in animals and humans. Canberra: Biotext.

Joersbo, M. 2001. Advances in the selection of transgenic plants using non-antibiotic marker genes. Physiol. Plant. 111: 269-272.

Joersbo, M., Donaldson, I., Kreiberg, J., Petersen, S.G., Brunstedt, J., and Okkels, F.T. 1998. Analysis of mannose selection used for transformation of sugar beet. Mol. Breeding 4: 111-117.

Joersbo, M., and Okkels, F.T. 1996. A novel principle for selection of transgenic plant cells: positive selection. Plant Cell Rep. 16: 219-221.

Johnston, S.A., Anziano, P.Q., Shark, K., Sanford, J.C., and Butow, R.A. 1988. Mitochondrial transformation in yeast by bombardment with microprojectiles. Science. 240: 1538-1541.

Kahn, M., Kolter, R., Thomas, C., Figurski, D., Meyer, R., Remaut, E., and Helinski, D.R. 1979. Plasmid cloning vehicles derived from plasmids ColE1, F, R6K and RK2. Meth. Enzymol. 68: 268-280.

Kao, H.M., Keller, W.A., Gleddie, S., and Brown, G.G. 1992. Synthesis of *Brassica oleracea/ Brassica napus* somatic hybrid plants with novel organelle DNA compositions. Theor. Appl. Genet. 83: 313-320.

Khachtourians, G.G. 1998. Agricultural use of antibiotics and the evolution and transfer of antibiotic-resistant bacteria. Canadian Med. Ass. J. 159: 1129-1136.

Kilby, N.J., Davies, G.J., Snaith, M.R., and Murray, J.A.H. 1995. FLP recombinase in transgenic plants - constitutive activity in stably transformed tobacco and generation of marked cell clones in Arabidopsis. Plant J. 8: 637-652.

Kivaisi, A.K., Dencamp, H., Lubberding, H.J., Boon, J.J., and Vogels, G.D. 1990. Generation of soluble lignin derived compounds during degradation of barley straw in an artificial rumen reactor. Appl. Microbiol. Biotechnol. 33: 93-98.

Komari, T., Hiei, Y., Saito, Y., Murai, N., and Kumashiro, T. 1996. Vectors carrying two separate T-DNAs for co-transformation of higher plants mediated by *Agrobacterium tumefaciens* and segregation of transformants free from selection markers. Plant J. 10: 165-174.

Kononov, M.E., Bassuner, B., and Gelvin, S.B. 1997. Integration of T-DNA binary vector 'backbone' sequences into the tobacco genome: Evidence for multiple complex patterns of integration. Plant J. 11: 945-957

Kuiper, H.A., Kleter, G.A., Noteborn, H.P.J.M., and Kok, E.J. 2001. Assessment of the food safety issues related to genetically modified foods. Plant J. 27: 503-528.

Kumar, S., and Fladung, M. 2001. Controlling transgene integration in plants. Trends Plant Sci. 6: 155-159.

Kunkel, T., Niu, Q.W., Chan, Y.S., and Chua, N.H. 1999. Inducible isopentenyl transferase as a high-efficiency marker for plant transformation. Nature Biotechnol. 17: 916-919.

Kunze, I., Ebneth, M., Heim, U., Geiger, M., Sonnewald, U., and Herbers, K. 2001. 2-Deoxyglucose resistance: a novel selection marker for plant transformation. Mol. Breeding 7: 221-227.

LaFayette, P.R., and Parrott, W.A. 2001. A non-antibiotic marker for amplification of plant transformation vectors in *E. coli*. Plant Cell Rep. 20: 338-342.

Last, R.L., Bissinger, P.H., Mahoney, D.J., Radwanski, E.R., and Fink, G.R. 1991. Tryptophan mutants in *Arabidopsis* - the consequences of duplicated tryptophan synthase beta genes. Plant Cell. 3: 345-358.

Leff, L.G., Dana, J.R., McArthur, J.V., and Shimkets, L.J. 1993. Detection of Tn5-like sequences in kanamycin-resistant stream bacteria and environmental DNA. Appl. Environ. Microbiol. 59: 417-421.

Li, Z.J., Hayashimoto, A., and Murai, N. 1992. A sulfonylurea herbicide resistance gene from *Arabidopsis thaliana* as a new selectable marker for production of fertile transgenic rice plants. Plant Physiol. 100: 662-668.

Livermore, D.M. 1995. Beta-lactamases in laboratory and clinical resistance. Clin. Microbiol. Rev. 8: 557-584.

Lorenz, M.G., Reipschlager, K., and Wackernagel, W. 1992. Plasmid transformation of naturally competent *Acinetobacter calcoaceticus* in nonsterile soil extract and groundwater. Arch. Microbiol.157: 355-360.

Luehrsen, K.R., and Walbot, V. 1991. Intron enhancement of gene expression and the splicing efficiency of introns in maize cells. Mol. Gen. Genet. 225: 81-93.

Lutz, K.A., Knapp, J.E., and Maliga, P. 2001. Expression of bar in the plastid genome confers herbicide resistance. Plant Physiol. 125: 1585-1590.

Maeser, S., and Kahmann, R. 1991. The Gin recombinase of phage Mu can catalyze site-specific recombination in plant protoplasts. Mol. Gen. Genet. 230: 170-176.

Maliga, P. 2001. Plastid engineering bears fruit - The tomato has been engineered to express in its plastids high levels of a recombinant protein. Nature Biotechnol. 19: 826-827.

Martineau, B., Voelker, T.A., and Sanders, R.A. 1994. On defining T-DNA. Plant Cell. 6: 1032-1033.

Matzke, M.A., Mette, M.F., and Matzke, A.J.M. 2000. Transgene silencing by the host genome defense: implications for the evolution of epigenetic control mechanisms in plants and vertebrates. Plant Mol. Biol. 43: 401-415.

McClintock, B. 1984. The significance of responses of the genome to challenge. Science. 226: 792-801

McKnight, T.D., Lillis, M.T., and Simpson, R.B. 1987. Segregation of genes transferred to one plant cell from 2 separate *Agrobacterium* strains. Plant Mol. Biol. 8: 439-445.

Mikkelsen, T.R., Andersen, B., and Jorgensen, R.B. 1996. The risk of crop transgene spread. Nature 380: 31.

Molinier, J., Himber, C., and Hahne, G. 2000. Use of green fluorescent protein for detection of transformed shoots and homozygous offspring. Plant Cell Rep. 19: 219-223.

Nap, J.-P., J., B., and Stiekema, W. 1992. Biosafety of kanamycin-resistant transgenic plants. Transgenic Res. 1: 239-249.

Nielsen, K.M., Bones, A.M., Smalla, K., and van Elsas, J.D. 1998. Horizontal gene transfer from transgenic plants to terrestrial bacteria - a rare event? Fems Microbiol. Rev. 22: 79-103.

Nilsson, B., Uhlen, M., Josephson, S., Gatenbeck, S., and Philipson, L. 1983. An improved positive selection plasmid vector constructed by oligonucleotide mediated mutagenesis. Nucleic Acids Res. 11: 8019-8030.

Onouchi, H., Nishihama, R., Kudo, M., Machida, Y., and Machida, C. 1995. Visualization of site-specific recombination catalyzed by a recombinase from *Zygosaccharomyces rouxii* in *Arabidopsis thaliana*. Mol. Gen. Genet. 247: 653-660.

Ow, D.W. 2001. The right chemistry for marker gene removal? Nature Biotechnol. 19: 115-116.

Palmer, J.D., and Shields, C.R. 1984. Tripartite structure of the *Brassica campestris* mitochondrial genome. Nature 307: 437-440.

Perl, A., Galili, S., Shaul, O., Bentzvi, I., and Galili, G. 1993. Bacterial dihydrodipicolinate synthase and desensitized aspartate kinase: 2 novel selectable markers for plant transformation. Bio/Technology. 11: 715-718.

Potrykus, I. 1991. Gene transfer to plants: assessment of published approaches and results. Ann. Rev. Plant Physiol. Plant Mol. Biol. 42: 205-225.

Prentki, P., and Krisch, H.M. 1984. *In vitro* insertional mutagenesis with a selectable DNA fragment. Gene. 29: 303-313.

Purrington, C.B., and Bergelson, J. 1997. Fitness consequences of genetically engineered herbicide and antibiotic resistance in *Arabidopsis thaliana*. Genetics. 145: 807-814.

Raleigh, E.A., Murray, N.E., Revel, H., Blumenthal, R.M., Westaway, D., Reith, A.D., Rigby, P.W.J., Elhai, J., and Hanahan, D. 1988. McrA and McrB restriction phenotypes of some *Escherichia coli* strains and implications for gene cloning. Nucleic Acids Res. 16: 1563-1575.

Randolph-Anderson, B., Boynton, J., Gillham, N., Harris, E., Johnson, A., Dorhtu, M.-P., and Matagne, R. 1993. Further characterisation of the respiratory deficient dum-1 mutation of *Chlamydomonas reinhardtii* and its use as a recipient for mitochondrial transformation. Mol. Gen. Genet. 236: 235-244.

Rayssiguier, C., Thaler, D.S., and Radman, M. 1989. The barrier to recombination between *Escherichia coli* and *Salmonella typhimurium* is disrupted in mismatch repair mutants. Nature. 342: 396-401.

Razin, B., and Carbon, J. 1977. Functional expression of cloned yeast DNA in *Escherichia coli*. Proc. Natl. Acad. Sci. USA. 74: 487-491.

Risseeuw, E., Offringa, R., Frankevandijk, M.E.I., and Hooykaas, P.J.J. 1995. Targeted recombination in plants using *Agrobacterium* coincides with additional rearrangements at the target locus. Plant J. 7: 109-119.

Russell, S.H., Hoopes, J.L., and Odell, J.T. 1992. Directed excision of a transgene from the plant genome. Mol. Gen. Genet. 234: 49-59.

Sambrook, J., Fritsch, E.F., and Maniatis, T. 1989. Molecular Cloning: a Laboratory Manual. Cold Spring Harbor 2nd Ed.

Schaefer, D.G. 2001. Gene targeting in *Physcomitrella patens*. Curr. Opin. Plant Biol. 4: 143-150.

Schubbert, R., Renz, D., Schmitz, B., and Doerfler, W. 1997. Foreign (M13) DNA ingested by mice reaches peripheral leukocytes, spleen, and liver via the intestinal wall mucosa and can be covalently linked to mouse DNA. Proc. Natl. Acad. Sci. USA. 94: 961-966.

Seed, B. 1983. Purification of genomic sequences from bacteriophage libraries by recombination and selection *in vivo*. Nucleic Acids Res. 11: 2427-2445.

Shah, D.M., Horsch, R.B., Klee, H.J., Kishore, G.M., Winter, J.A., Tumer, N.E., Hironaka, C.M., Sanders, P.R., Gasser, C.S., Aykent, S., Siegel, N.R., Rogers, S.G., and Fraley, R.T. 1986. Engineering herbicide tolerance in transgenic plants. Science. 233: 478-481.

Shen, P., and Huang, H.V. 1986. Homologous recombination in *Escherichia coli* - dependence on substrate length and homology. Genetics. 112: 441-457.

Shere, J.A., Bartlett, K.J., and Kaspar, C.W. 1998. Longitudinal study of *Escherichia coli* O157:H7 dissemination on four dairy farms in Wisconsin. Appl. Env. Microbiol. 64: 1390-1399.

Shoemaker, N.B., Vlamakis, H., Hayes, K., and Salyers, A.A. 2001. Evidence for extensive resistance gene transfer among *Bacteroides spp.* and among *Bacteroides* and other genera in the human colon. Appl. Environ. Microbiol. 67: 561-568.

Short, J.M., Fernandez, J.M., Sorge, J.A., and Huse, W.D. 1988. Lambda ZAP- a bacteriophage lamda vector with *in vivo* excision properties. Nucleic Acids Res. 16: 7583-7600.

Sidorov, V., Menczel, L., and Maliga, P. 1981. Isoleucine-requiring *Nicotiana* plant deficient in threonine deaminase. Nature. 294: 87-88.

Smalla, K., Vanoverbeek, L.S., Pukall, R., and Vanelsas, J.D. 1993. Prevalence of NptII and Tn5 in kanamycin-resistant bacteria from different environments. FEMS Microbiol. Ecol. 13: 47-58.

Smith, N., Kilpatrick, J.B., and Whitelam, G.C. 2001. Superfluous transgene integration in plants. Crit. Rev. Plant Sci. 20: 215-249.

Stalker, D.M., McBride, K.E., and Malyj, L.D. 1988. Herbicide resistance in transgenic plants expressing a bacterial detoxification gene. Science. 242: 419-423.

Stein, J.C., and Hansen, G. 1999. Mannose induces an endonuclease responsible for DNA laddering in plant cells. Plant Physiol. 121: 71-79.

Struhl, K., Cameron, J.R., and Davis, R.W. 1976. Functional genetic expression of eukaryotic DNA in *Escherichia coli*. Proc. Natl. Acad. Sci. USA. 73: 1471-1475.

Sugita, K., Kasahara, T., Matsunaga, E., and Ebinuma, H. 2000. A transformation vector for the production of marker-free transgenic plants containing a single copy transgene at high frequency. Plant J. 22: 461-469.

Sugita, K., Matsunaga, E., and Ebinuma, H. 1999. Effective selection system for generating marker-free transgenic plants independent of sexual crossing. Plant Cell Rep. 18: 941-947.

Svab, Z., Harper, E.C., Jones, J.D.G., and Maliga, P. 1990a. Aminoglycoside-3''-adenyltransferase confers resistance to spectinomycin and streptomycin in Nicotiana tabacum. Plant Mol. Biol. 14: 197-205.

Svab, Z., Hajdukiewicz, P., and Maliga, P. 1990b. Stable transformation of plastids in higher plants. Proc. Natl. Acad Sci. USA. 87: 8526-8530.

Svab, Z., and Maliga, P. 1993. High frequency plastid transformation in tobacco by selection for a chimeric aadA gene. Proc. Natl. Acad.Sci. USA. 90: 913-917.

Thimm, T., Hoffmann, A., Fritz, I., and Tebbe, C.C. 2001. Contribution of the earthworm *Lumbricus rubellus* (Annelida, Oligochaeta) to the establishment of plasmids in soil bacterial communities. Microbial Ecol. 41: 341-351.

Thomson, J.A. 2001. Horizontal transfer of DNA from GM crops to bacteria and to mammalian cells. J. Food Sci. 66: 188-193.

Thorpe, H.M., and Smith, M.C.M. 1998. *In vitro* site-specific integration of bacteriophage DNA catalyzed by a recombinase of the resolvase/invertase family. Proc. Natl. Acad. Sci. USA. 95: 5505-5510.

Twigg, A.J., and Sherratt, D. 1980. Trans-complementable copy number mutants of the plasmids ColE1. Nature. 283: 216-217.

Unseld, M., Marienfeld, J.R., Brandt, P., and Brennicke, A. 1997. The mitochondrial genome of *Arabidopsis thaliana* contains 57 genes in 366,924 nucleotides. Nature Genet. 15: 57-61.

Vancanneyt, G., Schmidt, R., Oconnorsanchez, A., Willmitzer, L., and Rochasosa, M. 1990. Construction of an intron-containing marker gene - splicing of the intron in transgenic plants and its use in monitoring early events in *Agrobacterium*-mediated plant transformation. Mol. Gen. Genet. 220: 245-250.

Vasquez, K.M., Marburger, K., Intody, Z., and Wilson, J.H. 2001. Manipulating the mammalian genome by homologous recombination. Proc. Natl. Acad. Sci. USA. 98: 8403-8410.

Vaucheret, H., Chabaud, M., Kronenberger, J., and Caboche, M. 1990. Functional complementation of tobacco and *Nicotiana plumbaginifolia* nitrate reductase deficient mutants by transformation with the wild-type alleles of the tobacco structural genes. Mol. Gen. Genet. 220: 468-474.

Vergunst, A.C., Schrammeijer, B., den Dulk-Ras, A., de Vlaam, C.M.T., Regensburg-Tuink, T.J.G., and Hooykaas, P.J.J. 2000. VirB/D4-dependent protein translocation from *Agrobacterium* into plant cells. Science. 290: 979-982.

Vollenhofer, S., Burg, K., Schmidt, J., and Kroath, H. 1999. Genetically modified organisms in food-screening and specific detection by polymerase chain reaction. J. Agricultur. Food Chem. 47: 5038-5043.

Walbot, V. 2000. A green chapter in the book of life. Nature. 408: 794-795.

Waldron, C., Murphy, E.B., Roberts, J.L., Gustafson, G.D., Armour, S.L., and Malcolm, S.K. 1985. Resistance to hygromycin B: a new marker for plant transformation studies. Plant Mol. Biol. 5: 103-108

WHO. 2001. WHO global strategy for containment of antimicrobial resistance: World Health Organization.

Wills, E.A., Roberts, I.S., Del Poeta, M., Rivera, J., Casadevall, A., Cox, G.M., and Perfect, J.R. 2001. Identification and characterization of the *Cryptococcus neoformans* phosphomannose isomerase-encoding gene, *MAN1*, and its impact on pathogenicity. Mol. Microbiol. 40: 610-620.

Yanisch-Perron, C., Vieira, J., and Messing, J. 1985. Improved M13 Phage Cloning vectors and host strains: nucleotide sequences of the M13mp18 and pUC19 vectors. Gene. 33: 103-119.

Ye, G.N., Hajdukiewicz, P.T.J., Broyles, D., Rodriguez, D., Xu, C.W., Nehra, N., and Staub, J.M. 2001. Plastid-expressed 5-enolpyruvylshikimate-3-phosphate synthase genes provide high level glyphosate tolerance in tobacco. Plant J. 25: 261-270.

Ye, G.N., Pang, S.Z., and Sanford, J.C. 1996. Tobacco (*Nicotiana tabacum*) nuclear transgenics with high copy number can express NPTII driven by the chloroplast psbA promoter. Plant Cell Rep. 15: 479-483.

Yoder, J.I., and Goldsbrough, A.P. 1994. Transformation systems for generating marker-free transgenic plants. Bio/Technology. 12: 263-267.

Zubko, E., Scutt, C., and Meyer, P. 2000. Intrachromosomal recombination between attP regions as a tool to remove selectable marker genes from tobacco transgenes. Nature Biotechnol.18: 442-445.

Zuo, J.R., Niu, Q.W., and Chua, N.H. 2000. An estrogen receptor-based transactivator XVE mediates highly inducible gene expression in transgenic plants. Plant J. 24: 265-273

Zuo, J.R., Niu, Q.W., Moller, S.G., and Chua, N.H. 2001. Chemical-regulated, site-specific DNA excision in transgenic plants. Nature Biotechnol. 19: 157-161.

From: *Transgenic Plants: Current Innovations and Future Trends*
Edited by: C. Neal Stewart, Jr.

Chapter 7

Site-Specific Recombination Systems and Their Uses for Targeted Gene Manipulation in Plant Systems

C. L. Baszczynski, W. J. Gordon-Kamm,
L. A. Lyznik, D. J. Peterson
and Z.-Y. Zhao

Abstract

Genetic transformation of the world's major crops has become routine, the result of significant technical advances over the past 10 years. Standard plant transformation methods do not provide for post-insertion manipulation of transgene sequences except through conventional breeding. Progress has been made in the development of new technologies, tools and methodologies to facilitate targeted approaches to gene integration or modification. Sequence-specific recombinases are being increasingly used to introduce or manipulate transgenes and as general tools for genetic manipulations. We focus in this chapter on the adaptation and application of naturally occurring site-specific recombination systems for use in plants, including relevant work from various species that has provided the basis for these new applications. In addition, we discuss some future directions in plants for these and other technologies.

Introduction

Plant transformation has been developed into a robust technology for studying gene expression and for the production of plants carrying new traits of interest. The vast majority of transformation work continues to rely on random integration, a process implicated in numerous problems associated with stability of transgene expression, variable expression between independent transgenic events and other position-related effects (Kooter *et al.*, 1999; van Leeuwen *et al.*, 2001). To better understand and address some of the concerns with random integration, as well as to better control the integration, expression or manipulation of transgenes, researchers have utilized and adapted naturally occurring site-specific recombination systems for use in both animals and plants. Some of these systems have been developed into tools useful for a wide variety of elegant genetic manipulations.

Investigations into recombinase-mediated sequence manipulation began in organisms most easily transformed, and has progressed into increasingly more recalcitrant species. The most widely used among the available site-specific recombinases originate from the Cre/*lox* system of bacteriophage P1 and the FLP/*FRT* system of the yeast 2μ plasmid (Cox, 1983). Cre and FLP are members of the tyrosine family of site-specific recombinases that includes the best-studied phage lambda integrase. They catalyze both integration and excision and share a similar catalytic mechanism of DNA recombination. The respective recombination sites are called *loxP* (locus of crossover) and *FRT* (FLP recombination target). The sites are composed of 13 bp inverted repeats providing binding sites for the recombinases and an 8 bp spacer region (see Figure 1A). Asymmetry of the spacer region determines the directionality of recombination sites. Two monomers bind to one recombination site bringing two of them together into a synaptic complex of four monomers and two recombination sites (Figure 1B). Conserved tyrosines at amino acid position 324 (or 343 in FLP monomers) provide hydroxyl groups for cleavage of the phosphodiester bonds (vertical arrows), leading to the formation of covalent linkages between monomers and DNA (only two are shown in Figure 1). The synaptic complex is resolved by subsequent rounds of transesterification reactions involving phosphotyrosine

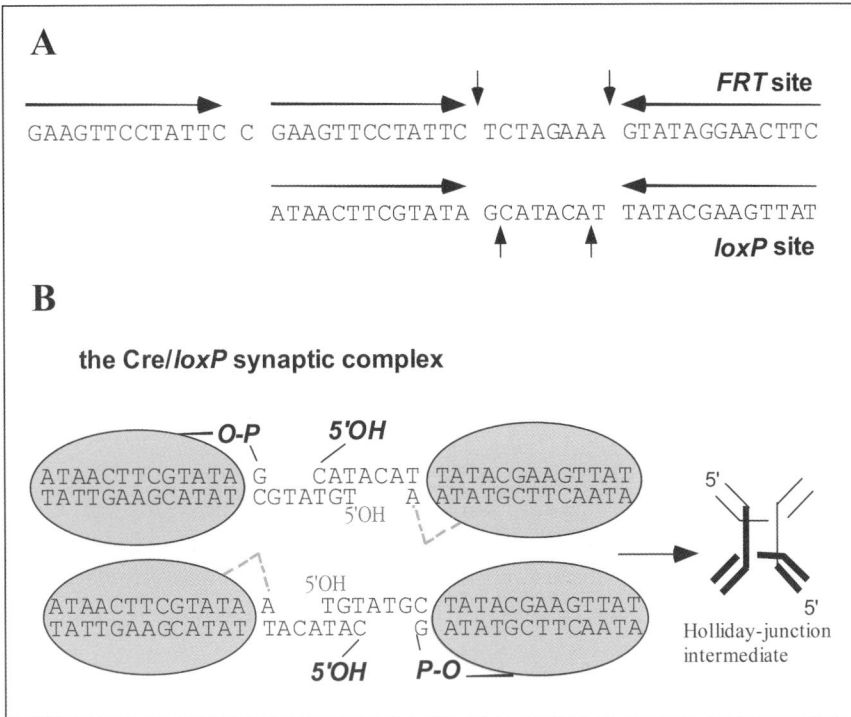

Figure 1. Schematic diagram of site-specific recombination target sites and a recombination complex. A. The recombination target site for either the Cre/*Lox* or FLP/*FRT* system consists of two 13 bp inverted repeats (long arrows) flanking an 8 bp spacer region. The FLP/*FRT* target site includes a third repeat which does not appear to be required for recombination. Small arrows show positions of recombinase cleavage. B. Synaptic complex formed during the Cre-mediated site-specific recombination reaction. Recombinase proteins bind to the repeats and catalyze cleavage as described in the text.

bonds and free hydroxyl groups of the newly formed 5' DNA strands (essentially a Holliday-junction intermediate is formed). There is an anti-parallel site arrangement of the Cre-*loxP* complex and the *cis* cleavage reaction catalyzed by Cre (the same Cre monomer both activates and cleaves a phosphodiester bond) (Guo *et al.*, 1997). In contrast, FLP recombinase cleaves in *trans* and the *FRT* sites are in a parallel orientation, in addition to other subtle structural differences in the complex (Huffman and Levene, 1999; Lee *et al.*, 1999).

Both systems are simple and well characterized, comprising the recombinase and its cognate target sites. Cre recombinase is functional in yeast (Sauer and Henderson, 1988), and both proteins (Cre and FLP) can recombine extrachromosomal substrates in mammalian (Sauer and Henderson, 1988; O'Gorman *et al.*, 1991) or plant cells (Dale and Ow, 1990;

Lyznik *et al.*, 1993). Chromosomal localization of recombination target sites does not prevent either recombinase binding or catalysis in heterologous organisms, including bacteria (Huang *et al.*, 1991), insects (Golic and Lindquist, 1989; Morris *et al.*, 1991; Konsolaki *et al.*, 1992), amphibian cells (Werdien *et al.*, 2001), mammalian cells (Sauer and Henderson, 1989; O'Gorman *et al.*, 1991), and plant cells (Odell *et al.*, 1990; Dale and Ow, 1991; Russell *et al.*, 1992; Lloyd and Davis, 1994; Lyznik *et al.*, 1995; Sonti *et al.*, 1995).

The systems have increasingly developed sophistication. Uses of site-specific recombination for transgenics fall into two categories. The first involves modification of genomically integrated transgene sequences *per se*. Precise changes can be made to transgenic loci, such as removal of selectable markers (Hohn *et al.*, 2001; Zuo *et al.*, 2001), gene activation or inactivation [for example, see (Kilby *et al.*, 2000; Hoff *et al.*, 2001)], or replacement of one transgenic sequence with another (Schlake and Bode, 1994; Bode *et al.*, 2000). The second type of modification encompasses a variety of genomic rearrangements. These range from small, local modifications at the transgenic locus, to full-scale chromosomal rearrangements. The type and extent depends on the creativity of the researcher. A major advantage of this second category that has not been fully realized is the impact of the chromosome environment on transgene expression. For breeding purposes the ability to manipulate chromosomal architecture could be of immense value, with a full range of controlled manipulations such as fragment exchange, deletions, inversions, translocations and duplications being possible. Exploration of the potential of these systems and future refinements for the modification of target sequences, coupled with the modulation of temporal and spatial expression of recombinase activity will contribute to further development of novel transgenic products and germplasm.

Using Site-Specific Recombinases for DNA Excision

A Robust Plant Transformation Tool
The original application of site-specific recombination systems in plants, as with yeast and mouse cells (Sauer and Henderson, 1988), demonstrated the use of Cre recombinase for gene excision (Dale and Ow, 1990; Odell *et al.*, 1990). In experiments performed in tobacco, Odell *et al.* (1990) also demonstrated that Cre activity could be delivered to plant cells by both re-transformation and cross-pollination. A functionally similar site-specific recombination system, the FLP/*FRT* system of yeast, was tested in plant

cells (maize and rice) with similar results (Lyznik *et al.*, 1993). Subsequent research, as documented in the following sections, has led to the development of both these systems into robust tools for transformation-based studies. Presently, genomic excisions can be achieved at 3-5% total transformation efficiency (without any selection for the recombination and transformation events) either by using Cre in mammalian cells (DiSanto *et al.*, 1995) or FLP in plant cells (Hodges and Lyznik, 1996).

Gene Removal and Controlled Activation of Genes

The use of site-specific recombination for exchange of foreign gene activities and/or removal of selectable marker genes followed the earlier experimentation on site-specific recombination systems. Cre-catalyzed excision was used in the first instance to remove a *loxP*-flanked hygromycin phosphotransferase gene from the tobacco genome, thereby demonstrating the feasibility of producing marker-free transgenic plants (Dale and Ow, 1991). In related studies, a mutant ALS gene, bounded by *loxP* sites, was used as a selectable marker gene for transformation of tobacco and *Arabidopsis*. Transgenic plants with β-glucuronidase (GUS) activity but no sulfonylurea resistance were generated using both cross-pollination and re-transformation strategies (Russell *et al.*, 1992). Linking two adjacent genes to the same promoter with the first gene flanked by recombination sites, enabled Cre-mediated excision of a firefly luciferase gene and the concomitant activation of a hygromycin phosphotransferase gene in tobacco (Bayley *et al.*, 1992). FLP recombinase was also used to eliminate the kanamycin resistance *neo* gene (removal of a selectable gene marker) and concurrently activate the *uidA* (*gusA*) gene in maize cells (Lyznik *et al.*, 1996). The efficiency of genomic recombinations was high enough in these experiments to eliminate the need for selection for recombination products, or for analysis of FLP expression. Marker-free transgenic tobacco seedlings also were produced using the FLP/*FRT* system (Lloyd and Davis, 1994; Bar *et al.*, 1996).

Examples of the production of marker-free transgenic plants (Gleave *et al.*, 1999; Sugita *et al.*, 2000), demonstrate that site-specific recombination systems are a highly reliable tool (see Chapter 6). The procedure has been simplified to a one-step transformation process by using a chemically inducible Cre/*lox* excision system in *Arabidopsis* (Zuo *et al.*, 2001). A similar strategy was used in tobacco with the R site-specific recombinase gene (Sugita *et al.*, 2000). The use of the Cre/*loxP* system for removal of selectable marker genes from transformed plastids in tobacco (Corneille *et al.*, 2001), demonstrates the versatility of recombinases in plant transformation studies.

Baszczynski et al.

Resolving Multiple-Copy Loci and Production of Hybrid Plants
Excision resulting from site-specific recombination systems has been used as a tool for trait enhancement and product development. Srivastava and co-workers used the Cre/*lox* system to eliminate multiple copies of an introduced foreign DNA sequence in wheat (Srivastava *et al.*, 1999) and later in maize (Srivastava and Ow, 2001). Transgenes were flanked by mutated (*lox511*) recombination sites in inverted orientations, and a second set of *loxP* sites were used to flank a selectable marker gene, *bar*. Cre recombinase was introduced by crossing transgenic parental plants. Recombination between the outermost excision sites in the progeny resolved the integrated molecules into a single copy of the transgene with concurrent excision of *bar*. As a result, single-copy, marker-free transgenic plants were produced. A similar outcome (e.g. removal of the selectable marker and simplification of the integration pattern) was produced by Sugita *et al.* (2000).

Excision may also be used to control gene expression. For example, a feasibility study using the FLP/*FRT* system for making plant hybrids was carried out in *Arabidopsis* (Luo *et al.*, 2000). Male-sterility was induced in this case by antisense expression of a tapetum-specific gene, *bcp1*, flanked by *FRT* sites. Upon cross-pollination with pollen from FLP-expressing plants, the male-sterility gene was removed, seeds were produced, and the progeny hybrid plants developed normally.

Temporal and Tissue-Specific Regulation of Gene Expression
Promoters responsive to environmental or tissue-specific factors can be used to regulate the activity of site-specific recombinases. Somatic mutations in *Drosophila* were effected by heat-shock inducible expression of FLP recombinase (Golic and Lindquist, 1989; Golic, 1991). Heat-shock promoters have been used also to induce activity of FLP recombinase in bacteria (Huang *et al.*, 1991), insect cells: *Aedes aegypti* and *Drosophila* (Morris *et al.*, 1991; Konsolaki *et al.*, 1992), and plant cells: maize and *Arabidopsis* (Kilby *et al.*, 1995; Lyznik *et al.*, 1995; Hoff *et al.*, 2001). In one of the more refined applications of the heat-shock inducible expression of FLP, Kilby *et al.* (2000) generated a high frequency of GUS sectoring events in transgenic *Arabidopsis* plants. Heat-shock activated FLP also has been used to study the function of floral transcription factors (*LEAFY and APETELA1*) (Sessions *et al.*, 2000), and a mosaic strategy using the Cre/*loxP* system was applied to study *AGAMOUS* gene function during *Arabidopsis* flower development (Sieburth *et al.*, 1998). In another example of FLP-mediated gene activation, Riou-Khamlichi *et al.* (1999), studied cytokinin-regulated cell division in

Arabidopsis using a heat shock induced recombinase to activate a constitutively-expressed CycD3 gene through excision of an intervening GUS reporter gene.

The Cre/*lox* system has been used to regulate gene expression in a tissue-specific manner in mouse cells by Lasko *et al.* (1992) and Orban *et al.* (1992). Also, Cre-mediated activation of GUS in developing tobacco embryos has been demonstrated (Odell *et al.*, 1994). The Cre recombinase coding sequence was fused to the promoters of the bean β-phaseolin and the soybean β-conglycinin genes. By crossing a plant containing a Cre expression cassette and a plant containing an inactivated *gusA* gene, a developmentally regulated pattern of GUS expression in tobacco seeds was observed in F1 progeny. Cre-mediated gene activation strategies have been utilized also to study *Agrobacterium*-mediated genetic transformation of *Arabidopsis* (De Buck *et al.*, 2000). Visual marker genes in incoming T-DNAs were excised using a Cre-mediated reaction, allowing differentiation between T-DNA transfer and T-DNA integration frequencies.

Using Recombinases for DNA Integration and Replacement

Integration at FRT or Lox Sites

Excisional recombination produces two DNA molecules containing unlinked recombination sites. The molecules may enter another round of recombination to form co-integrate molecules by intermolecular site-specific recombination. If such a reaction occurs between chromosomal DNA and an incoming transformation vector, site-specific integration takes place. In yeast or bacteriophage P1, natural control mechanisms allow for efficient excision and inversion reactions to take place through FLP- or Cre-mediated recombination, respectively. As a monomolecular reaction, excision is generally favored over integration. Following the success of a number of DNA integration studies in mammalian cells (Sauer and Henderson, 1989; O'Gorman *et al.*, 1991), Cre-mediated and FLP-mediated intermolecular site-specific recombination was demonstrated in tobacco and maize cells, respectively (Dale and Ow, 1990; Lyznik *et al.*, 1993). Since recombinases such as Cre and FLP mediate both excision and integration (the reaction is reversible), the stabilization of an integrated DNA sequence is a key issue in any genomic integration strategy. Albert *et al.* (1995) tested mutant *lox* sites (to reduce excision after integration) or the use of transient expression of the *cre* gene (to eliminate recombinase activity after integration) in tobacco. Site-specific recombination events were observed in both cases; however, the transient expression of *cre* produced more integration events.

A.

1. Target and substrate molecules combined.

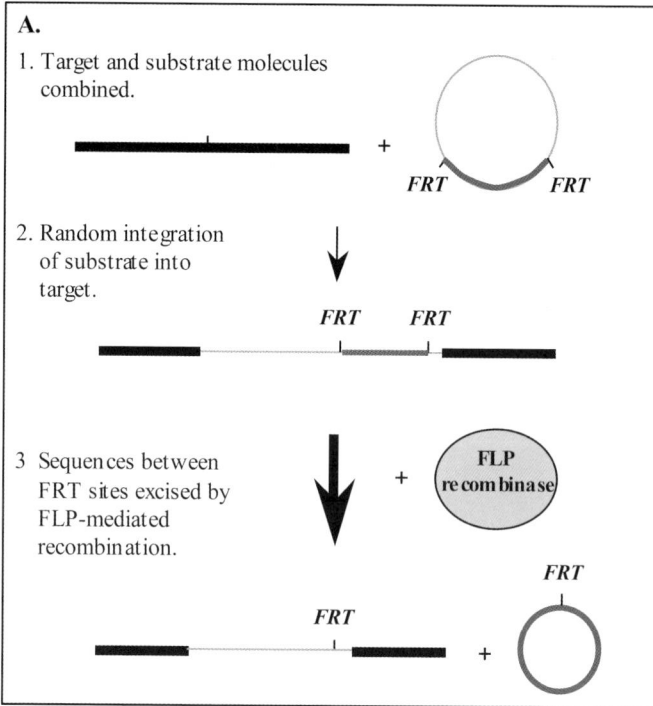

2. Random integration of substrate into target.

3. Sequences between FRT sites excised by FLP-mediated recombination.

B.

1. Target site is created in the genome.

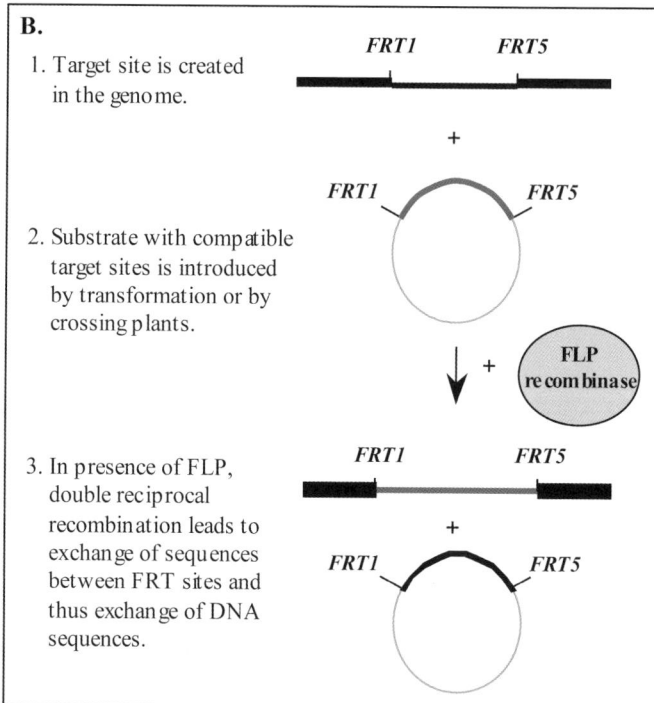

2. Substrate with compatible target sites is introduced by transformation or by crossing plants.

3. In presence of FLP, double reciprocal recombination leads to exchange of sequences between FRT sites and thus exchange of DNA sequences.

Figure 2. Comparison of FLP/*FRT* site-specific recombination with modifications to *FRT* target sites for applications in gene excision, integration or replacement. A. The gene to be excised (e.g. selectable marker) is flanked by two identical *FRT* sites on a plasmid construct used to transform plant tissues. When the marker is no longer needed for selection in culture or the field, FLP recombinase is introduced as DNA, RNA, protein or by crossing with a plant expressing FLP and events no longer carrying the marker are recovered. B. Genomic target sites containing two non-identical *FRT* sites flanking or adjacent to a gene of interest are created by standard transformation approaches. Introduction of a construct carrying a second gene flanked by the same pair of non-identical *FRT* sites in the presence of active FLP recombinase leads to targeted integration or replacement *via* double reciprocal recombination.

Sequence-specific recombinases also have been evaluated as a means of controlling integration of T-DNA during *Agrobacterium*-mediated transformation. In these studies, Cre/*lox* was used to gain more control over T-DNA integration during the plant transformation process (Vergunst and Hooykaas, 1998; Vergunst *et al.*, 1998). Two *loxP* sites were introduced into T-DNA in a direct repeat orientation to circularize a fragment of T-DNA before integration. T-DNA integration into a previously introduced *loxP* site located in the genome was monitored by activation of a silent *neo* target gene. Precise stable integration of T-DNA was demonstrated, although the efficiency was disappointingly low. The reported results demonstrate that while using site-specific recombination for T-DNA integration is possible, both systems are likely to require further refinements in order to become more compatible and practically useful.

Use of Mutant Target Sequences and Double Reciprocal Crossovers for Sequence Integration or Replacement
An interesting alternative strategy for site-specific integration was proposed by Schlake, Bode and co-workers (Schlake and Bode, 1994; Bode *et al.*, 2000). The strategy is termed "segmental genomic replacement", "double-reciprocal cross-over", or "recombinase-mediated cassette exchange" (RMCE). This procedure calls for two independent site-specific recombination events to take place between two independent sets of mutated or non-identical recombination sites flanking the DNA segment to be replaced. The outcome is the replacement of a chromosomal DNA fragment (target) with a vector DNA fragment (donor) flanked by the corresponding mutated recombination sites. Even in the presence of a recombinase protein, the product of this reaction is stabilized, if the recombined vector DNA is successfully eliminated following recombination. Using this strategy, experiments in mouse cells using Cre/*lox* with non-identical *lox* target sequences resulted in approximately a 3-fold increase in targeted integrations

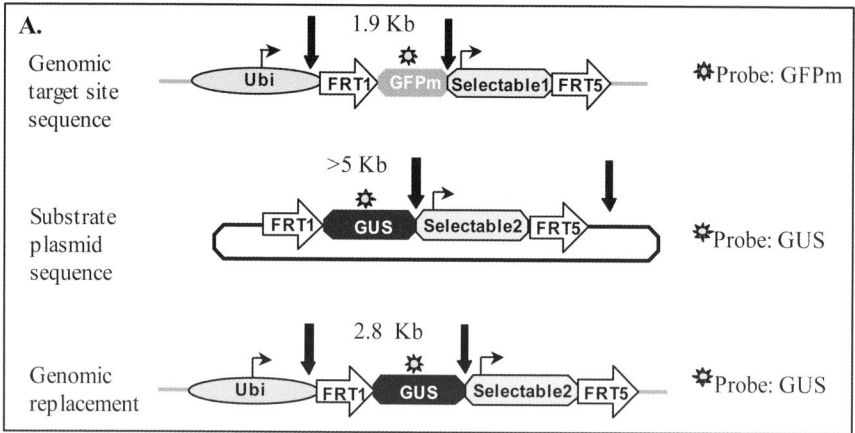

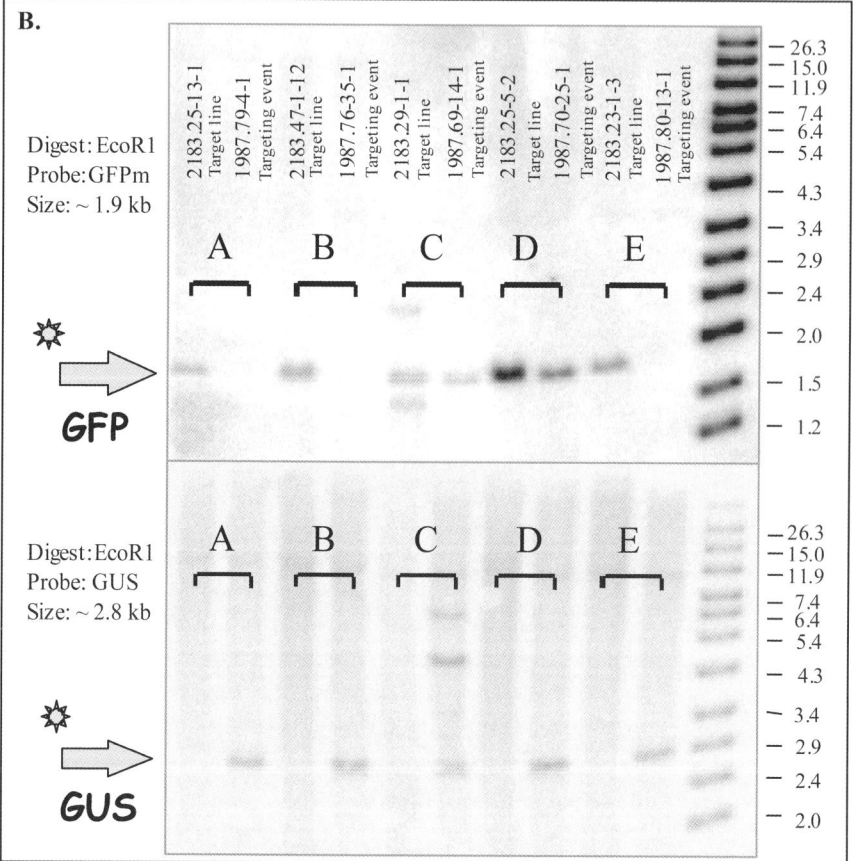

Figure 3. Demonstration of targeted gene replacement in maize using FLP/*FRT* site-specific recombination with modified target sites as described in the text. A. Schematic diagrams showing the original construct sequence used to generate a genomic target site for testing recombination, a second substrate plasmid construct used for the targeted replacement experiments and the predicted product of successful recombination to replace one reporter gene/selectable marker combination with another. Arrows indicate positions of *Eco*R1 restriction sites. Predicted sizes of hybridizing bands probed with GFP or GUS DNA probes are indicated. B. Southern blot analysis of *Eco*R1 digests from five independent targeted replacement events (A-E) using constructs as described above. In each pair, the left lane contains digested DNA from the original target line, while the right lane contains DNA from tissues following targeted replacement. Events A, B and E showed precise replacement while C and D had more complex integration. Molecular marker fragment sizes are indicated at right.

as compared to random integration events (Bethke and Sauer, 1997). In embryonic stem cells, no selection for recombination products was required to recover site-specific integration events (Soukharev *et al.*, 1999). Similarly, Feng *et al.* (1999) proposed a modification to the procedure (two recombining sites in an inverted orientation) that allowed recovery of integration events with no selection in transiently transformed MEL cells. The system can be further enhanced by providing more compatible mutated recombination sites (Kolb, 2001).

Over the last several years our laboratories have extended the observations of Schlake and Bode (1994) to develop and characterize a working targeted gene integration/replacement system for maize. This system utilizes novel and existing non-identical recombination *FRT* target sites, FLP recombinase (or derivatives such as FLPm, which is designed to work more efficiently in monocots), and novel accessory technologies to further extend the utility of the system. Essentially, a nucleotide sequence flanked by, or incorporating, two non-identical recombination target (*FRT*) sites is introduced into the plant's genome, establishing a target for insertion of new nucleotide sequences of interest. Once a stable plant or cultured tissue is established, a second construct or nucleotide sequence of interest (herein referred to as a transfer cassette), flanked by recombination target sites corresponding to those flanking the genomic target site, is introduced into the stable transformed plant or tissues. Concurrent expression of active FLP recombinase protein leads to a physical exchange of the nucleotide sequences between the non-identical *FRT* sites of the genomic target region and the transfer cassette. Functional recombinase (FLP or Cre) has been shown to be effective in catalyzing recombination reactions whether provided through transformation of plant cells with an expression cassette capable of expressing the recombinase in the plant, by transient expression of the recombinase (Gagneten *et al.*, 1997), by introducing messenger RNA for the recombinase (de Wit *et al.*, 1998) or the recombinase protein itself (Baubonis and Sauer,

1993; Jo *et al.*, 2001), or by crossing with a plant carrying an actively expressed or inducible recombinase (Kilby *et al.*, 2000). This approach is not limited to use of only one set of non-identical recombination target sites and the transformed plant may comprise multiple target sites from either or both the FLP/*FRT* or Cre/*lox* systems. In this manner, multiple manipulations of DNA sequences or genes at the target sites in the transformed plant are possible.

Figure 2 shows schematically how a single *FRT* site or two non-identical *FRT* sites can be used to conduct either excision or integration (or replacement) reactions, respectively. By selectively positioning the *FRT* sites in relation to DNA sequences to be introduced into plants, a variety of subsequent recombinase-mediated reactions can be carried out.

Figure 3 presents DNA construct designs and results from experiments conducted to demonstrate targeted replacement of genomic DNA sequences with new gene sequences provided in a transfer cassette. The genomic sequences targeted for recombination comprised a single copy insertion of a pair of genes (reporter plus selectable marker) previously introduced *via* *Agrobacterium*-mediated transformation of maize. Phenotypic and Southern analyses of multiple independent events resulting from these and other experiments confirmed the predicted targeted replacement of the originally introduced DNA sequences.

Apart from the obvious benefits of being able to integrate genes into the genome of plants, the approach described above provides a mechanism for introducing novel genes or DNA sequences into genomic locations that have been determined to be particularly beneficial for gene integration. It has been proposed that site-specific integration into pre-determined chromosomal locations could result in more uniform and predicable transgene expression (i.e. reducing position effects; see (Baur *et al.*, 2001; van Leeuwen *et al.*, 2001). This concept is being evaluated in our laboratories and others by creating multiple independent transformation events, characterizing the integration characteristics of these events and then evaluating expression of both the originally introduced genes as well as new genes retargeted into these characterized genomic sites. To this end, Day *et al.* (2000) examined the expression of targeted foreign genes in tobacco. A single-copy insertion of the *gusA* marker gene at a selected chromosomal site (*loxP*) produced a consistent spatial pattern of transgene expression in many re-transformation events. They also noted some unexpected variation in the *gusA* gene expression from insertions at the same chromosomal location.

The use of site-specific recombinases can also bring new insights to our understanding of transgene position effects. In a recent report, Cre-mediated recombination was used to switch the orientation of an integrated transgene cassette with a surprising result; transgene silencing was dramatically influenced by orientation (Feng *et al.*, 2001). Using recombinases in this manner can potentially help us understand the mechanisms of gene silencing by increasing the resolution of our observations. Also recently, the Cre/*lox* system was used to demonstrate a selective loss of methylation in *gusA* transgenes in *Arabidopsis* (De Buck and Depicker, 2001). Such results make it clear that useful genomic sites for re-targeting of transgenes will require careful characterization, including molecular structure, stability of transgene expression, and reproducibility at a given site or in a given orientation.

Unlike conventional transformation methods where transgene integration is random, often complex and unique each time, the approaches described here would provide the ability to repeatedly introduce sequences into the same genomic location. This capability should benefit promoter or gene expression evaluations, studies of transgene integration mechanisms, the characterization of different chromosomal regions, or the evaluation of impact of DNA sequence insertions on factors such as yield or agronomic performance.

Applications for Trait Stacking and Improved Breeding Efficiency

The results described above also provide an approach whereby integration of two or more genes can be targeted to the same genomic location, providing a mechanism for 'gene or trait stacking'. By selective positioning of non-identical *FRT* or *lox* sites adjacent to gene or trait sequences in constructs to be used to generate the initial transgenic events, subsequent transformation or crossing experiments can be conducted to bring a second trait adjacent to the first trait *via* recombination across the non-identical *FRT* or *lox* sites. The traits stacked in this way can then be maintained and managed as a closely linked pair of traits in conventional breeding programs. While conventional breeding approaches do provide ways to combine two to three traits by crossing and selection of lines carrying individually introduced and fixed traits, the time and resource requirements increase markedly with each trait to be added. Additionally, the disadvantage to the "breeding only" approach is that the combined traits most likely remain unlinked following crossing because of the random nature of integration of the original individual traits. With the targeted integration approach described above (also see schematic in Figure 4), the individual traits are brought together on the same chromosome and adjacent to each other by virtue of the positioning of the recombination target sites relative to the genes. The traits would segregate subsequently as a single linkage unit that can be more easily managed as part of breeding efforts.

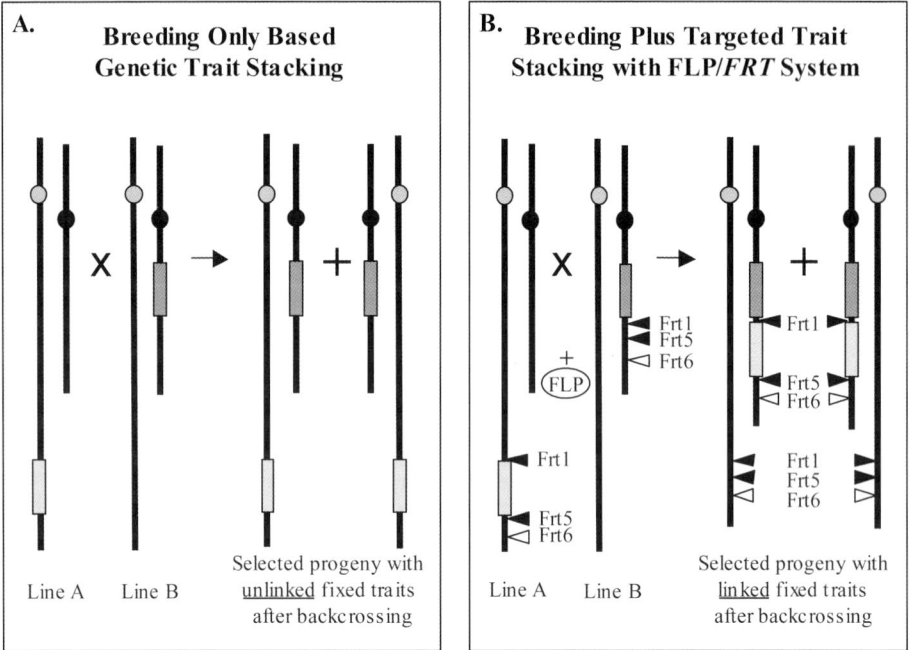

Figure 4. Schematic demonstrating how recombination target sites can be positioned relative to introduced genes or traits to facilitate subsequent linkage of the traits on chromosomes through stacking (panel B). Compare this to the non-linked traits in products arising from a conventional selection breeding approach for stacking (panel A).

Other alternatives for trait stacking exist. If the traits to be combined are known up front, then one can simply design DNA vectors that carry two or more genes for the traits of interest together on the same starting construct. However, in some cases all of the genes for traits to be combined may not be available at the start of the experiment, or a new or better gene may become available during production of the events that may necessitate substitution of the original for the new gene. Also, a breeder or researcher may wish to link a new trait or regulatory element to an already well-characterized event. Thus the described targeted integration/replacement strategy has many utilities. Regardless of the approach taken, it is important to characterize the expression of the individual and/or combined traits under many different conditions and in different genetic backgrounds. In commercial hybrid breeding programs, the behavior of the traits in the hybrid combinations, as well as the impact of integration, orientation and expression of the individual or combined genes on key factors such as yield and agronomic performance of the plants are important

On the Horizon for Plants

Chromosomal Engineering

Site-specific recombinases can effectively act on sites integrated into genomic structures to perform a wide-range of chromosomal modifications including translocations, insertions, deletions, and inversions. As opposed to localized transgene modifications discussed in the above sections, such large-scale chromosome modifications typically require positioning of target sequences in specific genomic configurations to affect the desired outcome. Examples exist in many species. The FLP system from yeast was originally used to demonstrate site-specific recombination between two homologous chromosomes in *Drosophila* (Gagneten *et al.*, 1997). Cre-mediated site-specific translocation was subsequently documented between non-homologous mouse chromosomes (Van Deursen *et al.*, 1995) and between tobacco chromosomes (Qin *et al.*, 1994). Chromosomal translocation between the *Dek* gene on chromosome 13 and the *Can* gene on chromosome 2 was observed in 1 out of 1200-2400 embryonic stem cells expressing Cre recombinase. In tobacco, chromosomal translocation resulted in activation of a promoterless hygromycin-resistance gene. Medberry *et al.* (1995) applied Cre/*lox* site-specific recombination to generate chromosomal deletions and inversions in tobacco. A modified *Ds* transposon was used to relocate one *lox* site away from the other one at the primary transgenic locus. Three chromosomal inversions and one deletion were recovered in the progeny of a cross with a Cre donor plant.

Chromosomal deletions, inversions and translocations can also be generated in a target plant containing a genomic *loxP* site that is randomly re-transformed with a T-DNA vector containing a second *loxP* site. Such rearrangements occurred at a frequency of 96% among identified Cre-mediated recombination events between the target *loxP* site and a *loxP* site of T-DNA (Vergunst *et al.*, 2000). The Cre/*lox* system has been tested also for the synthesis of hybrid chromosomes of different plant species (Koshinsky *et al.*, 2000). In this study, protoplast fusion was used to coalesce *Arabidopsis thaliana* and *Nicotiana tabacum* chromosomes. The interspecies transfer of a chromosomal fragment was detected in hygromycin-resistant calli; however, it was not maintained in regenerated plants except in a single T1 progeny event.

Gene Targeting and Site-Specific Recombination

The combination of homologous and site-specific recombinations provides a powerful tool for precise modifications of eukaryotic genomes. A basic paradigm of such procedures is to integrate a fragment of foreign DNA that contains site-specific recombination sites into a locus of interest *via* homologous recombination and, subsequently, to use a site-specific recombinase to make additional rearrangements including removal of a selectable marker gene from a targeted locus. Fiering *et al.* (1993) demonstrated the application of the FLP/*FRT* system for this purpose in human cells. First, the β-globin locus was disrupted by homologous recombination with the *neo*-containing targeting vector and, subsequently, the *neo* gene was excised using transient expression of FLP. As a result, the function of the mutated β-globin locus was restored. This procedure can be further simplified if the target site is a single copy gene located on the X chromosome, as for example the interleukin 2 receptor γ chain gene (Il-2Rγ) that is involved in differentiation of T- and NK-cells. Il-2Rγ-deficient male mice were obtained by targeting this locus with a *loxP*-flanked *neo*-resistance cassette and subsequent excision of the selectable marker together with a part of the gene by transient transformation of targeted cells with the pIC-Cre vector. Reported frequencies of Cre-mediated deletions were 4% (DiSanto *et al.*, 1995).

Gene Targeting in the Progeny of Transgenic Parents

Activation of site-specific recombination systems in the progeny of transgenic parents offers new possibilities for chromosomal engineering and gene targeting. Golic *et al.* (1997) demonstrated that FLP could catalyze DNA excision and re-integration into pre-selected chromosomal *FRT* target sites in *Drosophila* embryos. Such events were recovered from up to 5% of the progeny. Thus the procedure overcame the low-efficiency barriers of other transformation protocols. The method was subsequently extended into gene targeting applications in *Drosophila* (Rong and Golic, 2000). The FLP-excised DNA molecules were made highly recombinogenic by introduction of double-strand breaks within the homologous regions of the *yellow*[+] body color gene. The breaks were generated by a heat-inducible *I-SceI* gene (an endonuclease from yeast that recognizes and cuts a unique 18-base pair restriction site), producing about one gene-targeting event for every 500 progeny. Those observations were subsequently verified for another chromosomal locus (Rong and Golic, 2001).

With the remarkable exception of mouse cells and *Drosophila*, the combination of homologous recombination and site-specific recombination has not been tested in other organisms, partly because of the lack of suitable gene targeting model systems. However, the elements of site-specific recombination systems applicable to gene targeting have been investigated extensively in other higher eukaryotes, including plants, and research continues on alternative homologous recombination strategies for robust gene integration and modification in animal and plant systems.

Summary

Significant advances have been made over the last decade in plant transformation methodologies and strategies. Development or adaptation of systems, approaches and technologies for precise gene insertion, excision or modification, coupled (in many cases) with general transformation improvements, is expediting basic and applied research in plants as well as positively impacting the development of new traits for crop improvement. With the rapid progress in the area of plant genomics and the need to understand the structure, function, organization and behavior of the thousands of plant genes that influence biological processes and which provide a source of new trait opportunities, site-specific recombination and other tools are providing new and unique ways for studying the function, significance and utility of these genes.

References

Albert, H., Dale, E.C., Lee, E., and Ow, D.W. 1995. Site-specific integration of DNA into wild-type and mutant *lox* sites placed in the plant genome. Plant J. 7: 649-659.

Bar, M., Leshem, B., Gilboa, N., and Gidoni, D. 1996. Visual characterization of recombination at FRT-*gusA* loci in transgenic tobacco mediated by constitutive expression of the native FLP recombinase. Theor. Appl. Genet. 93: 407-413.

Baubonis, W. and Sauer, B. 1993. Genomic targeting with purified Cre recombinase. Nucleic Acids Res. 21: 2025-2029.

Baur, J.A., Zou, Y., Shay, J.W., and Wright, W.E. 2001. Telomere position effect in human cells. Science. 292: 2075-2077.

Bayley, C.C., Morgan, M., Dale, E.C., and Ow, D.W. 1992. Exchange of gene activity in transgenic plants catalyzed by the Cre-*lox* site-specific recombination system. Plant Mol. Biol. 18: 353-361.

Bethke, B. and Sauer, B. 1997. Segmental genomic replacement by Cre-mediated recombination: genotoxic stress activation of the *p53* promoter in single-copy transformants. Nucleic Acids Res. 25: 2828-2834.

Bode, J., Schlake, T., Iber, M., Schubeler, D., Seibler, J., Snezhkov, E., and Nikolaev, L. 2000. The transgeneticist's toolbox: novel methods for the targeted modification of eukaryotic genomes. Biol. Chem. 381: 801-813.

Corneille, S., Lutz, K., Svab, Z., and Maliga, P. 2001. Efficient elimination of selectable marker genes from the plastid genome by the CRE-*lox* site-specific recombination system. Plant J. 27: 171-178.

Cox, M.M. 1983. The FLP protein of the yeast 2-mm plasmid: expression of a eukaryotic genetic recombination system in *Escherichia coli*. Proc. Natl. Acad. Sci. USA. 80: 4223-4227.

Dale, E.C. and Ow, D.W. 1990. Intra- and intermolecular site-specific recombination in plant cells mediated by bacteriophage P1 recombinase. Gene. 91: 79-85.

Dale, E.C. and Ow, D.W. 1991. Gene transfer with subsequent removal of the selection gene from the host genome. Proc. Natl. Acad. Sci. USA. 88: 10558-10562.

Day, C.D., Lee, E., Kobayashi, J., Holappa, L.D., Albert, H., and Ow, D.W. 2000. Transgene integration into the same chromosome location can produce alleles that express at a predictable level, or alleles that are differentially silenced. Genes Dev. 14: 2869-2880.

De Buck, S., De Wilde, C., Van Montagu, M., and Depicker, A. 2000. Determination of the T-DNA transfer and the T-DNA integration frequencies upon cocultivation of *Arabidopsis thaliana* root explants. Mol. Plant Microbe Interact. 13: 658-665.

De Buck, S. and Depicker, A. 2001. Disruption of their palindromic arrangement leads to selective loss of DNA methylation in inversely repeated *gus* transgenes in *Arabidopsis*. Mol. Genet. Genomics. 265: 1060-1068.

de Wit, T., Drabek, D., and Grosveld, F. 1998. Microinjection of cre recombinase RNA induces site-specific recombination of a transgene in mouse oocytes. Nucleic Acids Res. 26: 676-678.

DiSanto, J.P., Muller, W., Guy-Grand, D., Fischer, A., and Rajewsky, K. 1995. Lymphoid development in mice with a targeted deletion of the interleukin 2 receptor γ chain. Proc. Natl. Acad. Sci. USA. 92: 377-381.

Feng, Y.Q., Lorincz, M.C., Fiering, S., Greally, J.M., and Bouhassira, E.E. 2001. Position effects are influenced by the orientation of a transgene with respect to flanking chromatin. Mol. Cell Biol. 21: 298-309.

Feng, Y.Q., Seibler, J., Alami, R., Eisen, A., Westerman, K.A., Leboulch, P., Fiering, S., and Bouhassira, E.E. 1999. Site-specific chromosomal integration in mammalian cells: highly efficient CRE recombinase-mediated cassette exchange. J. Mol. Biol. 292: 779-785.

Fiering, S., Kim, C.G., Epner, E.M., and Groudine, M. 1993. An "in-out" strategy using gene targeting and FLP recombinase for the functional dissection of complex DNA regulatory elements: analysis of the β-globin locus control region. Proc. Natl. Acad. Sci. USA 90: 8469-8473.

Gagneten, S., Le, Y., Miller, J., and Sauer, B. 1997. Brief expression of a GFP cre fusion gene in embryonic stem cells allows rapid retrieval of site-specific genomic deletions. Nucleic Acids Res. 25: 3326-3331.

Gleave, A.P., Mitra, D.S., Mudge, S.R., and Morris, B.A. 1999. Selectable marker-free transgenic plants without sexual crossing: transient expression of cre recombinase and use of a conditional lethal dominant gene. Plant Mol. Biol. 40: 223-235.

Golic, K.G. 1991. Site-specific recombination between homologous chromosomes in *Drosophila.* Science. 25: 958-961.

Golic, K.G. and Lindquist, S. 1989. The FLP recombinase of yeast catalyzes site-specific recombination in the *Drosophila* genome. Cell. 59: 499-509.

Golic, M.M., Rong, Y.S., Petersen, R.B., Lindquist, S.L., and Golic, K.G. 1997. FLP-mediated DNA mobilization to specific target sites in *Drosophila* chromosomes. Nucleic Acids Res. 25: 3665-3671.

Guo, F., Gopaul, D.N., and van Duyne, G.D. 1997. Structure of Cre recombinase complexed with DNA in a site-specific recombination synapse. Nature. 389: 40-46.

Hodges, T.K. and Lyznik, L.A. 1996. Genetic modifications of maize genome using DNA recombination reactions. Genetica Polonica. 37A: 36-49.

Hoff, T., Schnorr, K.M., and Mundy, J. 2001. A recombinase-mediated transcriptional induction system in transgenic plants. Plant Mol. Biol. 45: 41-49.

Hohn, B., Levy, A.A., and Puchta, H. 2001. Elimination of selection markers from transgenic plants. Curr. Opin. Biotechnol. 12: 139-143.

Huang, L.C., Wood, E.A., and Cox, M.M. 1991. A bacterial model system for chromosomal targeting. Nucleic Acids Res. 19: 443-448.

Huffman, K.E. and Levene, S.D. 1999. DNA-sequence asymmetry directs the alignment of recombination sites in the FLP synaptic complex. J. Mol. Biol. 286: 1-13.

Jo, D., Nashabi, A., Doxsee, C., Lin, Q., Unutmaz, D., Chen, J., and Ruley, H.E. 2001. Epigenetic regulation of gene structure and function with a cell-permeable Cre recombinase. Nat. Biotechnol. 19: 929-933.

Kilby, N.J., Davies, G.J., Snaith, M.R., and Murray, J.A. 1995. FLP recombinase in transgenic plants: constitutive activity in stably transformed tobacco and generation of marked cell clones in *Arabidopsis.* Plant J. 8: 637-652.

Kilby, N.J., Fyvie, M.J., Sessions, R.A., Davies, G.J., and Murray, J.A. 2000. Controlled induction of GUS marked clonal sectors in *Arabidopsis.* J. Exp. Bot. 51: 853-863.

Kolb, A.F. 2001. Selection-marker-free modification of the murine β-casein gene using a *lox2722* site. Anal. Biochem. 290: 260-271.

Konsolaki, M., Sanicola, M., Kozlova, T., Liu, V., Arca, B., Savakis, C., Gelbart, W.M., and Kafatos, F.C. 1992. FLP-mediated intermolecular recombination in the cytoplasm of *Drosophila* embryos. New Biol. 4: 551-557.

Kooter, J.M., Matzke, M.A., and Meyer, P. 1999. Listening to the silent genes: transgene silencing, gene regulation and pathogen control. Trends Plant Sci. 4: 340-347.

Koshinsky, H.A., Lee, E., and Ow, D.W. 2000. Cre-*lox* site-specific recombination between *Arabidopsis* and tobacco chromosomes. Plant J. 23: 715-722.

Lasko, M., Sauer, B., Mosinger, B., Jr., Lee, E.J., Manning, R.W., Yu, S.H., Mulder, K.L., and Westphal, H. 1992. Targeted oncogene activation by site-specific recombination in transgenic mice. Proc. Natl. Acad. Sci. USA. 89: 6232-6236.

Lee, J., Jayaram, M., and Grainge, I. 1999. Wild-type Flp recombinase cleaves DNA in trans. EMBO J. 18: 784-791.

Lloyd, A.M. and Davis, R.W. 1994. Functional expression of the yeast FLP/*FRT* site-specific recombination system in *Nicotiana tabacum*. Mol. Gen. Genet. 242: 653-657.

Luo, H., Lyznik, L.A., Gidoni, D., and Hodges, T.K. 2000. FLP-mediated recombination for use in hybrid plant production. Plant J. 23: 423-430.

Lyznik, L.A., Hirayama, L., Rao, K.V., Abad, A., and Hodges, T.K. 1995. Heat-inducible expression of *FLP* gene in maize cells. Plant J. 8: 177-186.

Lyznik, L.A., Mitchell, J.C., Hirayama, L., and Hodges, T.K. 1993. Activity of yeast FLP recombinase in maize and rice protoplasts. Nucleic Acids Res. 21: 969-975.

Lyznik, L.A., Rao, K.V., and Hodges, T.K. 1996. FLP-mediated recombination of *FRT* sites in the maize genome. Nucleic Acids Res. 24: 3784-3789.

Medberry, S.L., Dale, E., Qin, M., and Ow, D.W. 1995. Intra-chromosomal rearrangements generated by Cre-*lox* site-specific recombination. Nucleic Acids Res. 23: 485-490.

Morris, A.C., Schaub, T.L., and James, A.A. 1991. FLP-mediated recombination in the vector mosquito, *Aedes aegypti*. Nucleic Acids Res. 19: 5895-5900.

Odell, J., Caimi, P., Sauer, B., and Russell, S. 1990. Site-directed recombination in the genome of transgenic tobacco. Mol. Gen. Genet. 223: 369-378.

Odell, J.T., Hoopes, J.L., and Vermerris, W. 1994. Seed-specific gene activation mediated by the Cre/*lox* site-specific recombination system. Plant Physiol. 106: 447-458.

O'Gorman, S., Fox, D.T., and Wahl, G.M. 1991. Recombinase-mediated gene activation and site-specific integration in mammalian cells. Science. 251: 1351-1355.

Orban, P.C., Chui, D., and Marth, J.D. 1992. Tissue- and site-specific DNA recombination in transgenic mice. Proc. Natl. Acad. Sci. USA. 89: 6861-6865.

Qin, M., Bayley, C., Stockton, T., and Ow, D.W. 1994. Cre recombinase-mediated site-specific recombination between plant chromosomes. Proc. Natl. Acad. Sci. USA. 91: 1706-1710.

Riou-Khamlichi, C., Huntley, R., Jacqmard, A., and Murray, J.A. 1999. Cytokinin activation of *Arabidopsis* cell division through a D-type cyclin. Science. 283: 1541-1544.

Rong, Y.S. and Golic, K.G. 2000. Gene targeting by homologous recombination in *Drosophila*. Science. 288: 2013-2018.

Rong, Y.S. and Golic, K.G. 2001. A targeted gene knockout in *Drosophila*. Genetics. 157: 1307-1312.

Russell, S.H., Hoopes, J.L., and Odell, J.T. 1992. Directed excision of a transgene from the plant genome. Mol. Gen. Genet. 234: 49-59.

Sauer, B. and Henderson, N. 1988. Site-specific DNA recombination in mammalian cells by the Cre recombinase of bacteriophage P1. Proc. Natl. Acad. Sci. USA. 85: 5166-5170.

Sauer, B. and Henderson, N. 1989. Cre-stimulated recombination at *loxP*-containing DNA sequences placed into the mammalian genome. Nucleic Acids Res. 17: 147-161.

Schlake, T. and Bode, J. 1994. Use of mutated FLP recognition target (*FRT*) sites for the exchange of expression cassettes at defined chromosomal loci. Biochemistry. 33: 12746-12751.

Sessions, A., Yanofsky, M.F., and Weigel, D. 2000. Cell-cell signaling and movement by the floral transcription factors *LEAFY* and *APETALA1*. Science. 289: 779-782.

Sieburth, L.E., Drews, G.N., and Meyerowitz, E.M. 1998. Non-autonomy of AGAMOUS function in flower development: use of a Cre/*loxP* method for mosaic analysis in *Arabidopsis*. Development. 125: 4303-4312.

Sonti, R.V., Tissier, A.F., Wong, D., Viret, J.F., and Signer, E.R. 1995. Activity of the yeast FLP recombinase in *Arabidopsis*. Plant Mol. Biol. 28: 1127-1132.

Soukharev, S., Miller, J.L., and Sauer, B. 1999. Segmental genomic replacement in embryonic stem cells by double *lox* targeting. Nucleic Acids Res. 27: e21.

Srivastava, V., Anderson, O.D., and Ow, D.W. 1999. Single-copy transgenic wheat generated through the resolution of complex integration patterns. Proc. Natl. Acad. Sci. USA. 96: 11117-11121.

Srivastava, V. and Ow, D.W. 2001. Single-copy primary transformants of maize obtained through the co-introduction of a recombinase-expressing construct. Plant Mol. Biol. 46: 561-566.

Sugita, K., Kasahara, T., Matsunaga, E., and Ebinuma, H. 2000. A transformation vector for the production of marker-free transgenic plants containing a single copy transgene at high frequency. Plant J. 22: 461-469.

Van Deursen, J., Fornerod, M., Van Rees, B., and Grosveld, G. 1995. Cre-mediated site-specific translocation between nonhomologous mouse chromosomes. Proc. Natl. Acad. Sci. USA. 92: 7376-7380.

van Leeuwen, W., Ruttink, T., Borst-Vrenssen, A.W., van Der Plas, L.H., and van Der Krol, A.R. 2001. Characterization of position-induced spatial and temporal regulation of transgene promoter activity in plants. J. Exp. Bot. 52: 949-959.

Vergunst, A.C. and Hooykaas, P.J. 1998. Cre/lox-mediated site-specific integration of *Agrobacterium* T-DNA in *Arabidopsis thaliana* by transient expression of cre. Plant. Mol. Biol. 38: 393-406.

Vergunst, A.C., Jansen, L.E., and Hooykaas, P.J. 1998. Site-specific integration of *Agrobacterium* T-DNA in *Arabidopsis thaliana* mediated by Cre recombinase. Nucleic Acids Res. 26: 2729-2734.

Vergunst, A.C., Jansen, L.E.T., Fransz, P.F., de Jong, J.H., and Hooykaas, P.J.J. 2000. Cre/lox-mediated recombination in *Arabidopsis*: evidence for transmission of a translocation and a deletion event. Chromosoma. 109: 287-297.

Werdien, D., Peiler, G., and Ryffel, G.U. 2001. FLP and Cre recombinase function in Xenopus embryos. Nucleic Acids Res. 29: E53-53.

Zuo, J., Niu, Q.W., Moller, S.G., and Chua, N.H. 2001. Chemical-regulated, site-specific DNA excision in transgenic plants. Nat. Biotechnol. 19: 157-161.

From: *Transgenic Plants: Current Innovations and Future Trends*
Edited by: C. Neal Stewart, Jr.

Chapter 8

Transgenic Plants for Disease Resistance

Vipaporn Phuntumart

Abstract

The advances in biotechnology have enhanced new opportunities for crop improvement other than conventional plant breeding, which is limited by sexual barriers across species. Transgene-mediated resistance becomes a method of choice for plant disease resistance development. Different sources of transgenes have been proposed; these include plant resistance genes, pathogen-derived resistance genes, and antimicrobial proteins of plant and non-plant origins. Numerous disease resistance transgenes have been inserted into plant genomes and have been successful in protecting the plants from several diseases including nematodes. This chapter discusses the principles of plant-pathogen interaction, mechanisms of defense responses, signals involved in defense responses and the current status of transgenic plants that confer resistance to different kind of pathogens. Promising approaches for the production of transgenic plants with broad-spectrum resistance have also been discussed.

Introduction

Plant disease is one of the major problems for crop production. Integrated management for crop protection has been advanced greatly by using the combination of resistant varieties, application of pesticides, and conventional practices (i.e., crop rotation). Although conventional plant breeding to enhance disease resistance against several pathogens has been noticeably successful, it is a time-consuming process and has limited scope. Alternatively, numerous agrochemicals have also been developed. However, chemicals are expensive and pathogens often develop resistance to the chemicals after a certain period of time. In addition, concerns have arisen from the harmful effect of these chemicals on human health and environment. Therefore, advances in biotechnology that can contribute to further crop improvement seem like a good addition for integrated pest management. Genetic engineering of plants by the transformation of genes that confer resistant traits from any other species in greater genotypes has now been accepted as a consequence for disease resistance development. To engineer plant resistance against plant diseases, different sources of transgenes have been proposed including plant resistance genes, pathogen-derived resistance genes, and antimicrobial proteins of plant and non-plant origins. In this chapter, I discuss the principals of plant-pathogen interaction, mechanisms of the defense responses, and the current status of transgenic plants that confer resistance to different kind of pathogens.

Plant-Pathogen Interactions

During evolution, plants have developed defense mechanisms in response to pathogens, herbivores, and environmental stresses. These mechanisms consist of physical and chemical barriers, known as passive or preformed defenses, as well as active defense responses that are induced after perception of the stress or invasion. In preformed defense, physical barriers inhibit the penetration and the spread of pathogens through the plants, and chemical compounds act as inhibitors of hydrolytic enzymes produced by pathogens. The active defense responses include structural modifications and biochemical changes that lead to an increase in resistance against the pathogens. Depending on the plant-pathogen systems, a different combination of structural modification and biochemical reactions might be effective in the defense of the plant.

Assessment of resistance or pathogen virulence is based on disease reaction types. The interaction between plants and pathogens can be compatible, leading to disease, or incompatible, in which the plant is resistant.

Likewise, pathogens are either virulent or avirulent depending on their ability to successfully penetrate and invade their host. Invasion by a virulent pathogen leads to the development of symptoms and to a compatible interaction. In this case, the plant cannot detect that pathogen or the activated defense responses are ineffective or even suppressed. Avirulent pathogens fail to induce disease development thus leading to an incompatible interaction. In this case, the plant recognizes the pathogens and defense mechanisms are induced (Ohashi and Oshima, 1992; Goodman and Novacky, 1994; Agrios, 1997).

Genes Determining Host-Pathogen Specificity

In general, host resistance, pathogen virulence and disease reaction types are genetically determined. Host resistance is controlled by one or several resistance (*R*) genes. In the pathogen, the presence of one or more genes for pathogenicity, specificity, or virulence against a particular host is thought to be a key factor in disease development. Virulence genes of a pathogen are usually specific for a particular host. Some pathogens can attack several hosts because they either have many diverse genes for virulence or their virulence genes are less specific to those hosts. In many cases, the product of virulence genes are recognized by the product of a resistance gene in the plant and rapid host defense responses are triggered, leading to an incompatible interaction. In this case, the virulence genes are termed avirulence (*Avr*) genes and their product, elicitors. To explain the interaction of the *Avr* gene from pathogens and the *R* gene from the plant, it has been hypothesized that the *R* gene products encode receptors that bind *Avr* gene products and play a role as signaling intermediates in a signaling cascade that leads to the induction of resistance (Collmer, 1998).

Pathogen Avirulence Genes and Elicitors

Experimental evidence conclusively supports the notion that pathogens possess avirulence genes whose products are involved in determining host specificity. These avirulence factors include elicitors and *hrp* (*h*ypersensitive *r*eaction and *p*athogenicity) genes. Several reports show that the elicitors in the microbial culture extract can induce defense mechanisms in plants such as accumulation of phytoalexins, pathogenesis-related (PR) proteins or a hypersensitive reaction (HR). The elicitors are not only the products from pathogens but can also be plant-derived secondary elicitors such as cutin monomers that are released from plant cell wall during pathogen infection. Elicitors are chemically diverse but are thought to serve as ligands for the receptors encoded by *R* genes in plants. They include polysaccharides,

Table 1. Some elicitors of plant defense responses (Lucas, 1998)

Source	Common Name	Chemical Nature	Biological Activity
Fungi			
Monilinia fructicola	Monilicolin	Peptide	Induces phytoalexins
Phytophthora megasperma	-	Glucan	Induces phytoalexins
Phytophthora parasitica	Elicitin	Peptide	Induces necrosis
Rhynchosporium secalis	Necrosis-inducing peptides	Peptide	Induces necrosis
Cladosporium fulvum	avr elicitor	Peptide	Induces HR
Sacharomyces cerevisae	-	Glucan	Induces phytoalexins
Fungal cell wall	-	Chitosan	Induces phytoalexins
Bacteria			
Erwinia amylovora	Harpin	Protein	Induces HR
Pseudomonas syringae	Syringolide	Glycoside	Induces HR
Viruses			
Tobacco mosaic virus	Coat protein	Protein	Induces HR
Potato virus X	Coat protein	Protein	Induces resistance
Plants			
Endogenous	Oligosaccharins	Oligosaccharides	Induce host defense

glycoproteins and proteins (Jackson and Taylor, 1996; Collmer, 1998; Lucas, 1998). Some examples of elicitors are shown in Table 1.

Other avirulence factors of bacterial pathogens are *hrp* genes whose products are required for the delivery of the *Avr* gene-derived signal from the bacteria into the hosts, and are associated with both induction of HR on resistant plant and pathogenicity on susceptible plants (Jackson and Taylor, 1996). The *hrp* genes form a cluster and have been studied most extensively in *Erwinia amylovora, Pseudomonas syringae, Pseudomonas solanacearum* and *Xanthomonas campestris.* Sequence analysis of the *hrp* gene cluster in

Table 2. The products and suggested function of some *hrp* genes (Lucas, 1998)

Gene	Bacterial Source	Product	Suggested function
hrPS	*P.s. phaseolicola*	Regulatory protein	Activates *hrp* operon in response to plant signals
hrpH	*P.s. syringae*	Membrane protein	Component of protein secretory pathway
hrpC	*X.c. vesicatoria*	Membrane protein	Component of protein secretory pathway
HrpZ	*P.s. syringae*	Extracellular protein elicitor	Induces HR
HrpN	*E. amylovora*	Extracellular protein elicitor	Induces HR

X. campestris pv. *vesicatoria, E. amylovora* and *P. syringae* pv. *syringae* revealed that the genes are predicted to encode conserved components of a type III secretion system which are also conserved in the mammalian pathogens *Yersinia, Shigella* and *Salmonella* (Alfano and Collmer, 1996; Bonas and van den Ackerveken, 1999). The Hrp proteins have diverse functions but most likely seem to play a role in the secretion pathway required for pathogenicity. Some *hrp* genes and their possible function are shown in Table 2.

Plant Resistance Genes

As discussed earlier, plants possess *R* genes whose product is likely to encode receptors, which interact specifically with *Avr* gene products of the pathogen much in the same way as a receptor and ligand. Numerous disease resistance genes have been identified in plants (Table 3). The first *R* gene to be cloned and characterized was *Hm1* from maize, which provides resistance to *Cochliobolus carbonum* by inactivating the HC toxin produced by this fungus. In contrast, most of the R genes that have been cloned and characterized seem to have a common characteristic involved in signal transduction. The structural domains of *R* gene products comprise serine-threonine kinase, leucine-rich repeats (LRR) or nucleotide-binding sites (NBS). The *R* genes that contain NBS can be further subdivided into two classes, those containing the leucine zippers (LZ) and those containing domains similar to the *Drosophila Toll*/interleukin-1 receptors. However, the common motif of the *R* gene products is LRR, which is believed to determine specific recognition

Table 3. Some cloned plant disease resistance genes (adapted from Bent, 1996; Lucas, 1998)

R gene	Host	Pathogen	Structure	Location	Function
Hm1	Maize	*Cochliobolus carbonum*	Toxin reductase	Cytoplasmic	Toxin reductase
Pto	Tomato	*Pseudomonas syringae* pv. *tomato*	Protein kinase	Cytoplasmic	Kinase
RPS2	Arabidopsis	*P. syringae* pv. *tomato*	LRR, NBS, LZ	Membrane	Receptor
Xa21	Rice	*Xanthomonas campestris* pv. *oryzae*	LRR, protein kinase	Membrane	Receptor
N	Tobacco	Tobacco mosaic virus	LRR, NBS	Cytoplasmic	Receptor
Cf	Tomato	*Cladosporium fulvum*	LRR	Membrane	Receptor
L6	Flax	*Melampsora lini*	LRR, NBS	Cytoplasmic	Receptor

Class	R-gene	Structure				
1	*Pto, Fen*					Kinase
2	*Xa21, FLS2*		LRR		TM	Kinase
3	*Cf genes*		LRR		TM	
4	*Xa21*D		LRR			
5; 5.1	*Rps2, Rpm1,* and *Mi*	LZ	HD	NBS	LRR	
5.2	*N and L6*	TIR		NBS	LRR	

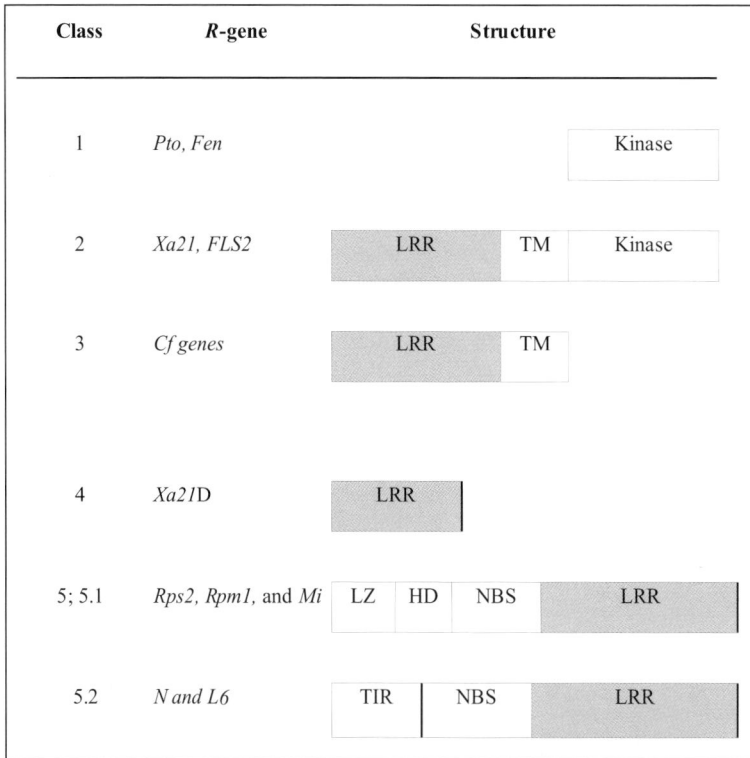

Figure 1. The five main classes of representative resistance genes. LRR= leucine-rich repeat; TM= transmembrane domain for *Xa21* or membrane anchor for the *Cf* class; LZ= leucine zipper; HD= hydrophobic domain; NBS= nucleotide- binding site; TIR= Toll/Interleukin-1 signaling domain (Richter and Ronald, 2000).

of elicitors by the receptor molecules (Bent, 1996; Richter and Ronald, 2000). The structure and classes of the cloned plant *R* genes are shown in Figure 1.

Plant Defense Mechanisms

Preformed Defense Mechanisms

The initial pathogen infection process usually occurs at the plant surface. Plant structural defenses can prevent the invasion of certain pathogens. These structural preformed barriers include wax, cuticle, and cell wall structures. In addition, other structural features such as leaf shape, vein pattern, leaf folding, the presence of hairs, size, shape, and number of stomata, and lenticels also influence the infection process of pathogens.

Preformed chemical defense substances are also involved in plant defense and comprise a wide diversity of secondary metabolites, which are toxic to pathogens and pests. They also include preformed inhibitors. Several studies have shown that preformed plant defense substances may play a role in host-pathogen interactions. For example, fungitoxic exudates that are released from leaves of tomato and sugarbeet inhibit the germination of spores of *Botrytis* and *Cercospora*, respectively (Agrios, 1997). Other examples of preformed chemical substances that are either present on the surface waxes or within host cells are phenolic glycosides or esters. These substances inhibit the activity of fungal hydrolytic enzymes (Lucas, 1998). In onion, resistance to smudge disease (*Colletotrichum cercinans*) is related to the presence of two phenolic compounds: catechol and protocatechuic acid. Apple and pear leaves contain toxic substances that inhibit fireblight bacterium (*Erwinia amylovora*). In potato tubers, resistance to common scab (*Streptomyces scabies*) is related to the presence of chlorogenic acid. Avenacin in oats and tomatines in tomato show antifungal activity against many fungi (reviewed in Osbourn, 1996; Agrios, 1997). The importance of the oat avenacin in the preformed defense of oat against the root-infecting fungi, *Gaeumannomyces graminis* var. *tritici* and other fungal pathogens was demonstrated using genetic evidence (Papadopoulou *et al.*, 1999). The resistance to infection by *Colletotrichum gleosporioides* of subtropical fruits is associated with the presence of 5-alkylated resorcinols (Prusky and Keen, 1993).

Induced Defense Mechanisms
Plants respond to pathogens either by modifying their cell wall structure or by inducing the synthesis of chemical compounds that lead to the inhibition of the invading pathogens. The biochemical processes induced upon pathogen infection include the hypersensitive reaction (HR), production of cell wall strengthening molecules, production of antimicrobial substances (PR proteins, phytoalexins and phenolic compounds), the production of active oxygen radicals and lipoxygenases (Goodman and Novacky, 1994; Agrios, 1997; Lucas, 1998). The induced defense mechanisms are very complex and are detailed below.

The Changes of Host Cell Wall
The induced structural defense involves morphological changes in the cell wall that, in turn, lead to increase resistance to penetration of fungal pathogens. These changes include the reinforcement of host cell walls with strengthening molecules, such as lignin, callose, esterified phenol and hydroxyproline-rich glycoprotein (HRGP). In addition, other barriers are also made, such as cork layers, abscission layers, formation of tylose and

deposition of gums. Amorphous, fibrillar materials that surround and trap bacteria, and prevent bacterial multiplication are also produced in certain cases (reviewed in Agrios, 1997; Lucas, 1998).

Hypersensitive Reaction (HR)

A central feature of the gene-for-gene plant disease resistance is the hypersensitive reaction (HR). HR results from a specific recognition event of avirulence factors or elicitors of pathogen by plant resistance genes. It is observed visually as small necrotic lesions in the infected region. The mechanism of HR is complex and consists of biochemical processes that lead to the death of the infected cells resulting in the restriction of the pathogen's invasion. It is accompanied by the formation of toxic antimicrobial compounds such as phytoalexins, active oxygen species (ROS), phenoxy radicals, and lipid radicals, together with the development of physical barriers (see above). In addition, HR is also accompanied by the increase of the level of degradative enzymes that may attack the microorganism such as PR proteins (reviewed in Goodman and Novacky, 1994; Mehdy, 1994; Jackson and Taylor, 1996; Agrios, 1997).

Pathogenesis-related Proteins

Pathogenesis-related (PR) proteins are produced and accumulated in local infected tissues as well as in uninfected systemic tissues during plant defense responses. The induction of PR proteins and acquired resistance after pathogen attack suggest that PR proteins may play a role in defense against pathogens (Ohashi and Oshima, 1992; Stintzi *et al.*, 1993; Goodman and Novacky, 1994). However, the biological function of some PR proteins is still not well known. The characterized PR proteins and their properties are summarized in Table 4 (van Loon and van Strien, 1999).

Phytoalexins

Phytoalexins are toxic antimicrobial compounds with low molecular weight, which are not present in healthy tissues but are produced upon pathogen infection, injury or stress. They are broad-spectrum inhibitors that active against a wide range of organisms. The induction of phytoalexins is associated with *de novo* synthesis of different biosynthetic enzymes. The resistance occurs when one or more phytoalexins reach a concentration sufficient to restrict pathogen development. Some of the better-studied phytoalexins include rishitin in tobacco, camalexin in *Arabidopsis*, phaseollin in bean,

Table 4. The families of PR proteins (adapted from van Loon and van Strien, 1999).

Family	Type member	Properties	References
PR-1	Tobacco PR-1a	unknown	van Loon *et al.*, 1987
PR-2	Tobacco PR-2	β-1,3-glucanase	van Loon *et al.*, 1987
PR-3	Tobacco P, Q	chitinase type I, II, IV, V, VI, VII	van Loon *et al.*, 1987
PR-4	Tobacco R	chitinase type I, II	van Loon *et al.*, 1987
PR-5	Tobacco S	thaumatin-like	Cornelissen *et al.*, 1986
PR-6	Tomato Inhibitor I	proteinase inhibitor	Green and Ryan, 1972
PR-7	Tomato P_{69}	endoproteinase	Vera and Conejero, 1988
PR-8	Cucumber chitinase	chitinase type III	Métraux *et al.*, 1988
PR-9	Tobacco 'lignin-forming peroxidase'	peroxidase	Lagrimini and Rothstein, 1987
PR-10	Parsley 'PR-1'	ribonuclease-like	Somssich and Hahlbrock, 1986
PR-11	Tobacco class V chitinase	chitinase, type I	Melchers *et al.*, 1994
PR-12	Radish Rs-AFP3	defensin	Terras *et al.*, 1992
PR-13	Arabidopsis THI2.1	thionin	Epple *et al.*, 1995
PR-14	Barley LTP4	lipid-transfer protein	Garcia-Olmedo *et al.*, 1995

pisatin in pea, glyceollin in soybean, alfalfa, and clover, pisatin in potato, gossypol in cotton and capsidiol in pepper (Lucas, 1998). In some plants, more than one phytoalexin is produced upon pathogen attack; for example, broad bean (*Vicia faba*) inoculated with *Botrytis cinerea* produced five different phytoalexins (reviewed in Agrios, 1997; Lucas, 1998).

Systemic Acquired Resistance

Plant Immunization
Immunization in human and animals is often activated by an application of inactivated pathogen, pathogen protein or other antigenic substances. This

resulted in the production of antibodies against the pathogens and thereby, protection occurred against the later infection of the pathogens. In contrast, plants do not produce antibodies in response to infection by pathogens, but rather, defense mechanisms are activated upon the signals transmitted from an attacking pathogen. Induced resistance occurs first around the site of infection, and it is called local acquired resistance. An interesting feature of plants is the subsequent development of resistance in sites distal from the local infection. This is called systemic acquired resistance (SAR). It is analogous to immunization and reflects the ability of the plant to protect themselves against a future infection. Thus, the resistance spreads out systemically to distal non-infected parts of the plant (reviewed in Schneider *et al.*, 1996; Agrios, 1997; Sticher *et al.*, 1997). The mechanisms involved in SAR include changes of the plant cell walls, an induction of PR proteins and an increase of peroxidase and lipoxygenase (LOX) activities.

Application of plant-growth promoting rhizobacteria (PGPR) to the roots of plants can induce resistance in the leaves or the stem, without causing any symptoms to the root that has been inoculated with the PGPR. Induced resistance triggered by PGPR has been termed induced systemic resistance (ISR) in order to differentiate it from SAR. ISR is mediated by a jasmonate- or ethylene- sensitive pathway and does not involved the expression of PR proteins (Pieterse *et al.*, 1998; van Loon *et al.*, 1998; van Wees *et al.*, 2000).

Inducers of Plant Resistance
Plants develop SAR upon the attack of insects, necrotizing pathogens, or treatment with certain chemicals. SAR acts nonspecifically throughout the plant and contributes to resistance against a broad spectrum of pathogens. In young cucumber plants, localized inoculation with either a fungus (*Colletotrichum lagenarium*) or a virus (tobacco necrosis virus, TNV) leads to a broad spectrum SAR to at least 13 diseases caused by fungi, bacteria and viruses (Dean and Kuc, 1986; Sticher *et al.*, 1997). Inoculation of cucumber plants with *Pseudomonas syringae* pv. *lachrymans* led to protection against *P.s.* pv. *lachrymans, P.s.* pv. *syringae, C. lagenarium* and TNV (Phuntumart, unpublished). Inoculation of spores of *Peronospora parasitica* pv. *tabaci*, the blue mold pathogen, into the stem of tobacco plants leads to SAR in the leaves against the same pathogen for 2-3 weeks after the primary inoculation (Cohen and Kuc, 1981). In Arabidopsis, SAR provides resistance against *P. parasitica, Pseudomonas syringae* pv. *tomato* DC3000 and turnip crinkle virus (Uknes *et al.*, 1992). As mentioned earlier, SAR can also be induced by certain chemicals. The known chemical inducers of SAR are not antimicrobial and include salicylic acid (SA), arachidonic acid, 2,6-

dichloroisonicotinic acid (INA) and benzo(1,2,3) thiadiazole-7-carbothionic acid S-methyl ester (BTH) (Schneider *et al.*, 1996; Sticher *et al.*, 1997). These chemicals may induce acquired resistance in plants at a level not causing necrosis and are effective when applied through the root, by foliar spray, or by stem injection. They induce the same set of *PR* genes as those induced systemically during SAR such as, PR-1, β-1,3-glucanase, chitinase and cysteine-rich protein (Ward *et al.*, 1991a; Agrios, 1997; Sticher *et al.*, 1997; Phuntumart, unpublished).

Genes Associated with SAR

Development of SAR is associated with the expression of various genes that belong to the PR family in the broadest sense. The classified *PR*-genes are shown in Table 4. Some *PR* genes, such as *PR-1* and *PR-8* have been used as marker genes since their induction is tightly correlated with the appearance of SAR in uninfected tissues (Métraux *et al.*, 1989; Ward *et al.*, 1991b; Uknes *et al.*, 1992; Phuntumart, unpublished). However, the expression level of *PR* genes can vary during SAR between different plant species.

SAR Signal Transduction
Salicylic Acid

Although the signal transduction triggering SAR is not well established, salicylic acid (SA) seems to play an important role in SAR signaling. This idea is based on several evidences such as (i) the level of SA increases after pathogen inoculation and it is correlated with SAR (Malamy *et al.*, 1990; Métraux *et al.*, 1990) (ii) exogenous application of SA can induce SAR and expression of genes associated with SAR (Ward *et al.*, 1991b; Uknes *et al.*, 1992), (iii) transgenic tobacco and *Arabidopsis thaliana* plants expressing the bacterial *NahG* gene encoding salicylate hydroxylase, an enzyme that catalyzes the conversion of SA to catechol, not only show a low level of SA but are unable to express SAR in response to viral, fungal or bacterial pathogens (Gaffney *et al.*, 1993; Delaney *et al.*, 1994). These results suggest that SA is required in the SAR signal transduction pathway. However, other evidence suggests that SA is not the long distance signal. First, in cucumber plants inoculated with *P. syringae*, the primary infected leaves can be removed before the accumulation of SA without affecting the SAR expression and the systemic increase of SA (Rasmussen *et al.*, 1991). Second, grafting experiment between *NahG* and wild-type tobacco plants show that a signal other than SA is responsible for systemic expression of SAR (Vernooij *et al.*, 1994). However, it is clear that SA must be present for SAR to occur.

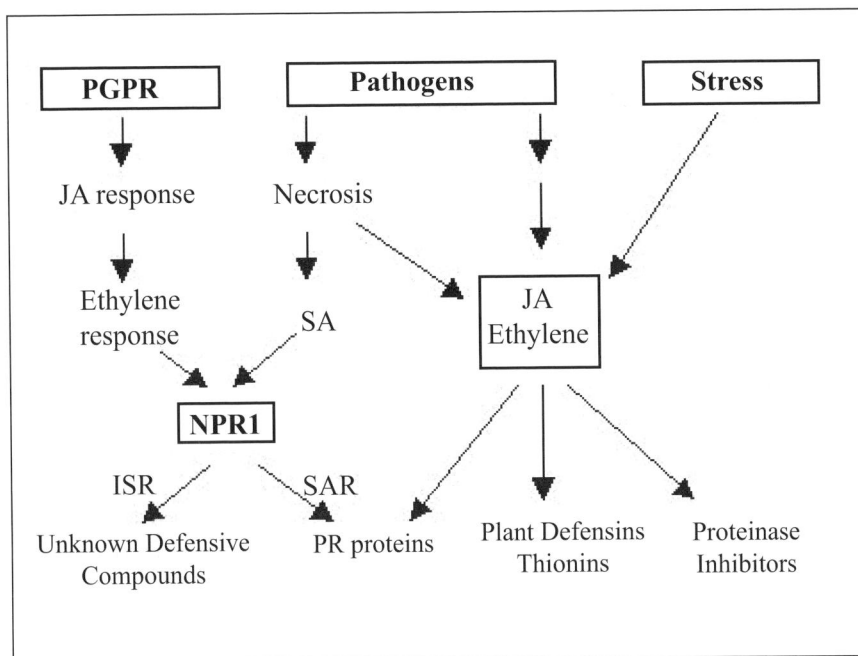

Figure 2. Proposed model showing systemic signals transduction and signal molecules associated with induced-resistance, JA=jasmonate, SA=salicylic acid, NPR1=*Arabidopsis* regulator of systemic acquired resistance, ISR= Induced systemic resistance, SAR= systemic acquired resistance (Pieterse and van Loon, 1999).

Several lines of evidence indicate that SA is an essential signal in SAR and it is required downstream of the long distance signal (Ryals *et al.*, 1996; Sticher *et al.*, 1997). A scheme of the signal transduction pathway of SA is shown in Figure 2.

Jasmonates
Jasmonic acid (JA) is derived from peroxidized linolenic acid, generated by lipoxygenase (LOX). JA and its volatile ester, methyl jasmonate (MeJA) are inducers of the expression of protease inhibitors (Farmer and Ryan, 1990; Farmer *et al.*, 1992) thionin (Andresen *et al.*, 1992; Epple *et al.*, 1995), defensin (Penninckx *et al.*, 1996), osmotin (Xu *et al.*, 1994) and proline-rich glycoprotein (Ryan, 1992). JA and MeJA were shown to induce resistance to *Phytophthora infestans* in potato (Cohen, 1993). Exogenous application of JA led to an accumulation of PR-1 in rice but did not induce resistance to the rice blast fungus, *Magnaporthe grisea* in the treated-leaves while partial protection occurred at the non-treated upper leaves (Schweizer *et al.*, 1995).

191

This suggests that JA may induce a systemic signal for SAR. The mutants of Arabidopsis, *fad3-2, fad7-2* and *fad8* that cannot accumulate jasmonate were found to be extremely susceptible to *Pythium mastophorum* (Drechs.), a root rot fungus, compared to wild-type plants. Inoculation with *Pythium* resulted in the induction of transcripts of three jasmonate-responsive defense genes (*LOX2, CHS, and PDF1.2*) in the wildtype but not in the jasmonate-deficient mutants. Exogenous application of methyl jasmonate to the mutants can reduce the disease to a level close to that of wild-type controls. These results indicate that jasmonate is important in the signal transduction pathway in defense to pathogens (Vijayan *et al.*, 1998). The proposed role of JA in signal transduction pathway is shown in Figure 2.

Lipoxygenase (LOX) is one of the important enzymes in the octadecanoid pathway leading to JA biosynthesis. The isolation and characterization of LOX in higher plants showed that it occurs as gene families, for example there are three isoforms in Arabidopsis and eight isoforms in soybean (Siedow, 1991; Fidantsef and Bostock, 1998). Exogenous application of INA or inoculation with *M. grisea* induced accumulation of LOX in rice and can protect rice seedlings from *M. grisea* (Schweizer *et al.*, 1995).

Ethylene
Ethylene, a plant hormone derived from methionine, is also produced upon wounding or infection by pathogens. It has been shown to enhance lignification and induce accumulation of hydroxyproline-rich glycoproteins (Boller, 1990), β-1,3-glucanase and chitinase (Abeles *et al.*, 1971). In ethylene-insensitive mutants of Arabidopsis plants, the expression of *PR* genes is similar to that in wild-type plants (Bent *et al.*, 1992). Thus, it is unlikely that ethylene is the systemic signal for SAR. Two experiments have shown that ethylene can lead to enhanced jasmonate level in tomato plants (Xu *et al.*, 1994; O'Donnell *et al.*, 1996). The induction of a defensin gene in Arabidopsis plants appears to require both JA and ethylene (Penninckx *et al.*, 1998). The proposed role of ethylene in the signal transduction pathway for induced resistance is shown in Figure 2.

Transgenic Plants for Fungal and Bacterial Resistance

Several genes that confer resistance to plant diseases caused by fungi and bacteria were found in different organisms. These include plant *PR* genes, genes encoding antimicrobial peptides, pathogen-derived genes and genes associated with defense responses (i.e. HR).

Transgenes That Encode Antimicrobial and Pathogenesis-related Proteins

Numerous antimicrobial proteins have been expressed in transgenic plants either constitutively or targeted to vacuoles or apoplasts. These include phytoalexins and PR proteins. Several lines of evidence demonstrated their role in the defense response against the pathogens when they are induced or constitutively expressed in transgenic plants (Table 5). The hydrolytic enzymes, chitinase and β-1,3 glucanase are PR proteins that are capable of degrading chitin and β-1,3 glucan, the major components of plant cell walls. Over-expression of chitinase in transgenic tobacco, canola (*Brassica napus*), and rice enhanced resistance to *Rhizoctonia solani,* a soilborne fungus (Broglie *et al.,* 1991; Benhamou *et al.,* 1993; Jach *et al.,* 1995; Datta *et al.,* 2001) and to *Fusarium thapsinum* in transgenic sorghum (Waniska *et al.,* 2001). Transgenic alfalfa over-expressing endogenous β-1,3 glucanase (*Aglu1*) showed reduction in disease severity when challenged with *Phytophthora megasperma* f.sp. *medicaginis* (Dixon *et al.,* 1996).

Combinations of two or more antimicrobial genes may provide significantly enhanced resistance to pathogens. A synergistic activity between the two enzymes of tobacco chitinase and β-1,3-glucanase genes, constitutively expressed in tomato plants conferred more resistance to a fungal pathogen, *Fusarium oxysporum,* than either gene alone (Jongedijk *et al.,* 1995). Similar results were observed during co-expression of a rice chitinase (*RCH10*) gene and an alfalfa glucanase (*Aglu1*) gene in transgenic tobacco to *Cercospora nicotianae.* However, the same combination of these genes did not show any significant resistance to plant pathogens when expressed in alfalfa (Dixon *et al.,* 1996). This may due to the fact that disease resistance in alfalfa required different combination of defense responses other than only the activities of these two hydrolytic enzymes. Co-expression of genes encode class II chitinase, class II glucanase and a type I ribosome-inactivating protein (RIP) from barley in transgenic tobacco displayed a greater protection against the *C. nicotianae* than either transgene alone (Zhu *et al.,* 1994). Transgenic carrots constitutively expressed tobacco class I chitinase and β-1,3 glucanase exhibited resistance to *Alternaria dauci, Alternaria radicina, Cercospora carotae* and *Erysiphe heracleï* (Melchers and Stuiver, 2000). Constitutively expressing chimeric chitinase genes (tomato/tobacco chitinase) in transgenic canola enhanced disease tolerance to *Cylindrosporium concentricum, Phoma lingam* and *Sclerotinia sclerotium* (Grison *et al.,* 1996). Transgenic tobacco and potato plants with the *ThEn-42,* a gene encoding chitin-degrading enzymes isolated from *Trichoderma harzianum* showed significantly increased resistance to the pathogens, *Alternaria alternata, A. solani, B. cinerea* and *R. solani* (Lorito *et al.,* 1998). Several reports showed

Table 5. Examples of transgenic crops for disease resistance

Plants	Transgene(s)	Enhanced resistance to	References
Tobacco	Chitinase	*Rhizoctonia solani*	Broglie *et al.*, 1991; Jach *et al.*, 1995
	RIP	*Rhizoctonia solani*	Logeman *et al.*, 1992
	PR-1a, SAR 8.2	*Peronospora tabacina,*	Alexander *et al.*, 1993
		Phytophthora parasitica,	Alexander *et al.*, 1993
		Pythium sp.	Alexander *et al.*, 1993
	Chitinase+Glucanase	*Cercospora nicotianae*	Zhu *et al.*, 1994
	Chitinase+Glucanase	*Rhizoctonia solani*	Jach *et al.*, 1995
	Chitinase+RIP	*Rhizoctonia solani*	Jach *et al.*, 1995
	AFP	*Alternaria longipes*	Terras *et al.*, 1995
	CPO-P	*Colletotrichum destructivum*	Rajasekaran *et al.*, 2000
	LTP	*Pseudomonas syringae*	Molina and Garcia-Olmedo, 1997
	expI	*Erwinia carotovora*	Maë *et al.*, 2001
	Virus CP	*TMV, ToMV*	Sanders *et al.*, 1992
	PAP	*PVX, PVY*	Lodge *et al.*, 1993
	hrmA	*Pseudomonas syringae pv. tabaci*	Shen *et al.*, 2000
		Phytophthora parasitica, ToVMV	Shen *et al.*, 2000
Arabidopsis	Thi2.1 (thionin)	*Fusarium oxysporum*	Epple *et al.*, 1997
	LTP	*Pseudomonas syringae*	Molina and Garcia-Olmedo, 1997
	NIM1/NPR1	*Peronospora parasitica*	Cao *et al.*, 1998
		Pseudomonas syringae	Cao *et al.*, 1998
Tomato	Chitinase+Glucanase	*Fusarium oxysporum*	Jongedijk *et al.*, 1995
	Bs2	*Xanthomonas campestris*	Tai *et al.*, 1999
	Tobacco N	TMV	Whitham *et al.*, 1996
	PAP	PVX, PVY	Lodge *et al.*, 1993
Potato	AP24	*Phytophthora infestans*	Liu *et al.*, 1994
	Glucose oxidase	*Phytophthora infestans,*	Wu *et al.*, 1995
		Verticillium dahliae,	Wu *et al.*, 1995
		Erwinia carotovora	Wu *et al.*, 1995
	AFP	*Verticillium*	Liang *et al.*, 1998
	Replicase	PLRV	Vazquez-Rovere *et al.*, 2001
Canola	Chitinase	*Rhizoctonia solani,*	Broglie *et al.*, 1991
		Cylindrosporium concentricum,	Grison *et al.*, 1996
		Sclerotinia sclerotium,	Grison *et al.*, 1996
		Phoma lingam	Grison *et al.*, 1996
Cucumber	Chitinase	*Botrytis cinerea*	Tabei *et al.*, 1998
Carrot	Chitinase+Glucanase	*Alternaria dauci,*	Melchers and Stuiver, 2000
		Alternaria radicina,	Melchers and Stuiver, 2000
		Cercospora carotae,	Melchers and Stuiver, 2000
		Erysiphe heracleï	Melchers and Stuiver, 2000
	Human lysozyme	*Alternaria dauci*	Takaichi M., and Oeda, K., 2000
		Erysiphe heracleï	Takaichi M.,and Oeda, K., 2000
Geranium	AFP	*Botrytis cinerea*	Bi *et al.*, 1999
Peanut	Chitinase	*Cercospora arachidicela*	Rohini and Rao, 2001

Broccoli	Chitinase	*Aternaria brassicicola*	Mora and Earle, 2001
Sweet potato	CP	SPFMV	Okada *et al.*, 2001
Papaya	RP	PRSV	Chen *et al.*, 2001
Grapevine	Chitinase	*Uncinula necator*	Yamamoto *et al.*, 2000
Sorghum	Chitinase	*Fusarium thapsinum*	Waniska *et al.*, 2001
Rice	Chitinase	*Rhizoctonia solani*	Datta *et al.*, 2001
		Verticillium dahliae	Tabaeizadeh *et al.*, 1999
		Magnaporthe grisea	Nishizawa *et al.*, 1999
	Thaumatin (PR-5)	*Rhizoctonia solani*	Datta *et al.*, 1999
	AFP (cecropin)	*Xanthomonas oryzae* pv. *oryzae*	Sharma *et al.*, 2000
	Xa21	*Xanthomonas oryzae* pv. *oryzae*	Tu *et al.*, 1998
Alfalfa	IOMT	*Phoma medicaginis*	He and Dixon, 2000
Wheat	AFP	*Fusarium sp.*	Liang *et al.*, 1998

that over-expression of chitinase are ineffective in some fungi that do not contain chitin such as oomycetes (i.e. *Pythium, Phytophthora*). Therefore, other types of PR or PR-like proteins that show antifungal activity include osmotins, defensins, thionins, hevein (a lectin from *Urtica dioica*) and lipid transfer proteins from various plant species have been used to generate disease resistant plants (Broekaert *et al.*, 1997). Constitutive expression of an endogenous PR1a in transgenic tobacco plants exhibited significant resistance to two oomycete pathogens, *Peronospora tabacina* and *Phytophthora parasitica* (Alexander *et al.*, 1993). Constitutive expression of a defensin from radish (*Rs-AFP2*) in tobacco resulted in enhanced resistance to *Alternaria longipes* (Terras *et al.*, 1995). Transgenic potato constitutively expressed tobacco osmotin, a PR-5 protein, significantly delayed a symptom caused by *Phytophthora infestans* (Liu *et al.*, 1994; Zhu *et al.*, 1996). Over-expression of alfalfa defensin (*alfAFP*) in transgenic potato showed resistance to *Verticillium dahliae* in both the greenhouse and under field conditions (Gao *et al.*, 2000). Increased protection of Arabidopsis plants from *Fusarium oxysporum* has been demonstrated in transgenic plants constitutively expressing an endogenous thionin gene (*Thi2.1, PR5*; Epple *et al.*, 1997). Transgenic potato expressing *pA13*, a gene encoding osmotin-like proteins showed an increased tolerance to the late-blight fungus *Phytophthora infestans* at different phases of infection, especially at an early phase of fungal infection (Zhu *et al.*, 1996).

Similar results have been observed in bacterial resistance transgenic plants. Expression of *PR-14*, a barley non-specific lipid-transfer protein enhanced resistance against *Pseudomonas syringae* when expressed in Arabidopsis and tobacco plants (Molina and Garcia-Olmedo, 1997).

Transgenic rice with a gene encoding cecropin B, an antimicrobial protein from *Bombyx mori*, driven by a signal peptide of chitinase gene from rice conferred resistance to bacterial leaf blight caused by *Xanthomonas oryzae* pv. *oryzae* under the greenhouse condition (Sharma *et al.*, 2000).

The other sources of promising plant antifungal peptides are ribosome-inactivating proteins (RIP) from plants and other organisms. RIPs are toxic N-glycosidases that modify ribosomes by specifically removing an adenosine residue located near the 3' terminus of a large subunit RNA, resulting in prevent the protein synthesis. RIPs are linked to plant defense by inhibiting protein translation of microorganisms an by their antimicrobial activity (Nielsen and Boston, 2001). Transgenic plants expressing different types of RIPs conferred resistance to a broad range of plant pathogens (Table 5). Tolerance to *R. solani* has been observed in transgenic tobacco constitutively expressing proRIP b-32 from maize (Maddolani *et al.*, 1997). The barley seed-RIP conferred resistance to *Erysiphe graminis* when over-expressed in transgenic wheat (Bieri *et al.*, 2000). However, abnormal development, such as, growth inhibition of transgenic plants over-expressing RIPs were often observed. (Lam *et al.*, 1996; Lodge *et al.*, 1993). It is thought that this might limit their antiviral activity efficiency. Thus, different types of inducible promoters have been exploited to drive different RIP genes, e.g., wound-inducible promoter controlled type I barley RIP exhibited resistance to *R. solani* when expressed in tobacco (Logemann *et al.*, 1992).

Transgenic Plants with Avirulence Genes

Transgenic tobacco plants expressing *hrmA*, a gene encoding avirulence against *Pseudomonas syringae* pv. *tabaci* (isolated from *Pseudomonas syringae* pv. *syringae)* conferred resistance to *P. parasitica* and *P. syringae* pv. *tabaci* (Shen *et al.*, 2000). Expression of *avrRpt2*, the avirulence gene of *P. syringae* pv. *tomato* in transgenic *Arabidopsis* plants have been reported to confer resistance to *P. syringae* pv. *tomato*, together with the induction of *PR1* gene (McNellis, *et al.*, 1998). Transgenic canola (*Brassica napus*) plants with avirulence (*Avr-9*) gene from *Claodosporium fulvum*, showed enhanced resistance to *Leptosphaeria maculans* together with the induction of genes associated with defense responses (i.e. *PR1, PR2* and *Cxc750*; Hennin *et al.*, 2001). The expression of bacterial ribonuclease barnase under the control of pathogen-inducible promoter (*prp-1*) from potato was found to reduce sporulation of *P. infestans* (Strittmatter *et al.*, 1995).

Transgenes Associated with Defense Responses

The *Arabidopsis NIM1/NPR1* gene, which is required in SA-mediated resistance pathway enhanced resistance to *Pseudomonas syringae* and *Peronospora parasitica* when expressed in transgenic plants (Cao *et al.*, 1998). Transgenic alfalfa over-expressing O-methyl transferase (IOMT) resulted in early production of the endogenous phytoalexin medicarpin and accordingly reduced disease severity caused by *Phoma medicaginis* (He and Dixon, 2000).

Transgenes those induce the very early steps of resistance are the potential genes to improve disease resistance in plants. Among them, the H_2O_2 – generating enzymes and proteins that involved in cell death were shown to function upstream of the reactive oxygen intermediates, O_2^-, and H_2O_2. Constitutive expression of the H_2O_2-generating enzyme, glucose oxidase, in transgenic potato displayed disease resistance to *Phytophthora infestans, Verticillium* wilt disease and *Erwinia carotovora* (Wu *et al.*, 1995). Transgenic rice over-expressing endogenous *OsRac1*, a gene that encodes a regulator of ROI and cell death, conferred resistance to a compatible race of *Magnaporthe grisea* as well as a compatible race of bacterial blight disease caused by *Xanthomonas oryzae* pv. *oryzae* (Ono *et al.*, 2001).

Other Transgenes for Disease Resistance

Over-production of chloroperoxidase (CPO-P) from *Pseudomonas pyrrocinia* in tobacco plants resulted in significantly increasing resistance against *Colletotrichum destructivum* (Rajasekaran *et al.*, 2000). Transgenic carrots constitutively expressing a human lysozyme gene exhibited resistance to *Alternaria dauci* and *Erysiphe heracleï* (Takaichi and Oeda, 2000). The N-oxoacyl-homoserine lactone (OHL) is a bacterial pheromone that plays a role in the signaling process during pathogenicity of *E. carotovora*. Expression of OHL biosynthesis, *expI* gene of *E. carotovora*, resulted in enhanced resistance to infection of *Erwinia carotovora*. It is believed that plants produced an endogenous active pheromone, which subsequently activated plant defense response (Maë *et al.*, 2001). Over-expression of a human lysozyme or *ttr* (tabtoxin resistance) gene in transgenic tobacco exhibited resistance against *Pseudomonas syringae* pv. *tabaci* (Nakajima *et al.*, 1997; Batchvarova *et al.*, 1998).

Transgenic Plants for Virus Resistance

Transgenic Approaches Including the Partial or Complete Sequences of Virus Genomes

Infection of a mild strain of virus often protects the plants from the subsequent infection with the same or related virus. This method is called cross-protection and has been used to protect crop plants in a variety of host/virus systems, such as citrus tristeza virus (CTV), cucumber mosaic virus (CMV), papaya ring spot virus (PRSV), and tobacco mosaic virus (TMV). The mechanisms of cross protection were not well understood until transgenic plant approach becomes a method of choice to produce virus disease resistant plants. This approach includes transgene-mediated virus resistance, co-suppression, antisense suppression, virus-induced gene silencing, and transcriptional gene silencing in plants (Waterhouse *et al.*, 2001).

Replicase-mediated resistance is one of the potential approaches to generate plants resistant to viral diseases. Transgenic plants produced by transformation of a complete or partial genome of viral replicases have been shown to confer resistance to many viruses (Baulcombe, 1994). This type of approach has been shown to exhibit resistance against tobacco mosaic virus (TMV; Golemboski *et al.*, 1990), potato virus X (PVX; Braun and Hemenway, 1992; Longstaff *et al.*, 1993), pea early browning virus (PEBV; MacFarlane and Davies, 1992), cucumber mosaic virus (CMV; Anderson *et al.*, 1992; Carr *et al.*, 1994; Wintermental *et al.*, 1997; Canto and Palukaitis, 1998), cymbidium ringspot virus (CymRSV; Rubino and Russo, 1995), potato leaf roll virus (PLRV; Vazquez-Rovere *et al.*, 2001) and papaya ringspot virus (PRSV; Chen *et al.*, 2001).

Transgenes that Encode Co-suppressors

Transgenic plants transformed with co-suppressor and virus resistance proteins are capable of degrading the invading virus RNA that homologous to the transgene. Resistance attributed to the high level of transcription rate of the transgene mRNA in the nucleus and a relatively low level in the cytoplasm (Dougherty and Parks, 1995). It has been proposed by Waterhouse *et al.* (2001) that the mRNA transcripts were recognized as foreign, led to the induction of sequence specific degradation of the transgene itself and also the other homologous or complementary RNA sequences by a nuclease in cytoplasm. This mechanism is known as post-transcriptional gene silencing (PTGS). Lindbo and Dougherty (1992) generated transgenic tobacco plants expressing untranslatable transcripts of tobacco etch virus coat protein containing the β-glucuronidase reporter gene (TEV/GUS). The plants were then challenged with wild type TEV or TEV/GUS. The transgenic plants showed resistance to the virus TEV/GUS but not to the wild type. This result indicates that the virus TEV/GUS was recognized by the plants and was specifically degraded in cytoplasm with the same mechanism as described above. Similar results have been observed in transgenic tomato expressing PVX/GUS (Angell and Baulcombe, 1997).

Transgenes that Encode GTP-binding Protein

Transgenic plants expressing GTP-binding proteins are also found to confer resistance to disease caused by virus. A high level of resistance against TMV of transgenic tobacco expressing *rgp1*, a gene encoding a Ras-related small GTP-biding protein has been observed. These transgenic plants showed an elevated level of salicylic acid in response to wounding as well as a significant increase in acidic PR protein and resulted in a significant resistance to TMV (Sano *et al.*, 1994).

Transformation of the R-gene and Genes That Encode Antiviral Proteins Some of the *R* gene and genes encoding antiviral proteins from plants enhance resistance to viral diseases when being expressed *in vivo*. The N gene from tobacco, which confers a 'gene for gene resistance' to tobacco mosaic virus (TMV) and to most other members of tobamovirus, showed enhanced resistance to TMV when expressed in tomato (Whitham *et al.*, 1996). Transgenic tobacco expressing dianthin (RIP from *Dianthus caryophyllus*) under the control of virion-sense promoter of African cassava mosaic virus (AFMV) showed resistance to Geminivirus infection (Hong *et al.*, 1996). Expression of pokeweed antiviral protein (PAP) in transgenic tobacco and tomato plants conferred a broad-spectrum virus disease resistance against PVX and PVY (Lodge *et al.*, 1993).

Transgenic Plants with Avirulence Genes

The transgene-mediated resistance by insertion of avirulence factors of pathogens is found to protect the plants from several diseases. Transgenic tobacco plants expressing *hrmA*, a gene that encodes avirulent factor from *Pseudomonas syringae* pv. *syringae* conferred resistance to tobacco vein mottling virus (ToVMV; Shen *et al.*, 2000). Expression of viral avirulence factors, coat proteins (CP) has been shown to confer resistance to the same virus that the CP was derived and sometimes to the closely related-viruses. Transgenic tobacco expressing CP of TMV or CP of tomato mosaic virus (ToMV) showed resistance to TMV (Powell-Abel *et al.*, 1986, Nelson *et al.* 1988, Sanders *et al.*, 1992). Coat protein-mediated resistance has also been reported in transgenic plants expressing cucumber mosaic virus (CMV;

Namba *et al.*, 1991; Quemada *et al.*, 1991), potato virus X (PVX; Hemenway *et al.*, 1988), potato virus Y (PVY; Perlak *et al.*, 1994), alfalfa mosaic virus (AlMV; Loeshc-Fries *et al.*, 1987; Tumer *et al.*, 1987), potato leaf roll virus (PLRV; Kaniewski *et al.*, 1993) and papaya ringspot virus (PRSV; Chiang *et al.*, 2001).

Transgenic Plants for Nematode Resistance

Nematodes are obligate parasites that obtain nutrients from living plant cells. The plant-parasitic nematodes can be divided into two main groups, the cyst nematodes of the genera *Heterodora* and *Globodera*, and the root-knot nematode of the genus *Meloidogyne*. In principle, nematodes infect a broad range of host plants and are able to survive under critical conditions in the soil for many years. The control of nematodes by nematode resistant breeding, nematocide fumigation and crop rotation has been employed. Due to the hazardous effect of the chemical to the environment, thus, the transgene-mediated resistance for nematode resistant plants has been developed. Transgenic plants for nematode resistance could be achieved by R-gene-mediated resistance, transformation of gene that results in disruption of the development of specialized feeding structure of nematodes and transformation of gene that interfere with the digestive system of nematodes. Resistance to nematode has been characterized as the reduction level of nematode reproduction. Several genes that confer resistance to nematode have been cloned but only few of them have been successfully transformed into different crop plants. These include R genes and genes encoding PR proteins (Grundler, 1996; Williamson and Hussey, 1999).

Examples of nematode resistance genes that have been cloned and well characterized are $Hs1^{pro-1}$, *Mi* and *Gpa2* (Williamson and Hussey, 1999). The $Hs1^{pro-1}$ was the first cloned R-gene that confers resistance to beet cyst nematode, *Heterodera schachtii* Schmidt, and was isolated from a wild sugar beet. It encodes a protein containing leucine rich region similar to those of disease resistance genes. However, $Hs1^{pro-1}$ does not fit to any well known classes of the plant R genea (Cai *et al.*, 1997; Ellis and Jones, 1998). Constitutive expression of $Hs1^{pro-1}$ in susceptible sugar beet *in vitro* led to enhance resistance to *H. schachtii* (Ellis and Jones, 1998). *Mi-1*, the NBS/LRR class of *R*-genes that isolated from tomato, conferred resistance to several species of root-knot nematodes and potato aphid, *Macrosiphum euphorbiae* (Milligan *et al.*, 1998; Rossi *et al.*, 1998). Transgenic tomato plants transformed with *Mi* conferred resistance against aphid (Rossi *et al.*, 1998). However, the effect of these transgenic plants with root knot nematode is still under investigation. Finally, the most recent nematode resistance gene to be cloned was *Gpa2*, the NBS/LRR class of *R* genes. It was isolated from the wild potato (*Solanum tuberosum* subsp. *andigena* CPC 1673) and confers resistance to some species of potato cyst nematode, *Globodera pallida* (Rouppe van der Voort *et al.*, 1999). Transgenic plants expressing *Gpa2* is currently being examined.

Similar to the defense mechanism against fungi and bacteria, several PRs are expressed in response to nematode infection. Transformation of proteinase inhibitors in transgenic plants led to the interruption of digestive proteinases of parasitic-nematodes. For instance, expression of rice cystein proteinase inhibitor in transgenic potato plants resulted in enhanced resistance against the cyst nematode, *G. papilllda* (Urwin *et al.*, 2001). Co-expression of genes encoding cystein proteinase inhibitor from rice and serine inhibitor from soybean or either gene alone conferred resistance to beet cyst nematode, *H. schachtii*, root-knot nematode, *Meloidegyne incognita* and reniform nematode, *Rotylenchulus reniformis* (Urwin *et al.*, 1997; Urwin *et al.*, 1998; Urwin *et al.*, 2000). Transgenic rice constitutively expressed endogenous cystein proteinase inhibitor provides a significant reduction in egg production of root-knot nematode, *M. incognita* (Vain *et al.*, 1998).

Future Perspectives of Transgenic Plants for Disease Resistance

The advent of biotechnology and the extensive information available on plant and microbial molecular genetic has made the production of transgenic plants for disease resistance feasible. Several genes that confer resistance to pathogens have been transformed into plants and have been successful in protecting the plants from several diseases. A transgenic approach seems especially useful in plants that have a limited amount of endogenous resistance genes to certain pathogens. However, several R genes do not function in heterologous systems and this phenomenon has been referred to as 'restricted taxonomic functionality'. For instance, a transient assay of the *Bs2* pepper gene (for an *AvrBs2*-dependent hypersensitive response) did not confer resistance when expressed in non-solanaceous plants (*Arabidopsis*, turnip, cucumber and broccoli) while it conferred resistance in some solanaceous plants (potato and eggplant; Tai *et al.*, 1999). The *RPS2* gene from *Arabidopsis* is also not able to function in transgenic tomato. In addition, transformation of a single transgene is not sufficient to protect plants from a wide-range of pathogens because of the rapid changes in the pathogen populations overcome the resistance provided by the transgene. Therefore, the multigenic defense will be highly desirable to enhance resistance in plants against a broad-spectrum of plant diseases. The combination of two or more genes that are involved in plant defense response could be an attractive approach to achieve broad-spectrum disease resistance in plants. For instance, transformation of genes that induce the very early steps of resistance mechanisms (i.e. genes encoding regulators or signal perception) together with genes that encode antimicrobial proteins and/or phytoalexins and/or genes that play a role in the biosynthesis pathway associated with the structural changes in the plant cell wall when challenged by pathogens. Another option to obtain transgenic disease resistance is to engineer proteins that are involved in signal transduction pathway, which, in turn, will induce active defense responses.

In a recent review, Stuiever and Custers (2001) suggested that one of the most promising approaches to generate broad-spectrum transgenic plants is to engineer elicitors from the pathogen under the control of a tightly inducible promoter, which conditionally triggers HR. This approach has been successfully demonstrated in transgenic tobacco and tomato plants transformed with a gene that encodes pathogen-derived elicitor under the control of pathogen-inducible promoter. These plants displayed an inhibition in the infection of tomato spotted wilt virus (TSWV).

As the genome sequencing projects and functional analysis of several genes are completed for both plants and pathogens, a better understanding of the function of *R* genes, along with their interactions with pathogens will be obtained. In addition, the function of avirulent factors will be better understood, which will facilitate development of more durable disease resistant crops.

References

Abeles, F.B., Bosshart, R.P., Forrence, L.E., and Habig, W.H. 1971. Preparation and purification of glucanase and chitinase from bean leaves. Plant Physiol. 47: 129-134.

Agrios, G.N. 1997. Plant pathology (fourth edition): Academic Press, San Diego, California, USA. 635 pp.

Alexander, D., Goodman, R.M., Gut-Rella, M., Glascock, C., Weymann, K., Friedrich, L., Maddox, D., Ahl-Goy, P., Luntz, T. Ward, E., and Ryals, J. 1993. Increased tolerance to two oomycete pathogens in transgenic tobacco expressing pathogenesis-related protein 1a. Proc. Natl. Acad. Sci. USA. 90: 7327-7331.

Alfano, J.R., and Collmer, A. 1996. Bacterial pathogens in plants: Life up against the wall. Plant Cell. 8: 1683-1698.

Anderson, J.M., Palukaitis, P., and Zaitlin, M. 1992. A Defective replicase gene induces resistance to cucumber mosaic virus in transgenic tobacco plants. Proc. Natl. Acad. Sci. USA. 89: 8759-8763.

Andresen, I., Becker, W., Schluter, K., Burges, J., Parthier, B., and Apel, K. 1992. The identification of leaf thionin as one of the main jasmonate-induced proteins of barley (*Hordeum vulgare*). Plant Mol. Biol. 19: 193-204.

Angell, S.M., and Baulcombe, D.C. 1997. Consistent gene silencing in transgenic plants expressing a replicating potato virus X RNA. EMBO J. 16: 3675-3684.

Batchvarova, R., Nikolaeva, V., Slavov, S., Bossolova, S., Valkov, V., Atanassova, S., Guelemerov, S., Atanassov, A., and Anzai, H. 1998. Transgenic tobacco cultivars resistant to *Pseudomonas syringae* pv. *tabaci*. Theor. Appl. Genet. 97: 986-989.

Baulcombe, D. 1994. Replicase-mediated resistance: a novel type of virus resistance in transgenic plants? Trends Microbiol. 2: 60-63.

Benhamou, N., Broglie, K., Broglie, R., and Chet, I. 1993. Antifungal effect of bean endochitinase on *Rhizoctonia solani*: ultrastructural changes and cytochemical aspects of chitin breakdown. Can. J. Microbiol. 39: 318-328.

Bent, A.F. 1996. Plant disease resistance genes: Function meets structure. Plant Cell. 8: 1757-1771.

Bent, A.F., Innes, R.W., Ecker, J.R., and Staskawicz, B.J. 1992. Disease development in ethylene-insensitive *Arabidopsis thaliana* infected with virulent and avirulent Pseudomonas and Xanthomonas pathogens. Mol. Plant-Microbe Interact. 5: 372-378.

Bi, Y.M., Cammue, B.P.A., Goodwin, P.H., KrishnaRaj, S., and Saxena, P.K. 1999. Resistance to *Botrytis cinerea* in scented geranium transformed with a gene encoding the antimicrobial protein Ace-AMP1. Plant Cell Rep.18: 835-840.

Bieri, S., Potrykus, I., and Futterer, J. 2000. Expression of active barley seed ribosome-inactivating protein in transgenic wheat. Theor. Appl. Genet. 100:755-763.

Boller, T. 1990. Ethylene and plant-pathogen interactions. In: Polyamines and Ethylene: Biochemistry, Physiology, and Interaction. Pennsylvania: Current Topics in Plant Physiology.Flores, H.E., Arteca, R.N., and Shannon, J.C., eds. American Soc. Plant Physiologists, MD, USA. p. 138-145.

Bonas, U., and van den Ackerveken, G. 1999. Gene-for-gene interactions: bacterial avirulence proteins specify plant disease resistance. Curr. Opin. Microbiol. 2: 94-98.

Braun, C.J., and Hemenway, C.L. 1992. Expression of amino-terminal portions or full-length replicase genes in transgenic plants confers resistance to potato virus X infection. Plant Cell. 4: 735-744.

Broekaert, W.F., Cammue, B.P.A., De Bolle, M.F.C., Thevissen, K., De Samblanx, G.W., and Osborn, R.W. 1997. Antimicrobial peptides from plants. Crit. Rev. Plant Sci., 16:297-323.

Broglie, K., Chet, I., Holliday, M., Cressman, R., Biddle, P., Knowlton, S., Mauvis, C.J., and Broglie, R. 1991. Transgenic plants with enhanced resistance to the fungal pathogen *Rhizoctonia solani*. Science. 254: 1194-1197.

Cai, D., Kleine, M., Kifle, S., Harloff, H.J., Sandal, N.N., Marcker, K.A., Klein-Lankhorst, R.M., Salentijn, E.M., Lange, W., Stiekema, W.J., Wyss, U., Grundler, F.M., and Jung, C. 1997. Positional cloning of a gene for nematode resistance in sugar beet. Science. 275: 832-834.

Canto, T., and Palukaitis, P. 1998. Transgenically expressed cucumber mosaic virus RNA 1 simultaneously complements replication of cucumber mosaic virus RNAs 2 and 3 and confers resistance to systemic infection. Virology. 25: 325-336.

Cao, H., Li, X., and Dong, X. 1998. Generation of broad-spectrum disease resistance by overexpression of an essential regulatory gene in systemic acquired resistance. Proc. Natl. Acad. Sci. USA. 26: 6531-6536.

Carr, J.P., Gal-On, A., Palukaitis, P., and Zaitlin, M.1994. Replicase-mediated resistance to cucumber mosaic virus in transgenic plants involves suppression of both virus replication in the inoculated leaves and long-distance movement. Virology. 199: 439-447.

Chiang, C.H., Wang, J.J., Jan, F.J., Yeh, S.D., Gonsalves, D. 2001. Comparative reactions of recombinant papaya ringspot viruses with chimeric coat protein (CP) genes and wild-type viruses on CP-transgenic papaya. J. Gen. Virol. 82: 2827-2836.

Chen, G., Ye, C.M., Huang, J.C., Yu, M., and Li, B.J. 2001. Cloning of the papaya ringspot virus (PRSV) replicase gene and generation of PRSV-resistant papayas through the introduction of the PRSV replicase gene. Plant Cell Rep. 20: 272-277.

Cohen, E. 1993. Chitin synthesis and degradation as targets for pesticide action. Arch. Insect Biochem. Physiol. 22: 245-261.

Cohen, Y., and Kuc, J. 1981. Evaluation of systemic acquired resistance to blue mold induced in tobacco leaves by prior stem inoculation with *Peronospora hyosciami* f.sp. *tabacina*. Phytopathol. 71: 783-787.

Collmer, A. 1998. Determinants of pathogenicity and avirulence in plant pathogenic bacteria. Curr. Opin. Plant Biol. 1: 329-335.

Cornelissen, B.J., van Huijsduijnen, H.R.A., and Bol, J.F. 1986. A tobacco mosaic virus-induced tobacco protein is homologous to the sweet-tasting protein thaumatin. Nature. 4: 531-532.

Datta, K., Tu, J., Oliva, N., Ona, I.I., Velazhahan, R., Mew, T.W., Muthukrishnan, S., and Datta, S.K. 2001. Enhanced resistance to sheath blight by constitutive expression of infection-related rice chitinase in transgenic elite indica rice cultivars. Plant Sci. 5: 405-414.

Datta, K., Velazhahan, R., Oliva, N., Mew, T., Muthukrishnan, S., and Datta, S. K. 1999. Over-expression of thaumatin-like protein (PR-5) gene in transgenic rice plants enhances environmentally friendly resistance to *Rhizoctonia solani* causing sheath blight disease. Theor. Appl. Genet. 98:1138-1145.

Dean, R.A., and Kuc, J. 1986. Induced systemic protection in cucumber: Effects of inoculum density on symptom development caused by *Colletotrichum lagenarium* in previously infected and uninfected plants. Phytopathol. 76: 186-189.

Delaney, T.P., Uknes, S., Vernooij, B., Friedrich, L., Weymann, K., Negrotto, D., Gaffney, T., Gutrella, M., Kessmann, H., Ward, E., and Ryals, J. 1994. A central role of salicylic acid in plant disease resistance. Science. 266: 1247-1250.

Dixon, R.A., Lamb, C.J., Masoud, S., Sewalt, V.J., Paiva, N.L. 1996. Metabolic engineering: prospects for crop improvement through the genetic manipulation of phenylpropanoid biosynthesis and defense responses. Gene. 7: 61-71.

Dougherty, W.G., and Parks, T.D. 1995. Transgenes and gene suppression: telling us something new? Curr. Opin. Cell Biol. 7: 399-405.

Ellis, J., and Jones, D. 1998. Structure and function of proteins controlling strain-specific pathogen resistance in plants. Curr. Opin. Plant Biol. 1: 288-293.

Epple, P., Apel, K., and Bohlmann, H. 1997. Overexpression of an endogenous thionin enhances resistance of *Arabidopsis* against *Fusarium oxysporum*. Plant Cell. 9: 509-520.

Epple, P., Apel, K., and Bohlmann, H. 1995. An *Arabidopsis thaliana* thionin gene is inducible via a signal transduction pathway different from that for pathogenesis-related proteins. Plant Physiol. 109: 813-820.

Farmer, E.E., Johnson, R.R., and Ryan, C.A. 1992. Regulation of expression of proteinase inhibitor genes by methyl jasmonate and jasmonic acid. Plant Physiol. 98: 995-1002.

Farmer, E.E., and Ryan, C.A. 1990. Interplant communication: Airborne methyl jasmonate induces synthesis of proteinase inhibitors in plant leaves. Proc. Natl. Acad. Sci. USA. 87: 7713-7716.

Fidantsef, A.L., and Bostock, R.M. 1998. Characterization of potato tuber lipoxygenase cDNAs and lipoxygenase expression in potato tubers and leaves. Physiologia Plantarum. 102: 257-271.

Gaffney, T., Friedrich, L., Vernooij, B., Negrotto, D., Nye, G., Uknes, S., Ward, E., Kessmann, H., and Ryals, J. 1993. Requirement of salicylic acid for the induction of systemic acquired resistance. Science 261: 754-756.

Gao, A.G., Hakimi, S.M., Mittanck, C.A., Wu, Y., Woerner, B.M., Stark, D.M., Shah, D.M., Liang, J., and Rommens, C.M. 2000. Fungal pathogen protection in potato by expression of a plant defensin peptide. Nat. Biotechnol. 18:1307-1310.

Garcia-Olmedo, F., Molina, A., Segura, A., and Moreno, M. 1995. The defensive role of nonspecific lipid-transfer proteins in plants. Trends in Microbiol. 3: 72-74.

Golemboski, D.B., Lomonossoff, G.P., Zaitlin, M. 1990. Plants transformed with a tobacco mosaic virus nonstructural gene sequence are resistant to the virus. Proc. Natl. Acad. Sci. USA. 87: 6311-6315.

Goodman, R.N., and Novacky, A.J. 1994. The Hypersensitive Reaction in Plants to Pathogens: A resistance phenomenon. APS Press, The American Phytopathological Society (St. Paul, Minnesota). p. 244.

Green, T.R., and Ryan, C.A. 1972. Wound-induced proteinase inhibitor in plant leaves: a possible defense mechanism against insects. Science. 175: 776-777.

Grison, R., Grezes-Besset, B., Schneider, M., Lucante, N., Olsen, L., Leguay, J.J., and Toppan, A. 1996. Field tolerance to fungal pathogens of *Brassica napus* constitutively expressing a chimeric chitinase gene. Nat. Biotechnol. 14: 643-646.

Grundler, F.M. 1996. Engineering resistance against plant-parasitic nematodes. Field Crops Res. 45: 99-109.

He, X.Z., and Dixon, R.A. 2000. Genetic manipulation of isoflavone 7-O-methyltransferase enhances biosynthesis of 4'-O-methylated isoflavonoid phytoalexins and disease resistance in alfalfa. Plant Cell. 12: 1689-1702.

Hemenway, L., Fang, R.X., Kaniewski, W.K., Chua, N.H., and Tumer, N.E. 1988. Analysis of the mechanism of protection in transgenic plants expressing the potato virus X coat protein or its antisense RNA. EMBO J. 7: 1273-1280.

Hennin, C., Hofte, M., and Diederichsen, E. 2001. Functional expression of Cf9 and Avr9 genes in *Brassica napus* induces enhanced resistance to *Leptosphaeria maculans*. Mol Plant-Microbe Interact. 14:1075-1085.

Hong, Y., Saunders, K., Hartley, M.R., and Stanley, J. 1996. Resistance to geminivirus infection by virus-induced expression of dianthin in transgenic plants. Virology. 220: 119-127.

Jach, G., Gornhardt, B., Mundy, J., Logemann, J., Pinsdorf, E., Leah, R., Schell, J., and Maas, C.1995. Enhanced quantitative resistance against fungal disease by combinatorial expression of different barley antifungal proteins in transgenic tobacco. Plant J. 8: 97-109.

Jackson, A.O., and Taylor, C.B. 1996. Plant-microbe interactions: life and death at the interface. Plant Cell. 8: 1651-1668.

Jongedijk, E., Tigelaar, H., van Roekel, J.S.C., Bres-Vloemans, S.A,. Dekker, I., van den Elzen, P.J.M., Cornelissen, B.J.C., and Melchers, L.S. 1995. Synergistic activity of chitinases and -1,3-glucanases enhances fungal resistance in transgenic tomato plants. Euphytica. 85: 173-180.

Kaniewski, W., Lawson, C., and Thomas, P. 1993. Agronomically useful resistance in Russet Burbank potato containing a PLRV *CP* gene. Abstract. IX International Congress of Virology, Glasgow, Scotland.

Lagrimini, L.M., and Rothstein, S. 1987. Molecular cloning of complementary DNA encoding the lignin-forming peroxidase from tobacco: molecular analysis and tissue-specific expression. Proc. Natl. Acad. Sci. USA. 84: 7542.

Lam, Y.H., Wong, Y.S., Wang, B., Wong, R.N.S., Yeung, H.W., and Shaw, P.C. 1996. Use of trichosanthin to reduce infection by turnip mosaic virus. Plant Science. 114: 111-117.

Liang, J., Wu, Y., Rosenberger, C., Hakimi, S., Castro S., and Berg, J. 1998. AFP genes confer disease resistance to transgenic potato and wheat plants. Abstract no.L-49. In: 5th International Workshop on Pathogenesis-related Proteins in Plants: Signalling Pathways and Biological Activities. Aussois, France.

Lindbo, J.A., and Dougherty, W.G. 1992. Untranslatable transcripts of the tobacco etch virus coat protein gene sequence can interfere with tobacco etch virus replication in transgenic plants and protoplasts. Virology. 189: 725-733.

Liu, D., Raghothama, K.G., Hasegawa, P.M., Bressan, R.A. 1994. Osmotin overexpression in potato delays development of disease symptoms. Proc. Natl. Acad. Sci. USA. 91: 1888-1892.

Lodge, J.K., Kaniewski, W.K., and Tumer, N.E. 1993. Broad-spectrum virus resistance in transgenic plants expressing pokeweed antiviral protein. Proc. Natl. Acad. Sci. USA. 15: 7089-7093.

Loesch-Fries, L.S., Merlo, D., Sinnen, T., Burhop, L., Hill, K., Krahn, K., Jarvis, N., Nelson, S., and Halk, E. 1987. Expression of alfalfa mosaic virus RNA 4 in transgenic plants confers virus resistance. EMBO J. 6: 1845-1852.

Logemann, J., Jach, G., Tommerup, M., Mundy, J., and Schell, J. 1992. Expression of a barley ribosome-inactivating protein leads to increased protection in transgenic plants. Bio/Technology. 10:305–308.

Longstaff, M., Brigneti, G., Boccard, F., Chapman, S., and Baulcombe, D. 1993. Extreme resistance to potato virus X infection in plants expressing a modified component of the putative viral replicase. EMBO J. 12: 379-386.

Lorito, M., Woo, S.L., Fernandez, I.G., Colucci, G., Harman, G.E., Pintor-Toro, J.A., Filippone, E., Muccifora, S., Lawrence, C.B., Zoina, A., Tuzun, S., and Scala, F. 1998. Genes from mycoparasitic fungi as a source for improving plant resistance to fungal pathogens. Proc. Natl. Acad. Sci. USA. 95: 7860-7865.

Lucas, J.A. 1998. Plant pathology and plant pathogens. J. A. Lucas, ed. Tokyo, Japan: Blackwell Science. 274 pp.

Maddaloni, M., Forlani, F., Balmas, V., Donini, G., Stasse, L., Corazza, L., and Motto, M. 1997. Tolerance to the fungal pathogen *Rhizoctonia solani* AG4 of transgenic tobacco expressing the maize ribosome-inactivating protein b-32. Transgenic Research. 6: 1-10.

MacFarlane, S.A., and Davies, J.W. 1992. Plants transformed with a region of the 201-kilodalton replicase gene from pea early browning virus RNA1 are resistant to virus infection. Proc. Natl. Acad. Sci. USA. 89: 5829-5833.

Mäe, A., Montesano, M., Koiv, V., and Palva, E.T. 2001. Transgenic plants producing the bacterial pheromone N-acyl-homoserine lactone exhibit enhanced resistance to the bacterial phytopathogen *Erwinia carotovora*. Mol. Plant-Microbe Interact. 14: 1035-1042.

Malamy, J., Carr, J.P., Klessig, D.F., and Raskin, I. 1990. Salicylic acid - a likely endogenous signal in the resistance response of tobacco to viral infection. Science. 250: 1002-1004.

McNellis, T.W., Mudgett, M.B., Li, K., Aoyama, T., Horvath, D., Chua, N.H., and Staskawicz, B.J. 1998. Glucocorticoid-inducible expression of a bacterial avirulence gene in transgenic Arabidopsis plants induces hypersensitive cell death. Plant J. 14: 247-257.

Mehdy, M.C. 1994. Active oxygen species in plant defense against pathogens. Plant Physiol. 105: 467-472.

Melchers, L.S., and Stuiver, M.H. 2000. Novel genes for disease-resistance breeding. Curr. Opin. Plant Biol. 3: 147-152.

Melchers, L.S., Apotheker-de Groot, M., van der Knaap, J., Ponstein, A.S., Selabuurlage, M.B., Bol, J.F., Cornelissen, B.J.C., van den Elzen, P.J.M., and Linthorst, H.J.M. 1994. A new class of tobacco chitinases homologous to bacterial exo-chitinases displays antifungal activity. Plant J. 5: 469-480.

Métraux, J.P., Burkardt, W., Moyer, M., Dichner, S., Middlesteadt, W., Pzyne, S., Carnes, M., and Ryals, J. 1989. Isolation of a complementary DNA encoding a chitinase with structural homology to a bifunctional lysozyme/ chitinase. Proc. Natl. Acad. Sci. USA. 86: 896-900.

Métraux, J.P., Signer, H., Ryals, J., Ward, E., Wyss-Benz, M., Gaudin, J., Raschdorf, K., Schmid, E., Blum, W., and Inverardi, B. 1990. Increase in salicylic acid at the onset of systemic acquired resistance in cucumber. Science. 250: 1004-1006.

Métraux, J.P., Streit, L., and Staub, T.A. 1988. A pathogenesis-related protein in cucumber is a chitinase. Physiol. Mol. Plant Pathol. 33: 1-9.

Milligan, S.B., Bodeau, J., Yaghoobi, J., Kaloshian, I., Zabel, P., and Williamson, V.M. 1998. The root knot nematode resistance gene Mi from tomato is a member of the leucine zipper, nucleotide binding, leucine-rich repeat family of plant genes. Plant Cell. 10: 1307-1319.

Molina, A., and Garcia-Olmedo, F. 1997. Enhanced tolerance to bacterial pathogens caused by the transgenic expression of barley lipid transfer protein LTP2. Plant J. 12: 669-675.

Mora, A., and Earle, E.D. 2001. Combination of *Trichoderma harzianum* endochitinase and a membrane-affecting fungicide on control of *Alternaria* leaf spot in transgenic broccoli plants. Appl. Microbiol. Biotechnol. 55: 306-310.

Nakajima, H., Muranaka, T., Ishige, F., Akutsu, K., and Oeda K. 1997. Fungal and bacterial disease resistance in transgenic plants expressing human lysozyme. Plant Cell Rep. 16: 674-679.

Namba, S., Ling, K., Gonsalves, C., Gonslaves, D., and Slightom, J.L. 1991. Expression of the gene encoding the coat protein of cucumber mosaic virus (CMV) strain WL appears to provide protection to tobacco plants against infection by several different CMV strains. Gene. 107: 181-188.

Nelson, R.S., McCormick, S.M., Delanney, X., Dube, P., Layton, J., Anderson, E.J., Kaniewska, M., Prosksch, R.K., Horsch, R.B., Rogers, S.G., Fraley, R.T., and Beachy, R.N. 1988. Virus tolerance, plant growth, and field performance of transgenic tomato plants expressing coat protein from tobacco mosaic virus. Bio/Technology. 6: 403-409.

Nielsen, K., and Boston, R.S. 2001. Ribosome-inactivating proteins: A plant perspective. Annu. Rev. Plant Physiol. Plant Mol. Biol. 52: 785-816.

Nishizawa T., Takano, R., and Muroga, K. 1999. Mapping a neutralizing epitope on the coat protein of striped jack nervous necrosis virus. J. Gen. Virol. 80: 3023-3027.

O'Donnell, P.J., Calvert, C., Leyser, H.M.O., and Bowles, D. J. 1996. Ethylene as a signal mediating the wound response of tomato plants. Science. 274: 1914-1917.

Ohashi, Y., and Oshima, M. 1992. Stress-induced expression of genes for pathogenesis-related proteins in plants. Plant Cell Physiol. 33: 819-826.

Okada, Y., Saito, A., Nishiguchi, M., Kimura, T., Mori, M., Hanada, K., Sakai, J., Miyazaki, C., Matsuda, Y., and Murata, T. 2001. Virus resistance in transgenic sweetpotato [*Ipomoea batatas* L. (Lam)] expressing the coat protein gene of sweet potato feathery mottle virus. Theor. Appl. Genet. 103: 743-751.

Ono, E., Wong, H.L., Kawasaki, T., Hasegawa, M., Kodama, O., and Shimamoto, K. 2001. Essential role of the small GTPase Rac in disease resistance of rice. Proc. Natl. Acad. Sci. USA. 98: 759-764.

Osbourn, A.E. 1996. Preformed antimicrobial compounds and plant defense against fungal attack. Plant Cell. 8: 1821-1831.

Papadopoulou, K., Melton, R.E., Leggett, M., Daniels, M.J., and Osbourn, A.E. 1999. Compromised disease resistance in saponin-deficient plants. Proc. Natl. Acad. Sci. USA. 22: 12923-12928.

Penninckx, I.A.M.A., Eggermont, K., Terras, F.R.G., Thomma, B.P.H.J., De Samblanx, G.W., Buchala, A., Métraux, J.P., Manners, J.M., and Broekaert, W.F. 1996. Pathogen-induced systemic activation of a plant defensin gene in *Arabidopsis* follows a salicylic acid-independent pathway. Plant Cell. 8: 2309-2323.

Penninckx, I.A.M.A., Thomma, B.P.H.J., Buchala, A., Métraux, J.P., and Broekaert, W.F. 1998. Concomitant activation of jasmonate and ethylene response pathways is required for induction of a plant defensin gene in *Arabidopsis*. Plant Cell. 10: 2103-2113.

Perlak, F., Kaniewski, W., Lawson, C., Vincent, M., and Feldman, J. 1994. Genetically improved potatoes: Their potential role in integrated pest management. In: Proceed. 3rd EFPP Conference. M. Manka, ed. J. Phytopathol. p. 451-454.

Pieterse, C.M.J., and van Loon, L.C. 1999. Salicylic acid-independent plant defence pathways. Trends Plant Sci. 4: 52-58.

Pieterse, C.M.J., van Wees, S.C.M., van Pelt, J.A., Knoester, M., Laan, R., Gerrits, N., Weisbeek, P.J., and van Loon, L.C. 1998. A novel signaling pathway controlling induced systemic resistance in *Arabidopsis*. Plant Cell. 10: 1571-1580.

Powell-Abel, P., Nelson, R.S., De, B., Hoffmann, N., Rogers, S.G., Fraley, R.T., and Beachy, R.N. 1986. Delay of disease development in transgenic plants that express the tobacco mosaic virus coat protein gene. Science. 232: 738-743.

Prusky, D., and Keen, N.T. 1993. Involvement of preformed antifungal compounds in the resistance of subtropical fruits to fungal decay. Plant Disease. 77: 114-119.

Quemada, H.D., Gonsalves, D., and Slightom, J.L. 1991. Expression of coat protein gene from cucumber mosaic virus strain C in tobacco: Protection against infections by CMV strains transmitted mechanically or by aphids. Phytopathol. 81: 794-802.

Rajasekaran, K., Cary, J.W., Jacks, T.J., Stromberg, K.D., and Cleveland, T.E. 2000. Inhibition of fungal growth *in planta* and *in vitro* by transgenic tobacco expressing a bacterial nonheme chloroperoxidase gene. Plant Cell Rep. 19: 333-338.

Rasmussen, J.B., Hammerschmidt, R., and Zook, M.N. 1991. Systemic induction of salicylic acid accumulation in cucumber after inoculation with *Pseudomonas syringae* pv. *syringae*. Plant Physiol. 97: 1342-1347.

Richter, T.E., and Ronald, P.C. 2000. The evolution of disease resistance genes. Plant Mol. Biol. 42: 195-204.

Rohini, V.K., and Sankara-Rao K. 2001. Transformation of peanut (*Arachis hypogaea* L.) with tobacco chitinase gene: variable response of transformants to leaf spot disease. Plant Sci. 160: 889-898.

Rossi, M., Goggin, F.L., Milligan, S.B., Kaloshian, I., Ullman, D.E., and Williamson, V.M. 1998. The nematode resistance gene *Mi* of tomato confers resistance against the potato aphid. Proc. Natl. Acad. Sci. USA. 95: 9750-9754.

Rouppe van der Voort, J.R., Kanyuka, K., van der Vossen, E., Bendahmane, A., Mooijman, P., Klein-Lankhorst, R., Stiekema, W., Baulcombe, D., and Bakker, J. 1999. Tight physical linkage of the nematode resistance gene *Gpa2* and the virus resistance gene *Rx* on a single segment introgressed from the wild species *Solanum tuberosum* subsp. *andigena* CPC1673 into cultivated potato. Mol. Plant-Microbe. Int. 12: 197-206.

Rubino, L., and Russo, M. 1995. Characterization of resistance to cymbidium ringspot virus in transgenic plants expressing a full-length viral replicase gene. Virology. 212: 240-243.

Ryals, J.A., Neuenschwander, U.H., Willits, M.G., Molina, A., Steiner, H.Y., and Hunt, M.D. 1996. Systemic acquired resistance. Plant Cell. 8: 1809-1819.

Ryan, C. A. 1992. The search for the proteinase inhibitor-inducing factor, PIIF. Plant Mol. Biol. 19: 123-133.

Sanders, P.R., Sammons, B., Kaniewski, W., Haley, L., Layton, J., LaVallee, B.J., Delannay, X., and Tumer, N.E. 1992. Field resistance of transgenic tomatoes expressing the tobacco mosaic virus or tomato mosaic virus coat protein genes. Phytopathol. 82: 683-690.

Sano, H., Seo, S., Orudgev, E., Youssefian, S., and Ishizuka, K. 1994. Expression of the gene for a small GTP-binding protein in transgenic tobacco elevates endogenous cytokinin levels, abnormally induces salicylic acid in response to wounding, and increases resistance to tobacco mosaic virus infection. Proc. Natl. Acad. Sci. USA. 25:10556-10560.

Schneider, M., Schweizer, P., Meuwly, P., and Métraux, J.P. 1996. Systemic acquired resistance in plants. International Review of Cytology - A Survey of Cell Biology. 168: 303-340.

Schweizer, P., Vallelian-Bindschedler, L., and Mösinger, E. 1995. Heat-induced resistance in barley to the powdery mildew fungus *Erysiphe graminis* f.sp. *hordei*. Physiol. Mol. Plant Pathol. 47: 51-66.

Sharma, A., Sharma, R., Imamura, M., Yamakawa, M., and Machii, H. 2000. Transgenic expression of cecropin B, an antibacterial peptide from *Bombyx mori*, confers enhanced resistance to bacterial leaf blight in rice. FEBS Lett. 484: 7-11.

Shen, S., Li, Q., He, S.Y., Barker, K.R., Li, D., and Hunt, A.G. 2000. Conversion of compatible plant-pathogen interactions into incompatible interactions by expression of the *Pseudomonas syringae* pv. *syringae* 61 *hrmA* gene in transgenic tobacco plants. Plant J. 23: 205-213.

Siedow, J.N. 1991. Plant lipoxygenase: structure and function. Annu. Rev. Plant Physiol. Plant Pathol. 42: 145-188.

Somssich, I.E., and Hahlbrock, K. 1986. Rapid activation by fungal elicitor of genes encoding pathogenesis-related proteins in cultured parsley cells. Proc. Natl. Acad. Sci. USA. 83: 2427-2430.

Sticher, L., Mauch-Mani, B., and Métraux, J.P. 1997. Systemic acquired resistance. Annu. Rev. Phytopathol. 35: 235-70.

Stintzi, A., Heitz, T., Prasad, V., Wiedemann-Merdinoglu, S., Kauffmann, S., Geoffroy, P., Legrand, M., and Fritig, B. 1993. Plant "pathogenesis-related" proteins and their role in defense against pathogens. Biochem. 75: 687-706.

Strittmatter, G., Janssens, J., Opsomer, C., and Botterman, J. 1995. Inhibition of fungal disease development in plants by engineering controlled cell death. Bio/Technology. 13: 1085-1088.

Stuiver, M.H., and Custers, J.H. 2001. Engineering disease resistance in plants. Nature. 14: 865-868.

Tabaeizadeh, Z., Agharbaoui, Z., Harrak, H., and Poysa, V. 1999. Transgenic tomato plants expressing a *Lycopersicon chilense* chitinase gene demonstrate improved resistance to *Verticillium dahliae* race 2. Plant Cell Rep. 19: 197-202.

Tabei, Y., Kitade, S., Nishizawa, Y., Kikuchi, N., Kayano, T., Hibi, T., and Akutsu, K. 1998. Transgenic cucumber plants harboring a rice chitinase gene exhibit enhanced resistance to gray mold (*Botrytis cinerea*). Plant Cell Rep. 17: 159-164.

Tai, T.H., Dahlbeck, D., Clark, E.T., Gajiwala, P., Pasion, R., Whalen, M.C., Stall, R.E., and Staskawicz, B.J. 1999. Expression of the *Bs2* pepper gene confers resistance to bacterial spot disease in tomato. Proc. Natl. Acad. Sci. USA. 96: 14153-14158.

Takaichi, M., and Oeda, K. 2000. Transgenic carrots with enhanced resistance against two major pathogens, *Erysiphe heraclei* and *Alternaria dauci*. Plant Sci. 25:135-144.

Terras, F.R.G., Schoofs, H.M.E., de Bolle, M.F.C., van Leuven, F., Rees, S.B., van der Leyden, J., Cammue, B.P.A., and Broekart, W.F. 1992. Analysis of two novel classes of plant antifungal proteins from radish (*Raphanus sativus* L.) seeds. J. Biol. Chem. 267: 15301-15309.

Terras, F.R.G., Eggermont, K., Kovaleva, V., Raikhel, N.V., Osborn, R.W., Kester, A., Rees, S.B., Torrekens, S., van Leuven, F., Vanderleyden, J., Cammue, B.P.A., and Broekaert, W.F. 1995. Small cysteine-rich antifungal proteins from radish (*Raphanus sativus* L.): their role in host defense. Plant Cell 7: 573-588.

Tu, J., Ona, I., Zhang, Q., Mew, T.W., Khush, G.S., and Datta, S.K. 1998. Transgenic rice variety 'IR72' with *Xa21* is resistant to bacterial blight. Theor. Appl. Genet. 97: 31-36.

Tumer, N.E., Hemenway, C., O'Connell, K., Cuozzo, M., Fang, R.X, Kaniewski, W., and Chua, N.H. 1987. Expression of coat protein genes in transgenic plants confers protection against alfalfa mosaic virus, cucumber mosaic virus and potato virus X. In: Plant Molecular Biology. D. von Wettstein and N-H. Chua, eds. Plenum Publishing Corporation. p. 351-356.

Uknes, S., Mauch-Mani, B., Moyer, M., Potter, S., Williams, S., Dincher, S., Chandler, D., Slusarenko, A., Ward, E., and Ryals, J. 1992. Acquired resistance in *Arabidopsis*. Plant Cell. 4: 645-656.

Urwin, P.E. 2001. Effective transgenic resistance to *Globodera pallida* in potato field trials. Mol. Breed. 8: 95-101.

Urwin, P.E., Levesley, A., McPherson, M.J., and Atkinson, H.J. 2000. Transgenic resistance to the nematode *Rotylenchulus reniformis* conferred by *Arabidopsis thaliana* plants expressing proteinase inhibitors. Mol. Breed. 6: 257-264.

Urwin, P.E., Lilley, C.J., McPherson, M.J., and Atkinson, H.J. 1997. Resistance to both cyst and root-knot nematodes conferred by transgenic *Arabidopsis* expressing a modified plant cystatin. Plant J. 12: 455-461.

Urwin, P.E., McPherson, M.J., and Atkinson, H.J. 1998. Enhanced transgenic plant resistance to nematodes by dual proteinase inhibitor constructs. Planta. 204: 472-479.

Vain, P., Worland, B., Clarke, M.C., Richard, G., Beavis, M., Liu, H., Kohli, A., Leech, M., Snape, J., Christou, P., and Atkinson, H. 1998. Expression of an engineered cysteine proteinase inhibitor (Oryzacystatin-I Delta D86) for nematode resistance in transgenic rice plants. Theor. Appl. Genet. 96: 266-271.

van Loon, L.C., Bakker, P.A.H.M., and Pieterse, C.M.J. 1998. Systemic resistance induced by rhizosphere bacteria. Annu. Rev. Phytopathol. 36: 453-483.

van Loon, L.C., Gerritsen, Y.A.M., and Ritter, C.E. 1987. Identification, purification and characterization of pathogenesis-related proteins from virus-infected Samsun NN tobacco leaves. Plant Mol. Biol. 9: 593-609.

van Loon, L.C., and van Strien, E. A. 1999. The families of pathogenesis-related proteins, their activities, and comparative analysis of PR-1 type proteins. Physiol. Mol. Plant Pathol. 55: 85-97.

van Wees, S.C., Luijendijk, M., Smoorenburg, I., van Loon, L.C., and Pieterse, C.M. 2000. Rhizobacteria-mediated induced systemic resistance (ISR) in *Arabidopsis* is not associated with a direct effect on expression of known defense-related genes but stimulates the expression of the jasmonate-inducible gene *Atvsp* upon challenge. Plant Mol. Biol. 4: 537-549.

Vazquez-Rovere, C., Asurmendi, S., and Hopp, H.E. 2001. Transgenic resistance in potato plants expressing potato leaf roll virus (PLRV) replicase gene sequences is RNA-mediated and suggests the involvement of post-transcriptional gene silencing. Arch Virol. 146:1337-13353.

Vera, P., and Conejero, V. 1988. Pathogenesis-related proteins of tomato-P-69 as an alkaline endoproteinase. Plant Physiol. 87: 58-63.

Vernooij, B., Friedrich, L., Morse, A., Reist, R., Kolditzjawhar, R., Ward, E., Uknes, S., Kessmann, H., and Ryals, J. 1994. Salicylic acid is not the translocated signal responsible for inducing systemic acquired resistance but is required in signal transduction. Plant Cell. 6: 959-965.

Vijayan, P., Shockey, J., Levesque, C.A., Cook, R.J., and Browse, J. 1998. A role for jasmonate in pathogen defense of *Arabidopsis*. Proc. Natl. Acad. Sci. USA. 95: 7209-7214.

Waniska, R.D., Venkatesha, R.T., Chandrashekar, A., Krishnaveni, S., Bejosano, F.P., Jeoung, J., Jayaraj, J., Muthukrishnan, S., and Liang, G.H. 2001. Antifungal proteins and other mechanisms in the control of sorghum stalk rot and grain mold. J. Agric. Food Chem. 49: 4732-4742.

Ward, E.R., Payne, G.B., Moyer, M.B., Williams, S.C., Dincher, S.S., Sharkey, K.C., Beck, J.J., Taylor, H.T., Ahl-Goy, P., Meins, F., and Ryals, J.A. 1991a. Differential regulation of β-1,3-glucanase messenger RNAs in response to pathogen infection. Plant Physiol. 96: 390-397.

Ward, E.R., Uknes, S.J., Williams, S.C., Dincher, S.S., Wiederhold, D.L., Alexander, D.C., Ahl-Goy, P., Métraux, J.P., and Ryals, J.A. 1991b. Coordinate gene activity in response to agents that induce systemic acquired resistance. Plant Cell. 3: 1085-1094.

Waterhouse, P.M., Wang, M.B., and Lough, T. 2001. Gene silencing as an adaptive defence against viruses. Nature. 411: 834-842.

Whitham, S., McCormick, S., and Baker, B. 1996. The N gene of tobacco confers resistance to tobacco mosaic virus in transgenic tomato. Proc. Natl. Acad. Sci. USA. 93: 8776-8781.

Williamson, V.M., and Hussey, R.S. 1999. Plant nematode resistance genes. Curr. Opin. Plant Biol. 2: 327-331.

Wintermantel, W.M., Banerjee, N., Oliver, J.C., Paolillo, D.J., and Zaitlin, M. 1997. Cucumber mosaic virus is restricted from entering minor veins in transgenic tobacco exhibiting replicase-mediated resistance. Virology. 231: 248-257.

Wu, G., Shortt, B.J., Lawrence, E.B., Levine, E.B., Fitzsimmons, C. and Shah, D.M. 1995. Disease resistance conferred by expression of a gene encoding H_2O_2-generating glucose oxidase in transgenic potato plants. Plant Cell. 7: 1357-1368.

Xu, Y., Chang, P.F.L., Liu, D., Narasimhan, M.L., Raghothama, K.G., Hasegawa, P.M., and Bressan, R.A. 1994. Plant defense genes are synergistically induced by ethylene and methyl jasmonate. Plant Cell. 6: 1077-1085.

Yamamoto, T., Iketani, H., Ieki, H., Nishizawa, Y., Notsuka, K., Hibi, T., Hayashi, T., and Matsuta, N. 2000. Transgenic grapevine plants expressing a rice chitinase with enhanced resistance to fungal pathogens. Plant Cell Rep. 19: 639-646.

Zhu, B., Chen, T.H.H., and Li, P.H., 1996. Analysis of late-blight disease resistance and freezing tolerance in transgenic potato plants expressing sense and antisense genes for an osmotin-like protein. Planta. 198: 70-77.

Zhu, Q., Maher, E.A., Masoud, S., Dixon, R.A., and Lamb, C.J. 1994. Enhanced protection against fungal attack by constitutive co-expression of chitinase and glucanase genes in transgenic tobacco. Bio/Technology 12: 807-812.

From: *Transgenic Plants: Current Innovations and Future Trends*
Edited by: C. Neal Stewart, Jr.

Chapter 9

Plant Biotechnology and Food Safety Evaluation

Harold Richards and Susan Hefle

Abstract

The age of genetically engineered foods is here. The promise of plant
biotechnology for agriculture and nutrition science is significant. Benefits
in the field are being realized already, from reduced pesticide use to increased
yields, and biotechnology food products designed to alleviate nutritional
deficiencies, protect from cancer, or reduce heart disease will shortly reach
the markets. However, there has been public confusion and concern over the
use of transgenic plants for crop and food improvement. This chapter will
seek to review food safety concerns in order to better understand the risks
involved and the benefits to food biotechnology. Food safety evaluations of
genetically modified (GM) products are needed to provide continued security
of the food supply as well as to maintain consumer confidence in the products
they purchase. Safety can be determined by establishment of substantial
equivalence of GM products to conventional food varieties, evaluation of
any potential toxicity, and identification of products that may be allergenic.
Combined, these methodologies will allow for the effective and safe
introduction of this technology into conventional food production.

In the Beginning...

It is important to understand that genetic modification of foods is not a new practice. In the earliest beginnings of agriculture, humans began selecting individual plants with desirable traits to pass on to the next generation of crops. This type of genetic selection has resulted, among other achievements, in the transformation of teosinte to domestic corn (Wang *et al.*, 1999). The process was laborious and took 10,000 years or more, but was effective genetic modification nonetheless. In the 1960s, crop hybrids were introduced and have become a staple in new crop production. This process of directed breeding programs, often involving distantly related species, was the foundation of the Green Revolution that has resulted in the ability to produce enough food to feed 6 billion people or more (Conway and Toenniessen, 1999). While effective, this technology is imprecise, resulting in the exchange of multiple traits and genes, and limited to sexually compatible species. Modern plant biotechnology is a refinement of that process, seeking to insert selected genes and traits from a broad and unlimited range of genetic backgrounds.

Plant biotechnology began in earnest when it was discovered that genes could be inserted into plant cells and the respective recombinant proteins were expressed (Horsch *et al.*, 1985; Klein *et al.*, 1988). Because plant cells are totipotent, that is, any one cell is capable of expressing the whole range of genetic material, it is possible to culture whole plants from as little as one cell. Combining these two techniques has allowed for the production of genetically modified plants and this technology has been refined and applied to a wide variety of crop species (Hansen and Wright, 1999). The earliest work focused on using genes that coded for marker proteins, which allowed for the process of transformation to be studied and optimized. As research progressed, functional gene expression was targeted. The first genetically engineered foods were designed for agricultural traits like pesticide resistance (Stewart *et al.*, 2000). Opposition to this technology initially began as environmentalist protests over concern for unpredictable and irreversible damage to the environment (Stewart *et al.*, 2000). As the controversy has expanded, concern has developed over the safety of these foods for human consumption.

What is Safe?

To discuss plant biotechnology and food safety, it is first important to establish what is considered safe. Many people make the mistaken assumption that "safe" means without risk. In fact, a level of risk is assumed when eating any food. Many plants contain endogenous compounds that are toxic at high doses. Furthermore, conventional agriculture relies on pesticides that can leave residue on food products and consumers of organic foods must accept a potential increase in exposure to insects, pathogens, and their by-products and toxins. Risk assessment is the process by which food safety regulatory agencies balance the likelihood and significance of risk with the benefits and need for a new product.

Establishing criteria for safety usually relies on defining several characteristics of a given food product or element. History of food use becomes an important factor for many food items. For instance, if a product has a history of safe human consumption in one part of the world, then it is assumed to be safe for consumption elsewhere (IFT Expert Report, 2000). For example, kiwi fruit, native of New Zealand and accepted as safe there, would not be subjected to rigorous safety tests in the USA. Substantial equivalence is another important safety concept (FAO/WHO, 2000). The principal is that if a traditional counterpart of the new food has a history of safe use, then safety studies should focus on the differences between the new food and its counterpart. Crops produced through conventional breeding practices would fall into this category. Hybrids that are comparable to established varieties are accepted despite the fact that they could contain numerous unique characteristics or components.

Toxicity standards are set for new compounds in foods, whether they are endogenous or exogenous. Any chemical application in agriculture, such as pesticides, is tested to establish safe intake levels. These tests usually involve animal models for toxicology studies. Animal toxicology studies must also be conducted for food additives, such as artificial sweeteners, to establish safe intake limits. Potential allergenicity of new food products must also be established due to the potential serious effects of an allergic response. This type of safety assessment can only be effectively applied when a specific compound for testing, such as food additives, can be identified. It is difficult to evaluate hybrids produced from conventional methods in this manner because of the wide variety of new compounds that could be in the food. Therefore, the standard of substantial equivalence is relied upon. If a product is significantly similar to a soybean, then people with soybean allergies should beware, but otherwise, the product is considered safe. The situation for evaluation is different for genetically modified (GM) food products because new crop varieties are modified in such a specific manner, often with only one gene product difference to allow for greater scrutiny.

The food products of biotechnology are rigorously evaluated for safety of human consumption based primarily on the above guidelines (FDA, 1992). They are scrutinized more so than food products produced from conventional means despite little scientific evidence that they pose any more risk. Still, the general populace, especially in Europe, is cautious and even distrustful of GM foods. In the USA, up until May 2000, safety evaluations of GM products slated for human consumption had been voluntary. The biotechnology companies had to ensure safety of their products and follow regulatory agency guidelines. Now these tests are required as part of a pre-market notification system (HHS, 2000). It remains to be seen if these more stringent requirements will bring more acceptance of the technology.

Substantial Equivalence

Given, by definition, that GM foods are novel, it is not possible to judge their safety by previous history of use. However, as some countries include GM foods into their diets, a case history may be compiled. Millions of tons of GM corn and soybeans have been consumed by the American populace since 1994. To date no health hazards have been identified (Stewart *et al.*, 2000), suggesting that GM food products are generally safe for human consumption across the globe. As more GM foods become staples in agriculture, a history of consumption will be established for a number of products. However, before that can occur, these products have to be identified as safe based on other criteria.

Establishing substantial equivalence will be the most useful evaluation of GM food safety. Determination of substantial equivalence is designed to identify the effects of genetic engineering on the food product that creates a difference in the product as a whole from a conventional counterpart (FAO/WHO, 2000). Factors that are considered include changes in the levels of essential nutrients, elevation of antinutritional factors, production of increased levels of inherent toxins, and changes in the levels of existent allergens (Miraglia *et al.*, 1998). Often the assessments involve animal feeding studies where groups are fed identical diets except for the replacement of a GM food for the traditional counterpart. Analyses can also focus on known potential hazards in an existing food or on a specific component of the new product. For instance, any new potato variety, conventional or GM, is evaluated for glycoalkaloid content, a known endogenous plant toxin (Zitnak and Johnston, 1970). Genetically modified food products lend themselves well to this type of analysis because they usually differ from their counterparts by only one trait or compound.

The primary goal of evaluation of GM foods based on substantial equivalence is to establish that no new hazards have been introduced into the food. That is, the risks you would normally associate with consumption of that food product are the same for the new variety. This concept has been an effective tool for determining the safety of new crop hybrids for decades, and should serve equally well, if not better, in the assessment of GM foods.

Potential Toxicity

Since the announcement and subsequent publication of the research of Ewen and Pusztai (1999), a furor over the risks of GM food consumption has raged in Europe. Based on the data from their experiments, they alleged that genetic engineering alone was responsible for intestinal damage and decreased immune function in a rat model feeding study. Despite no suggested mechanism for action, improper controls, unbalanced diets (Kuiper *et al.*, 1999), and similar studies with other GM foods suggesting evidence to the contrary (Hashimoto *et al.*, 1999; Hammond *et al.*, 1996; Harrison *et al*, 1996), the research is still widely cited by those opposed to GM foods. Regardless of politics, GM foods are considered by regulatory agencies in the US to be novel food products and therefore must be evaluated as such for safety. Although the data suggest that genetic engineering itself does not create a food safety risk, there remains concern that potential unintended effects could result in a complication. Until such a time that sufficient data has been collected to either validate or alleviate these concerns, toxicity studies should be conducted on GM products intended for human consumption.

Potential toxicity of GM foods can be evaluated on several different levels. To begin, the DNA itself can be considered. All food sources contain DNA, which humans have been consuming in large quantities since the beginning of our species (Beever and Kemp, 2000). It is therefore considered safe to eat and is not evaluated for safety. Since recombinant DNA represents a tiny fraction of the DNA consumed and is deemed to be identical in composition, it is also considered safe (Miraglia *et al.*, 1998). There have been unsubstantiated speculations that recombinant DNA could transfer from plants to microorganisms or humans (Ho *et al.*, 1999) creating a host of cataclysmic health disasters. There is, however, no scientific evidence to support these claims (Hodgson, 2000; Donaldson and May, 1999).

Most GM plants will produce specific recombinant proteins. These proteins are also evaluated for potential acute toxicity. In cases where the product is already an established part of the food supply or where toxicology standards already exist, those precedents are used to establish safety of the compound in the GM product. For example, overexpression of plant ferritin (Goto *et al.*, 1999) would not be considered toxic because all plants express this protein and there are no hazards associated with its over consumption. A protein from the soil bacterium *Bacillus thuringiensis* (Bt), which is a specific insecticide, has been approved for exogenous application in conventional and organic agriculture. Bt can be applied to crops in several fashions usually by direct application of *B. thuringiensis* or other bacteria recombinantly expressing the Bt protein. In addition, it has been shown to pose no additional dangers when expressed within the plant cells (Betz *et al.*, 2000). In cases where the protein has not been tested before, toxicology studies should be conducted on both the purified protein and the transgenic plants.

Often, transgenic proteins will be enzymes that result in the accumulation of secondary compounds that change the composition of the plant. In these cases, the potential risks of that biochemical change must be considered. As an example, plants have been engineered to express β-carotene (Ye *et al.*, 2000) and increased levels of α-tocopherol (Shintani and DellaPenna, 1998). In cases such as these, the resulting accumulation of new compounds or the over accumulation of native compounds that could be toxic at higher doses must be considered during toxicity studies. Also, enzymes that would result in the overproduction of natural defense compounds would necessitate toxicity evaluations. For instance, it is possible to engineer plants to express genistein and daidzein (Jung *et al.*, 2000), two soybean components thought to reduce the risk of certain cancers, but these compounds are biologically active and the effects of an increase level of consumption would need to be considered.

Nutritional factors must also be considered. If, for instance, a plant is modified for increase in iron accumulation, then the potential of iron toxicity from increased consumption must be explored. Conversely, increases in antinutritional factors that may limit nutrient bioavailability could also result in health risks. These concerns underline the need for substantial equivalence studies to establish a basis for safety evaluations.

It is difficult to justify the concerns and hysteria of GM food safety based on potential toxicity. Each evolution in agriculture has brought with it concerns for safety, and following the examples set during the Green Revolution, we are prepared to identify food products that would cause significant toxicity or otherwise compromised health. It is important to be aware of how long-term exposure of GM food products may impact human populations, as well as to look for subpopulations that may be particularly sensitive to certain products.

Assessing Allergenicity

The area of safety that has most dramatically affected concerns over genetically engineered foods is potential allergenicity. Food allergies occur in about 2% of the adult population and in 4-8% of children, and therefore are a significant health concern (Mendieta *et al.*, 1997, Boch *et al.*, 1978). Fortunately, the vast majority of food allergens occur in a select group of foods: cow's milk, wheat, tree nuts, peanuts, eggs, soybeans, fish, and crustaceans (FAO, 1995; Lehrer *et al.*, 1996; Taylor and Hefle, 2001). Complicating this issue is the fact that little is known about the causes behind the development of food allergies. It is known that food allergies are IgE-mediated immune responses, involving sensitization of mast cells and histamine degranulization (Yeung *et al.*, 2000; Bischoff *et al.*, 2000). The risk of concern is that transgenics could render an otherwise non-allergenic food allergenic, and that the populace would be surprisingly affected.

Currently, assessment of potential allergenicity relies on characterization of the transgene protein being synthesized in the GM food. This type of analysis for a new food product is made possible by the precision of biotechnology. Other new food products, such as those produced through hybridization, cannot be rigorously evaluated for potential allergenicity due to the numerous and unidentifiable compounds in the food products (IFT Expert Report, 2000). Rigorous testing of specific proteins ensures infinitesimal risk of a major allergen being introduced into the food supply.

The International Food Biotechnology Council (IFBC) and the Allergy and Immunology Institute of the International Life Sciences Institute (ILSI) developed a set of guidelines for allergenicity assessment (Metcalfe *et al.*, 1996). The criteria focus on comparing the characteristics of the recombinant protein to those of known allergens and involve following a decision tree (Figure 1). If the protein was derived from a source known to be allergenic, such as a peanut protein, then it is tested for binding of IgE in sensitized serum from individuals allergic to the source food (Yunginger and Adolphson, 1992). If no response is obtained, the procedure moves to a skin prick test, followed by a double-blind placebo-controlled food challenge (Taylor and Lehrer, 1996). If at any point along the testing, an immune response is triggered, the protein is labeled allergenic and not approved for the food supply unless products are labeled as containing the allergenic component.

The process is complicated if the allergenicity of the protein is unknown. The first step is to compare sequence similarity between the transgene protein and known food allergens. Despite sequencing over 300 known food allergens, no common allergen motif has been identified (Gendel, 1998a). Therefore, similarity comparisons attempt to identify potential IgE epitopes.

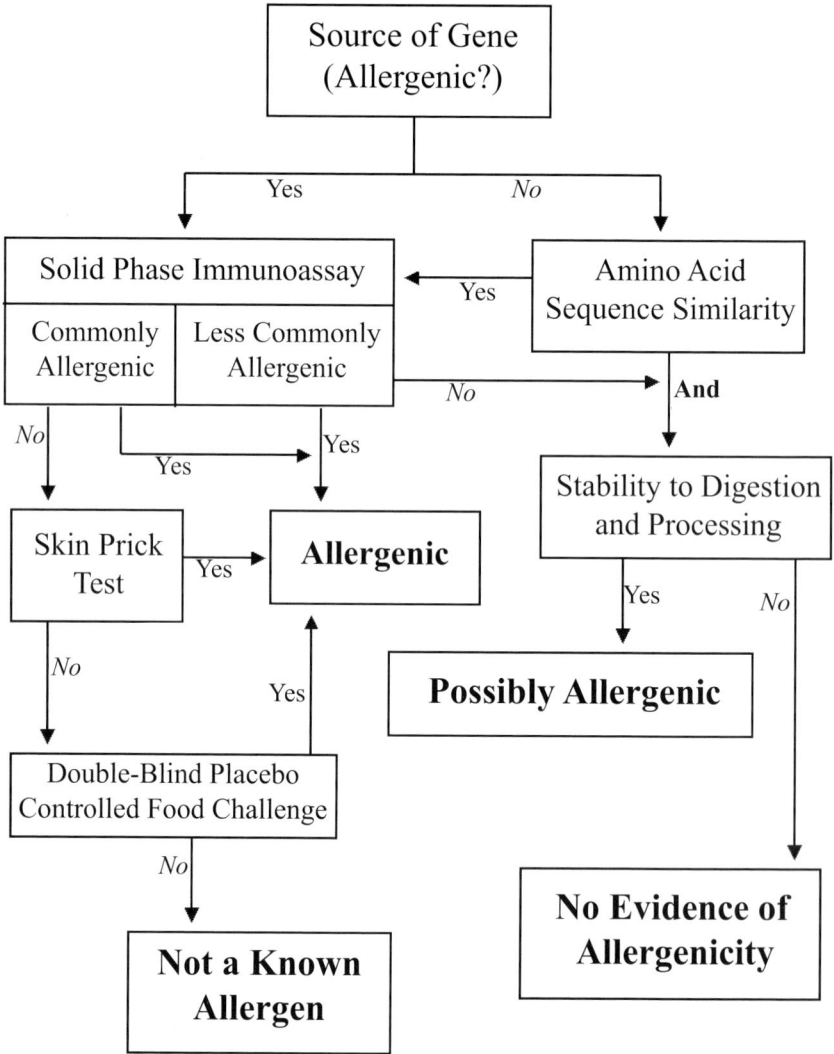

Figure 1. Assessment of the potential allergenicity of proteins used in biotechnology and resulting food products utilizes a decision-tree approach for evaluation. Proteins can be classified as allergenic, nonallergenic, low allergen probability, and possible allergenicity based on the immunological and physical characteristics of the protein. The figure is adapted from Metcalfe *et al.*, 1996.

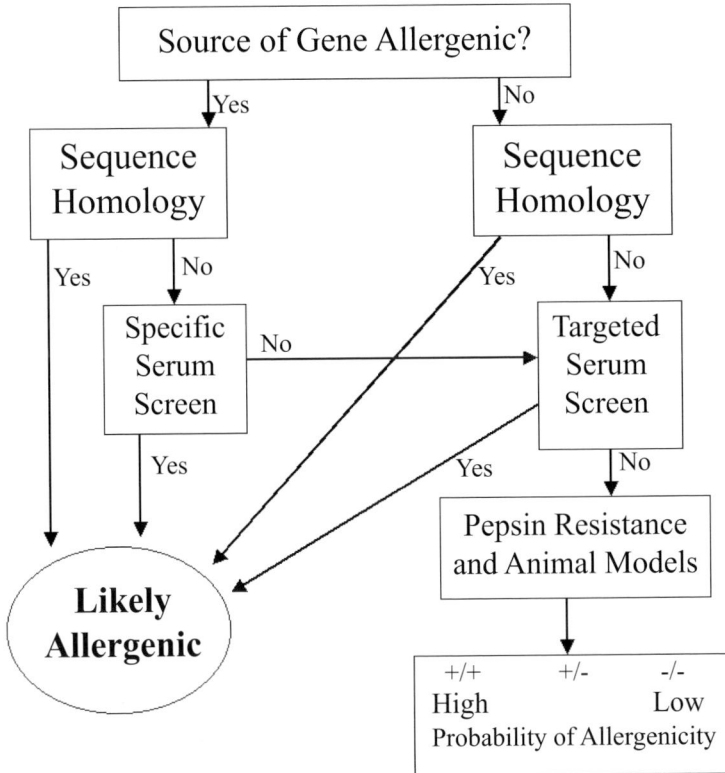

Figure 2. A panel convened by the FAO/WHO produced a revised decision tree. While similar to the previous tree, it entails a more stringent approval scheme. The tree incorporates the use of animal models to assesses potential human allergenicity as well as sequence homology now being indicative of allergenicity. In this scheme, proteins can be classified as likely allergenic or be rated as to a degree of allergenicity probability for high to low. This figure is adapted from FAO/WHO 2001 report.

If the recombinant protein has a significant identity over 8 amino acids, it is considered to have a potential epitope match. A standard of 8 sequential amino acids is used because it conforms to expectations of minimum epitope size and it is the number that is not significantly likely to occur due to chance (Gendel, 1998b). If an epitope match is found, then the protein is tested along the same lines as a protein from a known allergenic source. While this method may be effective in identifying linear existent epitopes, it will not allow for the identification of nonlinear or novel epitopes. A group convened by the Food and Agriculture Organization/World Health Organization in 2001 revised the previous decision tree (FAO/WHO 2001, Figure 2). This assessment scheme has not been broadly reviewed yet and contains some controversial recommendations, such as the use of animal models and the use of 6 amino acids instead of 8 for epitope matching. The model is currently unworkable as there are no validated animal models for human allergy study and a 6 amino acid match significantly complicates the bioinformatics assessment because of the increased likelihood of matches by chance.

If no sequence similarity is detected, then the next assessment is to determine the recombinant protein's stability to digestion. Most food allergens are stable to digestion (Astwood *et al.*, 1996), a characteristic that is important because allergens must reach the gut immune cells in the small intestine fairly intact to elicit and immune response (Bischoff *et al.*, 2000). While not an exclusive characteristic, stability to digestion can be used as an indicator of potential allergenicity. Transgenic proteins that are readily digestible are classified as having no evidence of allergenicity and are approved for consumption. Proteins that are stable to digestion are classified as possibly allergenic and are not approved until further tests can be conducted.

Given the few proteins that are food allergens (Taylor, 1997), it is most likely that transgenic proteins expressed in GM food will pose no risk. However, if a protein is an allergen in its original source food, then it will be an allergen in the GM product. An example of this involved expression of a Brazil nut protein in soybeans (Nordlee *et al.*, 1996). When the transgenic soybeans were tested against serum from Brazil nut-sensitive individuals, IgE from these individuals bound to the soybean extract. In addition, three of the Brazil nut sensitive individuals registered positive skin tests to the transgenic soybean while giving negative skin tests to nontransgenic soybean, indicating that the transgenic soybean could elicit an allergenic response. After the assessment was completed, the variety was not marketed as the result of allergy concerns. Therefore it is important to realize the potential risks and to recognize that a system is in place to safe guard against the inclusion of known allergens in GM foods.

The assessment scheme has a few notable problems. First, it is designed based on knowledge of known food allergens and cannot confidently identify new sources of food allergies. It should be remembered that allergies are the result of a hyperfunctional immune system and not a problematic protein. Allergies are specific to individuals, and the vast majority of proteins cause no response at all. Only a single protein or a select group in any one food item cause allergenic responses. Therefore, it is unlikely that a recombinant protein would be introduced that would become a major food allergen. To address the issue of allergenicity effectively, research will need to focus on the response itself to find treatments to alleviate the symptoms of those who suffer from allergies. Until then, the assessment protocols that exist should be utilized to minimize exposure of the food supply to potential allergenic proteins.

The second problem, and a more likely occurrence, is that proteins that are not allergenic will be restricted from the food supply. An excellent example of this is the recent StarLink^{TM} corn scare. StarLink^{TM} corn was genetically modified to express a certain insecticidal protein (Bt Cry9C). Cry9C is similar to other Bt proteins produced in transgenic plants, in that it specifically kills lepidopteron species (caterpillars). Unlike other Bt proteins, it is resistant to digestion and is digested at a slower rate than other Bt proteins (Aventis, 2000). Therefore, the EPA did not approve this corn for human consumption. It was, however, approved for animal feed, where allergenicity was not a concern. Subsequently, a consumer watchdog group conducted an analysis of several food products and Starlink™ corn was found in several taco shell products. Since Starlink™ was only approved for animal feed, this violation of food quality standards resulted in the products being removed from retail shelves.

But is CryC9 really an allergen risk? The allergenicity assessment was based on the protein's resistance to digestion and heat, but there are many commonly consumed proteins that are resistant or stable to digestion and heat that are not allergens (Astwood *et al.*, 1997; Strobel, 1997). In fact, no single characteristic of allergens is, by itself, indicative of allergenicity (Stanley and Bannon, 1999). Therefore, allergenicity assessment should consider even more criteria and examine each protein on a case-by-case basis.

There are several characteristics of allergens that may also be considered in allergenicity assessment. Most food allergens are small (10-40 kDa) and tend to be glycosylated (Hefle, 1996). The CryC9 protein does not share either of these characteristics (Aventis, 2000), suggesting that while resistant to digestion and heat, perhaps it is not itself an allergen. One of the more significant criteria that may be useful in identifying risk of allergenicity is the level of specific protein production in the food. Native proteins that have been identified to date as food allergens are expressed at levels greater than 1% of the total protein (Metcalfe *et al.*, 1996; Fuchs and Astwood, 1996; Taylor, 1992; Hefle, 1996). In fact, it has been demonstrated that, even for sensitive individuals, there is a minimum dose for initiation of a hypersensitivity response (Moneret-Vautrin *et al.*, 1998). That being the case, if a transgenic protein that shares few characteristics with known allergens is expressed below this threshold level, it is unlikely to be an allergenicity risk to the general population. The vast majority of transgene proteins, including CryC9, will be expressed below 1% of the total protein. Based on this information, Aventis applied to have the regulation of Starlink™ revised to set a tolerance level of human consumption. However, an independent scientific advisory panel convened by the U.S. Environmental Protection Agency found that Starlink™ still poses as 'medium' allergenicity risk, and recommended that it not be approved for human consumption at any level. While contamination of Starlink™ corn in products for human consumption was a regulatory nightmare, no evidence was discovered to suggest that Starlink[TM] corn poses an allergy concern. Individuals that claimed to have developed allergic reactions to Starlink[TM] did not possess immunoglobulin E (IgE) to the recombinant protein in Starlink[TM], as demonstrated by the Centers for Disease Control and the Food and Drug Administration (FIFRA Report, 2001). IgE is a fundamental component in the allergy response and must be present for the development of an allergic reaction.

Conclusions

So where does that leave the issue? Plant biotechnology brings tremendous promise for the improvement of diet, safety, and health, but also brings concerns and risks. These concerns must be evaluated and balanced with existing risks so that the technology can be effectively utilized and the benefits realized. There have always been individuals that are resistant to new technologies and advances, but fluoride in the drinking water did not poison people, irradiation of meat does not cause illness, and microwave ovens do not cause cancer or reduce the quality of foods. These are only a few examples of how unfounded fears empowered those who sought to limit the use of technologies that brought increased convenience and safety to the consumers. Plant biotechnology is no different. The hysteria surrounding GM foods and human health is mostly based on rejection of genetic engineering, but safety evaluations should focus on the safety of the food product and not on the methodology used to obtain the product. The evidence suggests that GM foods approved for human consumption will be beneficial and that safety assessment schemes are in place and are being utilized to protect consumers.

References

Astwood, J., Fuchs, R., and Lavrik, P.B. 1997. Food biotech and genetic engineering. In: Food Allergy: Adverse Reactions to Food and Food Additives. D.D. Metcalfe, H.A. Sampson, and R.A. Simon, eds. Blackwell Science, Cambridge, MA. p. 65-92.

Astwood, J.D., Leach, J.N., and Fuchs, R.L. 1996. Stability of food allergens to digestion *in vitro*. Nature Biotech. 14: 1269-1273.

Aventis. 2000. Response to docket control number OPP-00678.

Beever, D.E., and Kemp C.F. 2000. Safety issues associated with the DNA in animal feed derived from genetically modified crops. A review of scientific and regulatory procedures. Nutr. Abstr. Reviews, Series B: Livestock Feeds and Feeding. 70: 175-182.

Betz, F.S., Hammond, B.G., and Fuchs, R.L. 2000. Safety and advantages of *Bacillus thuringiensis*-protected plants to control insect pests. Reg. Tox. Pharm. 32: 156-173.

Bischoff, S.C, Mayer, J.H., and Manns, M.P. 2000. Allergy and the gut. Int. Arch. Allergy Immunol. 121: 270-283.

Bock, S.A., Lee, W.Y., Remigio, L., and May, C.D. 1978. Studies of hypersensitivity reactions to foods in infants and children. J. Allergy Clin. Immunol. 62: 327-334.

Conway, G., and Toenniessen, G. 1999. Feeding the world in the twenty-first century. Nature. 402 (suppl. 2): C55-C58.

Donaldson, L., and, May, R. 1999. Health implications of genetically modified foods. Depart. of Health (www.doh.gov.uk/pub/docs/gmfood.pdf).

Ewen, S.W.B., and Pusztai, A. 1999. Effects of diets containing genetically modified potatoes expressing *Galanthus nivalis* lectin on rat small intestine. Lancet. 354: 1353-1355.

FAO. 1995. Report on the FAO Technical Consultation on Food Allergies, Rome, Nov. 13-14. Food and Agric. Org. of the UN/WHO, Rome.

FAO/WHO. 2001. Evaluation of Allergenicity of genetically modified foods. Report of the joint FAO/WHO expert consultation on allergenicity of foods derived from biotechnology. Food and Agric. Org. of the UN/WHO, Rome.

FAO/WHO. 2000. Safety aspects of genetically modified foods of plant origin. Report of a Joint FAO/WHO Expert Consultation. Food and Agriculture Organization of the UN and WHO, Geneva, Switzerland.

FDA. 1992. Statement of policy: Foods derived from new plant varieties. Food and Drug Admin., Fed. Reg. 57: 22984.

FIFRA Report. 2001. Scientific advisory panel meeting: Assessment of additional scientific information concerning Starlink[TM] corn. Arlington, VA July 17-18. http:/www.epa.gov/scipoly/sap.

Fuchs, R.L., and Astwood, J.D. 1996. Allergenicity assessment of food derived from genetically modified plants. Food Technol. 50: 83.

Gendel, S.M. 1998a. Sequence databases for assessing the potential allergenicity of proteins used in transgenic foods. Adv. Food Nutr. Res. 42: 63-92.

Gendel, S.M. 1998b. The use of amino acid sequence alignments to assess potential allergenicity of proteins used in genetically modified foods. Adv. Food Nutr. Res. 42:45-62

Goto, F., Yoshihara, T., Shigemoto, N., Toki, S., and Takaiwa, F. 1999. Iron fortification of rice seed by the soybean ferritin gene. Nature Biotech. 17: 282-286.

Hammond, B.G., Vicini, J.L., Hartnell, G.F., Naylor, M.W., Knight, C.D., Robinson, E.H., Fuchs, R.L., and Padgette, S.R. 1996. The feeding value of soybeans fed to rats, chickens, catfish, and dairy cattle is not altered by genetic incorporation of glyphosate tolerance. J. Nutr. 126: 717-727.

Hansen, G., and Wright, M.S. 1999. Recent advances in the transformation of plants. Trends Plant Sci. 4: 226-231.

Harrison, L.A., Bailey M.R., Naylor M.W., Ream, J.E., Hammond, B.G., Nida, D.B., Burnette, B.L., Nickson, T.E. Mitsky, T.A., Taylor, M.L., Fuchs, R.L., and Padgette, S.R. 1996. The expressed protein in glyphosate-tolerant soybean, 5-enolpyruvyl-shikimate-3-phosphate synthase from *Agrobacterium* sp. Strain CP4, is rapidly digested *in vitro* and is not toxic to acutely gavaged mice. J. Nutr. 126: 728-740.

Hashimoto, W., Momma, K., Yoon, H.J., Ozawa, S., Ohkawa, Y., Ishige, T., Kito, M. Utsumi, S., and Murata, K. 1999. Safety assessment of transgenic potatoes with soybean glycinin by feeding studies in rats. Biosci. Biotechnol. Biochem. 63: 1942-1946.

Hefle, S. 1996. The chemistry and biology of food allergens. Food Technol. 50: 86-92.

HHS. 2000. FDA to strengthen premarket review of bioengineered foods. Press Release, May 3, US Dept of Health and Human Services, Washington, DC.

Ho, M-W., Ryan, A., and Cummins, J. 1999. Cauliflower mosaic viral promoter- A recipe for disaster? Micro. Ecol. Health Disease. 11: 194-197.

Hodgson, J. 2000. Scientists avert new GMO crisis. Nature Biotech. 18: 13.

Horsch, R.B., Fry, J.E., Hoffman, N.L., Eichholtz, D., Rogers, S.G., and Fraley, R.T. 1985. A simple and general method for transferring genes into plants. Science. 227: 1229-1231.

IFT Expert Report on Biotechnology and Foods 2000. Human food safety evaluation of rDNA biotechnology-derived foods. Food Technol. 54: 53-61.

Jung, W., Yu, O., Lau, S.M.C., O'Keefe, D.P., Odell, J., Fader, G., and McGonigle, B. 2000. Identification and expression of isoflavone synthase, the key enzyme for biosynthesis of isoflavones in legumes. Nature Biotech.18: 208-212.

Klein, T.M., Fromm, M., Weissinger, A., Tomes, W., Schaaf, S., Sletting, M., and Sanford, J. 1988. Transfer of foreign genes into intact plant cells with high-velocity microprojectiles. Proc. Natl. Acad. Sci. USA. 85: 4305-4309.

Kuiper, H.A., Noteborn, P.J.M., and Peijnenburg, A.C.M. 1999. Adequacy of methods for testing safety of genetically modified foods. Lancet. 354: 1315-1316.

Lehrer, S.B., Horner, W.E., and Reese, G. 1996. Why are some proteins allergenic? Implications for biotechnology. Crit. Rev. Food Sci. Nutr. 36: 553-564.

Metcalfe, D.D., Astwood, J.D., Towsend, R., Sampson, H.A., Taylor, S.L., and Fuchs, R.L. 1996. Assessment of the allergenic potential of food derived from genetically engineered crop plants. Crit. Rev. Food Sci. Nutr. 36: S165-S186.

Mendieta, N.L.R., Nagy, A.M., and Lints, F.A. 1997. The potential allergenicity of novel foods. J. Sci. Food Agric. 75: 405-411.

Miraglia, M., Onori, R., Brera, C., and Cava, E. 1998. Safety assessment of genetically modified food products: an evaluation of developed approaches and methodologies. Microchem. J. 59: 154-159.

Moneret-Vautrin, D.A., Rance, F., Kanny, G., Olsewski, A., Gueant, J.L., Dutau, G., and Guerin, L. 1998. Food allergy to peanuts in France— evaluation of 142 observations. Clin. Exp. Allergy. 28: 1113-1119.

Nordlee, J.A., Taylor, S.T., Townsend, J.A., Thomas, L.A., and Bush, R.K. 1996. Identification of a Brazil-nut allergen in transgenic soybeans. New Eng. J. Med. 334: 688-692.

Shintani, D., and DellaPenna, D. 1998. Elevating the vitamin E content of plants through metabolic engineering. Science. 282: 98-100.

Stanley, J.S., and Bannon, G.A. 1999. Biochemistry of food allergens. Clin. Rev. Allergy Immunol. 17: 279-291.

Stewart, C.N., Richards, H.A., and Halfhill, M.D. 2000. Transgenic plants and biosafety: science, misconceptions, and public perceptions. BioTechniques 29: 832-843.

Strobel, S. 1997. Oral tolerance: immune responses to food antigems. In: Food Allergy: Adverse Reactions to Food and Food Additives. D.D. Metcalfe, H.A. Sampson, and R.A. Simon, eds, p. 107-135.

Taylor, S.L. and Hefle, S.L. 2001. Food allergies and other food sensitivities. Food Technol. 55: 68-83.

Taylor, S.L. 1997. Food from genetically modified organisms and potential for food allergy. Envron. Toxicol. Pharmacol. 4: 121-126.

Taylor, S.L., and Lehrer, S.B. 1996. Principles and characteristics of food allergens. Crit. Rev. Food Sci. Nutr. 36: S91-S118.

Taylor, S.L. 1992. Chemistry and detection of food allergens. Food Technol. 46: 146-152.

Wang, R.L., Stec, A., Hey, J., Lukens, L., and Doebley, J. 1999. The limits of selection during maize domestication. Nature. 398: 236-239.

Ye, X., Al-Babili, S., Kloti, A., Zhang, J., Lucca, P., Beyer, P., and Potrykus, I. 2000. Engineering the provitamin A (β-carotene) biosynthetic pathway into (carotenoid-free) rice endosperm. Science. 287: 303-305.

Yeung, J.M., Applebaum, R.S., and Hildwine, R. 2000. Criteria to determine food allergen priority. J. Food. Prot. 63: 982-986.

Yunginger, J.W., and Adolphson, C.R. 1992. Standardization of allergens. In: Manual of Clinical Immunology, Am. Soc. Microbiol., Washington, DC. p. 678-684.

Zitnak, A., and Johnston, G.R. 1970. Glycoalkaloid content of B5141-6 potatoes. Am. Potato J. 47: 256-260.

From: *Transgenic Plants: Current Innovations and Future Trends*
Edited by: C. Neal Stewart, Jr.

Chapter 10

Plant-Based Vaccines

James E. Carter III, Nak-Won Choi,
Cheree Rivers-Khalid,
and William H.R. Langridge

Abstract

The increasing number of new incurable infectious diseases and the re-emergence of antibiotic resistant pathogens over the past several decades, in combination with the expected increase in world population from six to ten billion by the middle of this century, underscores the urgent need for new methods of vaccination which are more effective, easier to administer and less costly to produce. The advent of recombinant DNA technology and plant transformation methods coupled with the ability to regenerate plants from single cells has made possible the synthesis of antigens and autoantigens in edible plants. In this review, we describe recent developments in plant biotechnology where production of mucosal vaccines in edible plants for effective and inexpensive protection against infectious and autoimmune diseases will soon be feasible.

Introduction

Antibiotics have played a major role in the prevention and spread of infectious diseases. In 1945, penicillin became available to the public and pneumonia deaths in the United States declined from 68 to 27 per 100,000 within 5 years (Chase, 1982). Based on the control of a wide variety of bacterial pathogens, the medical community embraced antibiotics as a panacea for the treatment and prevention of infectious diseases. Thus, for the last half of the 20th century, the successful use of antibiotics and parenteral vaccination has created an illusion that the spread of infectious disease was under control. Unfortunately, this optimistic picture has been dampened by changing political climates, burgeoning world populations, increased international travel and immigration, greater dependence on a global food economy and expansion of international commerce, which has intimately connected countries that in the past were isolated from one another. In the past 28 years, the increased connections between nations has provided novel opportunities for the emergence of at least twenty-nine previously unknown and largely incurable infectious diseases, such as HIV, Ebola virus, Hanta virus, and West Nile virus. Further, the abuse of antibiotics coupled with the slow development of new antibiotics has resulted in the re-emergence of a variety of more virulent common pathogen strains.

During the past several years, an increasing number of infectious disease outbreaks have occurred in both industrialized and developing countries. Multi-drug resistant tuberculosis, acute micrococcal infections, and Hanta virus, as well as food and waterborne outbreaks of salmonella, cholera, and digestive tract illnesses caused by enteropathogenic and enterotoxic *Escherichia coli* have escalated (Satcher, 1995). International accounts of infectious disease include meningococcal meningitis in Zimbabwe and the U.K., human *Pfiesteria* infection in the U.S., dengue fever in Vietnam, tick-borne encephalitis in Saudi Arabia, and mosquito-borne encephalitis in the U.S. (Morse, 1995). More recently, the spread of infectious diseases has been further exacerbated by the emergence of bioterrorism. The continuous appearance of new lethal diseases and the re-emergence of more virulent commonly known pathogens is alarming, since neither parenteral immunization programs nor the application of antibiotics have been able to impede their global dissemination.

To appreciate why traditional injectable vaccines have so far failed to provide complete global protection against the persistence and continued appearance of lethal infectious diseases, we must understand more about the nature of the mammalian immune response. While traditional parenteral immunization methods generate a humoral immune response, they fail to generate significant levels of mucosal immunity. Mucosal immune responses in the gastrointestinal, respiratory, and urogenital tracts form the first line of defense against invasion by the majority of infectious diseases. Pathogens can proliferate in the body's tissues prior to stimulation of antibody-secreting memory cells generated by parenteral vaccination. Thus, infected individuals may display symptoms of illness before the body's mechanisms of humoral immunity begin to mount a significant immune response against the invader. Mucosal immunization has the distinct advantage of generating secretory antibodies (sIgA) in addition to humoral antibodies (IgG). The secretory antibodies are designed to survive the harsh environment of the gut and other mucosal surfaces. These secreted antibodies can intercept pathogens prior to their attachment to the mucosal surface, thereby preventing infection before it can become established. The impact of this "Ounce of prevention is worth a pound of cure" strategy that the body's mucosal immune system has devised to overcome infectious disease is not utilized by injected vaccines.

It is clear that problems associated with parenteral immunization such as refrigeration, availability of trained medical personnel for dissemination, lack of a protective mucosal immune response, and needle stick injuries that spread disease must be eliminated before the relentless progression of new and re-emerging infectious diseases can be halted. To accomplish this goal, we must take advantage of new technologies that have the ability to revolutionize immunization programs beyond their current status. Based on their protective efficacy in combination with their inexpensive cost, durability and global availability, recombinant plant-based mucosal vaccines are emerging as a feasible source of inexpensive, easy to apply method of immunization. These vaccines can be made available to everyone. Farmers in rural areas of economically developing nations can grow indigenous vaccine-containing plants through use of local agricultural techniques and available agricultural machinery. Further, plant-based vaccine programs eliminate requirements for refrigerated storage, trained medical personnel, and the use of needles and syringes for administration, which in addition to causing pain, can cause needle stick injury and disease spread.

Poverty and political unrest have been largely responsible for the lack of effective vaccination programs in developing nations. To make global vaccination programs more effective, basic changes are required in the way vaccines are produced, distributed and administered. In 1990, the World Health Organization (WHO) launched an initiative establishing goals for the development of vaccines that are safe, inexpensive, easily (orally) administered and widely accessible (Mitchell *et al.*, 1993). To meet these needs, several laboratories were able to show that transformed edible plants could express bacterial and virus antigens (Moffat, 1995). Further, the antigen

Table 1. Plant-based vaccine patents for infectious and autoimmune diseases.

US/PCT Patent Number	Title	Applicant	Date Issued
Vaccines For Infectious Diseases			
1. 5,484,719	Vaccines produced and administered through edible plants.	Lam & Arntzen	1/96
2. 5,612,487	Anti-viral vaccines expressed in plants.	Lam & Arntzen	3/97
3. 5,654,184	Oral immunization by transgenic plants.	Curtiss & Cardineau	7/97
4. 5,679,880	Oral immunization by transgenic plants.	Curtiss & Cardineau	10/97
5. 5,686,079	Oral immunization by transgenic plants.	Curtiss & Cardineau	11/97
6. 6,034,298	Vaccines expressed in plants.	Lam, Arntzen & Mason	3/00
7. 6,194,560	Oral immunization with transgenic plants.	Arntzen, Mason & Haq	2/01
8. D12273-2 (Prov.)	Production of a cholera toxin B subunit-rotavirus NSP4 enterotoxin fusion protein in potato.	Langridge & Arakawa	1/00
9. 6,261,561	Method of stimulating an immune response by administration of host organisms that express intimin alone or as a fusion protein with one or more other antigens.	Stewart, McKee, O'Brien, and Wachtel	7/01
Vaccines For Autoimmune Diseases			
10. WO95/08347	Methods and products for controlling immune responses in mammals.	Jevnikar	Filed, 9/94
11. WO1977US0013634	Tolerogenic antigens via edible plants or plant-derived products.	Agrivax, Inc.	2/98
12. 60/082,688	Methods and substances for preventing and treating autoimmune disease.	Langridge & Arakawa	Filed, 4/98

levels synthesized in the plant were able to generate detectable antibody titers in immunized animals (Mason *et al.*, 1992) and even provide protection against pathogen challenge (Arakawa *et al.*, 1997). These results indicated that transgenic plants are feasible production and delivery systems for providing protective immunization against infectious diseases (Castanon *et al.*, 1999). Based on applications for international and U.S. patents, laboratories pioneering the development of edible plant vaccines for the prevention and treatment of infectious and autoimmune diseases are presented in Table 1. In this review, we investigate recent advances in the development of experimental plant based vaccines. Finally, we focus on the future of plant-based vaccines for the prevention and treatment of infectious and autoimmune diseases.

Plant Vaccine Protection Against Infectious Disease

When pathogenic bacteria and viruses are passed on through food, water or by human contact, they may colonize or invade eukaryotic organisms predominantly through mucosal surfaces. Vaccines for protection against these infections must stimulate the mucosal immune system to produce secretory IgA (sIgA) at mucosal surfaces, especially the gut and respiratory epithelia (McGhee and Kiyono, 1993; Castro and Arntzen, 1993). Where organisms such as *Vibrio cholerae* and enterotoxigenic *Escherichia coli* (ETEC) are limited to infection of the epithelium, a strong mucosal immune response is produced. However, when the organism moves from the mucosal surface into the bloodstream, a humoral response is required to generate protective immunity. Therefore, it is necessary that mucosal vaccines generate local immunity at mucosal surfaces to prevent the spread of the pathogen into the circulation (Brennan *et al.*, 1999a).

The slow release of antigens from plant cells during digestion protects them from low pH degradation and proteolytic cleavage in the gut. The increased longevity of the antigen provides it with a greater chance to reach receptors on microfold (M) cells in the intestinal epithelium (Richter and Kipp, 1999). Following uptake into M cells, the antigen is transported across the basolateral membrane, where it is taken up by antigen presenting cells (APC)—macrophages, dendritic cells, and B cells. The APCs present digested fragments of the antigen to T helper lymphocytes that specifically recognize the antigen. The activated T cells secrete a variety of cytokines that stimulate inflammatory responses in cytotoxic macrophages and lymphocytes (CTL), which eliminate the pathogen.

Two strategies for delivery of plant-based vaccines have emerged based on methods used to obtain antigen gene expression in plants. A variety of methods are being used for transfer and stable integration of foreign genes into plant cells. Biological methods of gene transfer mediated by *Agrobacterium tumefaciens* are useful for high-frequency stable transformation of dicotyledenous plants (Nester *et al.*, 1984; Shahin and Simpson, 1986; Zambryski, 1992). Chemical and physical methods, such as calcium phosphate-polyethylene glycol (PEG), dimethylsulfoxide (DMSO), microinjection, electroporation and microprojectile bombardment are low-frequency transformation methods useful for all other plant cell types (Davey *et al.*, 1989; Klein *et al.*, 1987; Newell, 2000). The second approach is based on transient gene expression of antigens in plants mediated by engineered plant viruses. This strategy, exemplified by tobacco mosaic virus (TMV) can make use of two methods for generation of antigens in plant viruses: (i) subgenomic virus promotor-driven antigen gene expression and (ii) fusion of antigen epitopes to the virus coat protein.

Stable Transformation by *Agrobacterium tumefaciens*-Mediated Gene Transfer

Agrobacterium-mediated gene transfer is the most widely used biological technique for transformation of higher plants (Schell, 1987). During plant cell transformation, the T-DNA region (20 kb) containing foreign genes is transferred and integrated in the plant nuclear genome. *Agrobacterium* has three elements that are essential for the transfer of the T-DNA from Ti plasmids to plant cells: The virulence (*vir*) genes (Stachel and Nester, 1986), T-DNA border sequences (Zambryski *et al.*, 1982), and *chv*A and *chv*B genes or chromosomal virulence genes (Douglas *et al.*, 1985). The chromosomal *chv* loci control attachment of the bacterium to the plant cell. Activation of *vir* gene expression results in the generation of site-specific nicks in T-region border sequences (Yanofsky *et al.*, 1986; Wang *et al.*, 1984; Peralta and Ream, 1985) and leads to T-DNA transfer (Stachel *et al.*, 1987). The T (transfer) region on Ti plasmids is flanked by right and left 25-bp T-region border sequences essential for efficient T-DNA transfer (Joos *et al.*, 1983; Wang *et al.*, 1984). The gene of interest inserted between the T-DNA borders is efficiently co-transferred and stably maintained within the plant genome. T-DNA encoded genes 1, 2, and 4 are involved in the control of neoplastic growth in transformed plants (Willmitzer *et al.*, 1982). Genes 1 and 2 encode auxin synthesis in root-sprouting tissues, and gene 4 encodes cytokinin synthesis in shoot-inducing teratomas (Yamamoto *et al.*, 1987). Deletion of these genes has resulted in *Agrobacterium* strains which can transfer the

T-DNA but which do not cause tumors in transformed plants. By addition of selectable marker genes to the T-DNA, transformed plants can be obtained. The transformed plant cells are allowed to differentiate into normal plantlets on selective growth media and are grown into mature plants in the greenhouse. A list of infectious diseases for which plant based vaccines have demonstrated various levels of protective efficacy is presented in Table 2.

Alternative Methods for Vaccine Production

Although *Agrobacterium tumefaciens*-mediated nuclear transformation has been widely used for stable plant expression of foreign antigens for vaccine purposes, most monocotyledonous plants are presently beyond the host range of *A. tumefaciens* and are difficult to transform by this method. The efficiency of biological transformation methods has lead to increased interest in the development of alternatives to circumvent current problems associated with *A. tumefaciens* transformation methods. Plastid genome engineering has recently received considerable attention as a method for plant transformation based on the relatively high yields of engineered protein that are possible with this method and the ability to target the gene of interest to specific sites on the plastid chromosome (see Chapter 5).

Plastid transformation methods for introduction of the *Bacillus thuringiensis* lepidopteran protoxin gene into the chloroplast chromosomes of tobacco plants using microprojectile methods was shown to be effective (McBride *et al.*, 1995). The Bt protoxin was expressed at levels up to 2-3% of the total soluble leaf protein. The relatively high level of transgene expression in comparison to nuclear transformation was attributed to the simplicity of the plastid genome, which has ~50 chloroplasts per leaf cell and 60-100 chromosomes per chloroplast (Daniell *et al.*, 1990). Each leaf cell could therefore contain 3,000-5,000 gene copies. This high gene copy number has the potential for producing high levels of foreign protein. The *B. thuringiensis* delta endotoxin protein and aadA spectinomycin resistance proteins were synthesized at levels exceeding 45% of the total soluble protein of tobacco and tomato, respectively (de Costa, *et al.*, 2001; Ruf *et al.*, 2001). Thus, plastid transformation may present an opportunity for significant improvement in plant-based vaccine transmission to intestinal epithelial cells for presentation to the gut-associated lymphoid tissues.

Table 2. Diseases for which plant-produced vaccines have been constructed.

Disease	Pathogen	Plant used*	Reference
Autoimmune deficiency syndrome (AIDS)	Human immunodeficiency virus	Cowpea (V) Tobacco (V) Cowpea (V) Tobacco (V) Tobacco (V)	Porta et al., 1994 Sugiyama et al., 1995 McLain et al., 1996 Yusibov et al., 1997 Joelson et al., 1997
B-cell lymphoma	-	Tobacco (V/A)	McCormick et al., 1999
Bronchiolitis, pneumonia	Respiratory syncytial virus	Apple leaf (S) Tomato (S) Tobacco (V)	Sandhu et al., 1999 Sandhu et al., 2000 Belanger et al., 2000
Cholera	Vibrio cholerae	Potato (S) Tobacco (C)	Arakawa et al., 1997 Daniell et al., 2001
Colon cancer	-	Tobacco (A)	Verch et al., 1998
Colon, breast, and other epithelial tumors	-	Tobacco (A) Rice (A), Wheat (A)	Vaquero et al., 1999 Stoger et al., 2000
Common cold	Human rhinovirus 14	Cowpea (V)	Porta et al., 1994
Cytomegalovirus disease	Human cytomegalovirus	Tobacco (S)	Tackaberry et al., 1999
Dental caries (cavities)	Streptococcus mutans	Tobacco (A)	Ma et al., 1998
Diabetes, type 1A (autoimmune)	-	Tobacco (S), Potato (S) Potato (S)	Ma et al., 1997 Arakawa et al., 1998b, 1999

Disease	Target pathogen	Plant	References
"Travellers diarrhea"	Enterotoxigenic *E. coli*	Tobacco (S)	Haq *et al.*, 1995
		Potato (S)	Haq *et al.*, 1995
		Potato (S)	Tacket *et al.*, 1998
		Potato (S)	Mason *et al.*, 1998
		Potato (S)	Lauterslager *et al.*, 2001
		Corn (S)	Streatfield *et al.*, 2001
Diarrhea (human)	*V. cholerae, E. coli*, rotavirus, Norwalk virus	Potato (S)	Yu and Landridge, 2001
		Tobacco (S)	Mason *et al.*, 1996
		Potato (S)	Tacket *et al.*, 2000
Diarrhea (dogs)	Canine parvovirus	Tobacco (V)	Fernandez-Fernandez *et al.*, 1998
		Arabidopsis (S)	Gil *et al.*, 2001
Diarrhea (mink)	Mink enteritis virus	Cowpea (V)	Dalsgaard *et al.*, 1997
Flu	Influenza virus	Tobacco (V)	Sugiyama *et al.*, 1995
Food poisoning	*Staphylococcus aureus*	Cowpea (V), Tobacco (V)	Brennan *et al.*, 1999b
Foot-and-mouth disease	FMD virus	Cowpea (V)	Usha *et al.*, 1993
		Arabidopsis (S)	Carrillo *et al.*, 1998
		Alfalfa (S)	Wigdorovitz *et al.*, 1999a
		Tobacco (V)	Wigdorovitz *et al.*, 1999b
Gastroenteritis (pigs)	Swine-transmissible gastroenteritis virus	*Arabidopsis* (S)	Gomez *et al.*, 1998
		Potato (S)	Gomez *et al.*, 2000
		Tobacco (S)	Tuboly *et al.*, 2000
		Corn (S)	Streatfield *et al.*, 2001
Genital herpes	Herpes simplex virus-2	Soybean (A)	Zeitlin *et al.*, 1998

Table 2. Continued

Disease	Pathogen	Plant used*	Reference
Hemorrhagic disease (rabbit)	Rabbit hemorrhagic disease virus	Potato (S) Tobacco (V)	Castanon et al., 1999 Fernandez-Fernandez et al., 2001
Hepatitis B	Hepatitis B virus	Tobacco (S) Tobacco (S) Lupine (S) Lettuce (S) Potato (S)	Mason et al., 1992 Thanavala et al., 1995 Kapusta et al., 1999 Kapusta et al., 1999 Richter et al., 2000
Hepatitis C	Hepatitis C virus	Tobacco (V)	Nemchinov et al., 2000
Hepatitis (mouse)	Murine hepatitis virus	Tobacco (V)	Koo et al., 1999
Malaria	*Plasmodium* protozoa	Tobacco (V)	Turpen et al., 1995
Measles	Measles virus	Tobacco (S)	Huang et al., 2001
Pneumonia, meningitis, etc.	*Pseudomonas aeruginosa*	Cowpea (V)	Brennan et al., 1999a
Rabies	Rabies virus	Tomato (S) Tobacco (V) Tobacco (V), Spinach (V)	McGarvey et al., 1995 Yusibov et al., 1997 Modelska et al., 1998
Shipping fever (cattle)	*Mannheimia haemolytica* A1	White clover (S)	Lee et al., 2001

*Method used: S = stable (nuclear) transformation, V = engineered plant virus, A = plant-produced antibodies, C = chloroplast transformation.

Plant Vaccines Protect Against Bacterial Pathogens

Spa A Protein

One of the first described antigens expressed in transgenic plants was the *Streptococcus mutans* cell surface-adhesion protein (Spa A) (Curtiss and Cardineau, 1990). Spa A is a 185 kDa adhesion protein that binds *S. mutans* cells to the surface of the tooth enamel, initiating the onset of tooth decay. Data from oral immunogenicity studies showed that Spa A produced in *E. coli* stimulated the production of anti-Spa A sIgA in saliva. Tobacco plants were engineered to produce Spa A protein at levels up to 0.02% of the total protein in leaf samples. Further, the plant synthesized Spa A was found to be effective in the reduction of dental caries in orally immunized patients.

Vibrio cholerae and E. coli Enterotoxin B Subunits

Both LT from enterotoxigenic *E. coli* (ETEC) and CT from *Vibrio cholerae* are a major cause of lethal diarrhea. The heat-labile enterotoxin (LT) is a multimeric protein that is 80% identical to cholera toxin (CT), based on amino acid sequence analysis. Each holotoxin molecule of LT is composed of one A subunit (Mr of 27 kDa) and five B subunits (11.6 kDa) that form a pentameric membrane binding structure. LT-B subunits, when assembled, can bind to the GM1 gangliosides present on epithelial cell surfaces and allow the toxic LT-A to enter into the cells. LT-A and CT-A are potent oral immunogens that can be used in small amounts as adjuvants to stimulate a stronger immune response (Clements *et al.*, 1988). Antibodies against one B subunit can block the effect of either toxin because these toxins cross-react immunologically. Immunization against CT or LT gives some protection against both disease agents. Several mutants of CT and LT have been used as mucosal adjuvants (Dickinson and Clements, 1996). A and B subunits, which are coordinately expressed to form the holotoxin in plants, could enhance the vaccine value of other, less immunogenic, plant-expressed vaccine antigens expressed in the same tissues (Holmgren *et al.*, 1993).

LT-B synthesized in potato tubers resulted in very low levels of expression—up to 0.01% of the total protein (approximately 10-fold less in tobacco leaves). Modification of the recombinant protein with an endoplasmic reticulum retention signal (amino acids SEKDEL) at its carboxyl terminus resulted in a threefold increase over initial expression found without the SEKDEL segment. The recombinant protein in the transgenic tubers produced 3-4 µg LT-B/g fresh tuber weight. The expression of the B subunit of CT was also reported in transgenic potato plants (Arakawa *et al.*, 1997). The expression of CT-B with an ER retention signal at its carboxyl terminus generated 0.3% of total soluble plant protein. More recently, Mason *et al.* designed a synthetic LT-B gene with both dicot and monocot codon preferences, A-T richness, putative polyadenylation and cryptic splicing signals, and potential RNA destabilizing sequences which resulted in potato tuber expression of 10-20 μg per gram fresh weight (Mason *et al.*, 1996; 1998).

Plant Vaccines Against Viral Pathogens

Hepatitis B Virus (HBV)

The first attempt at production of a plant-produced vaccine was the synthesis of hepatitis B surface antigen (HBsAg) in tobacco and potato plants. The HbsAg was shown to assemble into virus-like particles (VLPs) that are similar in antigenicity to the yeast derived commercial vaccine that is considered to be prohibitively expensive for use in many developing countries (Mason *et al.*, 1992; Thanavala *et al.*, 1995). The expression of HBsAg was synthesized at levels equal to 0.01% of the total soluble protein in tobacco. The HBsAg content in plant extracts was approximately 3% of the total soluble protein. Furthermore, recombinant HBsAg (rHBsAg) extracted from tobacco leaves assembled as a VLP with an average size of 22 *n*m, important because the particle form of HBsAg retained conformational antigens required for immunogenicity (Cabral *et al.*, 1978). The rHBsAg of a crude extract from plants was injected to mice for immunization studies. The T-cell epitope expressed by the tobacco-derived rHBsAg was identified based on identification of anti-HBsAg serum and mucosal antibodies essential for prevention of hepatitis B infection in the serum of orally immunized subjects (Thanavala *et al.*, 1995).

Norwalk Virus (NV)

Norwalk virus causes acute epidemic gastro-enteritis in humans. Virus-like particles (VLPs) of assembled Norwalk virus capsid protein (NVCP) were produced and purified from baculovirus infected insect cell cultures. The VLP's have demonstrated oral immunogenicity for protection against Norwalk virus infection in mice (Ball *et al.*, 1996). Dose response experiments indicated that mice require at least 200 μg of NVCP to stimulate a significant immunity without the addition of adjuvants. In these studies, the mice consumed four doses of 4.0 g each of tuber containing 10-20 μg NVCP/g fresh tuber weight (Mason *et al.*, 1996). VLPs were produced in tobacco and potato (Ball *et al.*, 1996). NVCP does not have a signal sequence and Norwalk virions are produced in the cytoplasm of infected cells.

Recombinant NVCP generated in tobacco leaves was expressed at levels of up to 0.23% of the total soluble protein (TSP) and 0.37% TSP in potato tubers.

Rabies Virus (RV)

The rabies virus glycoprotein (G-protein) was expressed in transgenic tomato leaf and fruit tissue (McGarvey *et al.*, 1995). The native G-protein from denatured rabies virus has a molecular mass of 66 kDa. G-protein produced in plants was shown to contain immunological epitopes. However, recombinant G-protein was expressed at very low levels, in the range of 0.001% of TSP (1-10 ng/mg).

Foot-and-Mouth-Disease Virus (FMDV)

The VP1 of FMDV containing critical epitopes for the induction of neutralizing antibodies was expressed in *Arabidopsis thaliana*. Oral immunization with a crude extract of transformed plants showed that immunized mice developed specific serum antibodies to intact FMDV particles and to a synthetic peptide derived from the sequence of VP1 (Carrillo *et al.*, 1998). In addition, it was shown that the antigen produced in transgenic plants could stimulate a protective immune response in immunized mice.

Swine Transmissible Gastroenteritis Virus (TGEV)

Transmissible gastroenteritis is caused by TGEV, a multi-subunit, enveloped, single-stranded RNA virus (Laude *et al.*, 1990). TGE is a highly contagious enteric disease that is characterized by vomiting, severe diarrhea and high mortality in piglets. TGEV is composed of three structural proteins. Protein M is an integral membrane protein, N is a phosphoprotein encapsulating the viral RNA genome, and S is a large surface glycoprotein. Expression of recombinant TGEV S protein in transgenic corn was found to induce protective immune responses at mucosal surfaces of the intestine through activation of S-IgA (Streatfield *et al.*, 2001).

Production of Vaccines by Altered Plant Viruses

Transient Expression Using Viral Vectors

Genetically altered plant viruses have been constructed for expression of vaccine antigen proteins at a higher level. Two approaches to the generation of recombinant viruses have been applied. In the first approach, the pathogen antigen gene is expressed as a soluble protein by sub-genomic promotor-driven gene expression and second, small antigenic peptides encoding pathogen epitopes are fused to the viral coat proteins prior to assembly and formation of infectious virus particles, which then display the antigens on the virus surface.

Tobacco Mosaic Virus (TMV)

TMV is a rod shaped plant RNA virus composed of 2,100 copies of TMV coat protein monomers. After plant infection, the virus is harvested and can amount to as much as 50% of plant dry weight. Antigenic peptides or proteins can be fused to the coat protein monomer amino- or carboxyl-terminus or inserted at a structurally unimportant site within the coat protein. Those constructs are designed to have a leaky termination codon, separating the cp from the DNA encoding antigen. Thus, these viral vectors are able to produce both native and recombinant coat proteins, resulting in a functional virus that contains a coat protein displaying the antigen.

Malarial epitopes were expressed in TMV using this technique (Turpen *et al.*, 1995). The sporozoite B-cell epitope AGDR was inserted into a surface loop of the TMV coat protein and the epitope QGPGAP was inserted at the carboxyl terminus of TMV coat protein. After infection of plants, recombinant viruses expressing coat proteins with three copies of AGDR or two copies of QGPGAP were recovered in high yields: 0.4-1.2 mg/g antigen protein per gram fresh weight of recombinant plant tissue. The malaria epitopes were recognized by anti-malarial monoclonal antibodies in ELISA and western blot assay. These results reinforce the condition that the malaria epitopes were displayed on the surface of the viral particles.

Epitopes from *H. influenzae* haemagglutinin and human immunodeficiency virus (HIV) envelope protein were expressed as carboxyl-terminal fusions of TMV coat protein (Sugiyama *et al.*, 1995). In each case, the foreign DNA was cloned downstream of the TMV coat protein gene using a leaky stop codon. Virus particles purified from infected plants were recognized by antiserum specific for their epitopes in western blot assay, demonstrating the presence of pathogen epitopes on the viral surface.

Cowpea Mosaic Virus (CPMV)

CPMV is a positive-strand RNA virus containing 60 copies of large and small subunits. Viral particles are arranged with icosohedral symmetry. A viral vector was modified for insertion of foreign sequences, ensuring that no CPMV-specific sequences were deleted (Porta *et al.*, 1994). Through this modification, epitopes from VP1 of FMDV, VP1 of human rhinovirus-14 and gp41 of HIV-1 were stably expressed by modified CPMV. The viral particles purified from infected plants were recognized by antisera specific for their epitopes and mice orally or parenterally immunized with these particles demonstrated immunogenicity.

The VP2 capsid protein of mink enteritis virus (MEV) contains a 17 amino acid linear epitope (Dalsgaard *et al.*, 1997). MEV virus was synthesized in infected black-eyed bean plants (*Vigna unguiculata*), resulting in yields of 1.0-1.2 mg of MEV per gram fresh weight of plant tissue. Antibody responses to MEV VP2 epitope were generated in immunized animals in a dose-dependent manner and protection against infection was found with reduced viral shedding. Virus neutralizing antibodies were detected in the antiserum in 9 of 12 animals that received the recombinant viral vaccine. The results demonstrate that plant virus-based vaccines can protect animals from infectious diseases.

Tomato Bushy Stunt Virus (TBSV)
A 13 amino acid V3 loop peptide from HIV-1 gp120 was fused to the carboxyl terminus of the TBSV cp (Joelson *et al.*, 1997). Virus particles purified from tobacco plants displayed 180 copies of the antigen per virus. Mice immunized with the TBSV particles demonstrated a specific primary antibody response to the peptide and a strong antibody response to TBSV.

Alfalfa Mosaic Virus (AMV)
Whereas other plant viruses are restricted in their ability to accommodate large antigen proteins in the structure of their capsid, the coat protein of AMV is quite flexible and can form particles of different sizes and shapes, depending on the length of the encapsulated RNA. Both a 40 amino acid sequence derived from Rabies virus glycoprotein and a 47 amino acid sequence from the V3 loop of an HIV-1 MN isolate were expressed in recombinant virus particles (Yusibov *et al.*, 1997). Immunogenicity was demonstrated in mice that generated antisera capable of neutralizing 90% of rabies virus and up to 80% neutralization of an HIV-1 MN isolate.

Potato Virus X (PVX)
Potato virus X is composed of several thousand copies of a coat protein to which antigenic molecules can be displayed on. A 27 kDa marker protein was fused to the amino terminus of the 25kDa PVX cp without affecting

viral assembly (Cruz *et al.*, 1996). This system may rely on purification of the recombinant antigen, as the modified virus is prone to precipitation and is resistant to re-solubilization.

Plant Vaccines for Suppression of Autoimmune Disease

About 3-5% of the world population suffers from an autoimmune disease (Marrack *et al.*, 2001). Despite extensive research into the events that trigger autoimmunity and associated susceptibility factors, significant gaps in the knowledge of immunologic mechanisms leading to autoimmune diseases remain. The use of experimental animal models of autoimmune diseases and extensive gene mapping in families with a history of autoimmunity has helped to resolve some of the complex immune reactions involved in the loss of self-tolerance. Several hypotheses regarding autoimmune susceptibility have arisen from these studies that continue to gain support. Many individuals have increased susceptibility to autoimmunity because of haplotype differences in MHC genes that result in self-antigen presentation to the immune system. In immune mediated diabetes—IMD (also known as Type 1A, juvenile onset diabetes or insulin-dependent diabetes mellitus), variations in the HLA-DQ beta genes determine disease susceptibility, while additional alleles incur significant resistance to IMD (Todd *et al.*, 1987). There is also increasing evidence that people may be at risk of developing an autoimmune disease simply through exposure to certain pathogens, although this concept has not yet been clearly determined. Molecular mimicry, epitope spreading, and direct bystander activation by pathogen epitopes with similar structure or binding capacity to human proteins can lead to cross-reactive antibody or T cell receptor formation that can trigger an immune attack against self-tissues (Olson *et al.*, 2001). In IMD, 13 different viruses have been reported thus far that may cause or lead to diabetes in humans and various animal models (Jun and Yoon, 2001). However, recent reports demonstrate that in some cases, molecular mimicry by pathogen proteins may not directly lead to autoimmune disease and that additional mechanisms leading to perpetuation of immunopathogenesis must also be involved (Davies, 1997; Cainelli *et al.*, 2000; Benoist and Mathis, 2001; Schloot *et al.*, 2001).

Research into the etiology and molecular mechanisms of autoimmunity has revealed potential strategies for which the breakdown of self-tolerance leading to autoimmunity may be delayed or prevented. By shifting the intricate balance between T_H1 and T_H2 CD4$^+$ lymphocytes to bolster T_H2 activation and the concomitant release of "suppressor" cytokines, restoration of immune tolerance is possible. A major role of T_H2 cells is to down-regulate

Table 3. Suppression of autoimmune diseases by oral administration of autoantigens

Autoimmune Disease	Autoantigen*	Produced in Plant	Reference
Anti-phospholipid syndrome	Beta2-glycoprotein I^M	—	Blank et al., 1998
Autoimmune ear disease	Type II collagen cyanogen bromide peptide 11^M	—	Kim et al., 2001
Colitis (inflammatory bowel disease)	Haptenized colonic proteins^M Colon epithelial cell proteins^R	— —	Neurath et al., 1996 Dasgupta et al., 2001
Diabetes, type 1A	Insulin^M Glutamic Acid Decarboxylase^M Insulin-CTB^M, GAD-CTB^M	— Tobacco, Potato Potato	Zhang et al., 1991 Ma et al., 1997 Arakawa et al., 1998b, 1999
Multiple sclerosis	Myelin basic protein (MBP)^R MBP^R Myelin antigens^G MBP-CTB^R	— — — —	Higgins and Weiner, 1988 Bitar Whitacre, 1988 Brod et al., 1991 Sun et al., 2000
Myasthenia gravis	Acetylcholine receptor^R	—	Wang et al., 1993
Rheumatoid arthritis	Type II collagen^R	—	Zhang et al., 1990
Thyroiditis	Porcine thyroglobulin^M Porcine thyroid^H	— —	Peterson and Braley-Mullen, 1995 Lee et al., 1998
Uveitis	Retinal S antigen$^{R, H}$	—	Nussenblatt et al., 1990, 1996

*Superscript denotes animal tested in experimental autoimmune disease model:
M = mouse, R = Rat, G = guinea pig, H = human

T_H1 stimulation of cytotoxic macrophage and $CD8^+$ cytotoxic T lymphocyte (CTL) destruction of cells. Studies utilizing animal models of autoimmune diseases have shown that oral administration of autoantigens can induce T_H2 cell proliferation and cytokine secretion, inducing immunotolerance (Table 3). Since 100% suppression of insulitis and hyperglycemia has not yet been achieved in non-obese diabetic (NOD) mice, a mammalian model for autoimmune diabetes, further experiments will be required before IMD can be suppressed in humans.

Oral immunization with autoantigens poses several challenges to effective autoimmune suppression. First, large amounts of autoantigen are generally required. This constraint may be due to digestive degradation of much of the protein. Co-delivery of autoantigens with mucosal adjuvants has been shown to boost the tolerization response to fed autoantigens (Nagler-Anderson *et al.*, 1986; Lider *et al.*, 1989; Bergerot *et al.*, 1997; Hartmann *et al.*, 1997; Arakawa *et al.*, 1998b, 1999). Not all individuals may respond to the same autoantigen to the same degree. Autoimmune diseases can frequently be triggered by one or several different autoantigens, and certain antigens may trigger a lapse in tolerance at various stages throughout disease progression. Examples of autoimmune diseases propagated by a variety of autoantigens include Type 1 diabetes (pancreatic β-cell autoantigens—glutamic acid decarboxylase, insulin, IA-2, hsp 60, etc.) and systemic lupus erythematosus (nuclear antigens—dsDNA, histones, ribonucleoproteins, snRNP's, etc.) (Bach and Chatenoud, 2001; Hoffman, 2001). Prevention or suppression of these diseases may require co-delivery of several autoantigens given at precise times during development of the disease. Further, while it is relatively easy to identify people with autoimmune disease for tolerance/suppression therapy, a preventative approach will require screening individuals genetically at risk for developing an autoimmune response. If pathogens prove to be major cause of autoimmune disease induction, immunization against these pathogens may be necessary. Finally, several reports have suggested the possibility for disease induction when oral autoantigens are given (Genain *et al.*, 1996; Bellmann *et al.*, 1998; Blanas and Heath, 1999) leading to the possibility that accidental ingestion of autoantigens with adjuvants by non-autoimmune individuals could potentially lead to disease onset, although this possibility has not yet been confirmed.

One novel strategy currently used to overcome the previous obstacles is the production of autoantigens in transgenic plants. Large quantities of foreign protein can be produced in transformed plants—over 10 kg of protein per acre in some species, which can be synthesized at 1/10-1/50 the cost of production in transgenic mammalian, yeast or bacterial cell cultures (Kusnadi *et al.*, 1997; Gavilondo and Larrick, 2000). The majority of this cost is attributed to the process of purification, which can be eliminated if transformed plant tissues can be consumed directly. The first report of an autoantigen being generated by plants was in 1997, when the IMD pancreatic autoantigen glutamic acid decarboxylase (GAD) was synthesized in tobacco and potato at a level of 0.4% of the total soluble protein (Ma *et al.*, 1997). Although GAD expression was low, it was sufficient enough to suppress the onset of diabetes when fed to mice genetically destined to develop the disease. Additional studies revealed that low doses of autoantigens are better at inducing tolerance through the immunological mechanism of active suppression, whereas high autoantigen doses can incur immune "paralysis" (anergy) with little or no active suppression (Friedman and Weiner, 1994). To make autoantigens more immunogenic, gene fusions were made with the diabetic autoantigen insulin and the non-toxic B subunit of cholera toxin (CTB). CTB was shown to be a strong mucosal adjuvant and carrier molecule when co-delivered or chemically fused with antigens (Czerkinsky *et al.*, 1996). Chemically conjugated insulin-CTB fusion proteins were shown to be effective in diabetes suppression when fed to non-obese diabetic (NOD) mice using as little as one dose containing 500–5,000 times less autoantigen than unconjugated insulin given orally (Bergerot *et al.*, 1997). Edible plants were used to generate diabetes autoantigens genetically fused to CTB. The plant-synthesized insulin-CTB and GAD-CTB fusion proteins were found to be more effective in the reduction of diabetes symptoms than were insulin or GAD alone when fed to NOD mice (100-fold and 1,000 fold less required, respectively) (Arakawa *et al.*, 1998b, 1999) .

Additional methods currently under investigation for improvement of oral tolerance immunotherapy include the use of alternative enterocyte-binding proteins (work in progress in our laboratory). Another immunotolerization approach currently under investigation in our laboratory is to introduce genes encoding more than one autoantigen into a plant. Eating these will provide exposure to a broader spectrum of tolerizing autoantigens, which may be especially useful for the suppression of autoimmune diseases in which several self-proteins act as autoantigens. A proactive strategy to enhance autoimmune disease prevention could be to supply the population with mucosal vaccines that protect against pathogens suspected of eliciting self-reactive antibodies. This approach would require widely available and inexpensive subunit vaccines. Plant vaccines would be ideal for mass immunization programs. Immunogenic epitopes would be selected that provide protective immunity against the offending viruses or bacterial pathogens, but that were not cross-reactive with antibodies to self-proteins. However, prophylactic vaccination in humans may have to wait until more is known about the mechanisms of autoimmune disease immunopathogenesis and the safety of these plant-based mucosal vaccines is determined.

The Future of Plant-Based Vaccines

To provide immune protection against infectious and autoimmune diseases to the anticipated 10 billion people who will occupy our planet by 2050 will require development of simple, easy to prepare and deliver immunization against multiple infectious and autoimmune diseases (Yu and Langridge, 2001). Transgenic plants can produce foreign proteins almost 24 hours a day and could therefore be one of the most inexpensive sources of antigen protein and potentially one of the cheapest sources of vaccine proteins. Manipulation of antigen encoding genes at both the transcriptional and translational levels may provide increased levels of recombinant protein antigens in transformed plants. The use of constitutively expressed promoters, adjustment of codon usage to favor plant gene expression and removal of intron sequences may favor higher levels of antigen and autoantigen expression in plants (Mor *et al.*, 1998; Ma and Vine, 1999). However, there may be additional limitations to how much foreign protein can be synthesized or accumulated in transformed plant cells. Therefore, solutions to increased production of antigen proteins in plants may include improvements in the targeting of antigen molecules to enterocytes or to lymphocytes of the mucosal immune system, which may reduce the levels of plant-produced antigen proteins required for successful oral immunization (Arakawa *et al.*, 1998b).

Plant synthesized cholera toxin B subunits are effective carrier molecules for targeting antigens to receptors in the intestinal epithelial cell membrane (Arakawa, 1998a). Cholera toxin A subunits that have been genetically altered to eliminate toxicity while retaining strong adjuvant activity may also be employed as homologous carriers to further enhance the mucosal immune response. Antigen variability in individual transformed plants could be reduced through minimal processing of the plant material. Lyophilization and compression of ground plant tissues into tablets or capsules could provide consistent amounts of autoantigen for therapeutic treatment. Foreign proteins are post-translationally modified in plant cells following synthesis in a way similar to that found in animal cells (McGarvey *et al.*, 1995). Correct protein folding, glycosylation, and proteolysis may greatly affect antigen stability and immunogenicity. Small but significant variations may occur between post-translational modification in plant and animal cells. These differences will underscore continued efforts to determine the structure and function of plant-produced antigen proteins. To assure public acceptance, education concerning the biological and environmental safety of plant delivered vaccines will be essential (Nordlee *et al.*, 1996; Danner, 1997). For a review of the potentials and hazards of genetically modified plants, see McHughen,

2000. Current research efforts have clearly demonstrated the feasibility of plant based vaccines for protective mucosal immunization against infectious and autoimmune diseases. Increased understanding of the mechanisms regulating mucosal immunity will improve antigen and autoantigen targeting to enterocytes and lymphocytes of the mucosal immune system, which may be sufficient to raise vaccine efficacy to levels where edible vaccines will begin to replace parenteral immunization in animals and humans.

References

Arakawa, T., Chong, D.K., and Langridge, W.H. 1998a. Efficacy of a food plant-based oral cholera toxin B subunit vaccine. Nat. Biotechnol. 16: 292-297.

Arakawa, T., Chong, D.K., Merritt, J.L., and Langridge, W.H. 1997. Expression of cholera toxin B subunit oligomers in transgenic potato plants. Transgenic Res. 6: 403-413.

Arakawa, T., Yu, J., Chong, D.K., Hough, J., Engen, P.C., and Langridge, W.H. 1998b. A plant-based cholera toxin B subunit-insulin fusion protein protects against the development of autoimmune diabetes. Nat. Biotechnol. 16: 934-938.

Arakawa, T., Yu, J., Chong, D.K.X., Hough, J., Engen, P.C., and Langridge, W.H.R. 1999. Suppression of autoimmune diabetes by a plant-delivered cholera toxin B subunit-human glutamate decarboxylate fusion protein. Transgenics. 3: 51-60.

Bach, J.F., and Chatenoud, L. 2001. Tolerance to islet autoantigens in type 1 diabetes. Annu. Rev. Immunol. 19: 131-161.

Ball, J.M., Hardy, M.K., Conner, M.E., and Opekun, A.A. 1996. Recombinant Norwalk virus-like particles as an oral vaccine. Arch. Virol. Suppl. 12: 243-249.

Belanger, H., Fleysh, N., Cox, S., Bartman, G., Deka, D., Trudel, M., Koprowski, H., and Yusibov, V. 2000. Human respiratory syncytial virus vaccine antigen produced in plants. FASEB J. 14: 2323-2328.

Bellmann, K., Kolb, H., Rastegar, S., Jee, P., and Scott, F.W. 1998. Potential risk of oral insulin with adjuvant for the prevention of Type I diabetes: A protocol effective in NOD mice may exacerbate disease in BB rats. Diabetologia. 41: 844-847.

Benoist, C., and Mathis, D. 2001. Autoimmunity provoked by infection: How good is the case for T cell epitope mimicry? Nat. Immunol. 2: 797-801.

Bergerot, I., Ploix, C., Petersen, J., Moulin, V., Rask, C., Fabien, N., Lindblad, M., Mayer, A., Czerkinsky, C., Holmgren, J., and Thivolet, C. 1997. A cholera toxoid-insulin conjugate as an oral vaccine against spontaneous autoimmune diabetes. Proc. Natl. Acad. Sci. USA. 94: 4610-4614.

Bitar, D.M., and Whitacre, C.C. 1988. Suppression of experimental autoimmune encephalomyelitis by the oral administration of myelin basic protein. Cell Immunol. 112: 364-370.

Blanas, E., and Heath, W.R. 1999. Oral administration of antigen can lead to the onset of autoimmune disease. Int. Rev. Immunol. 18: 217-228.

Blank, M., George, J., Barak, V., Tincani, A., Koike, T., and Shoenfeld, Y. 1998. Oral tolerance to low dose beta 2-glycoprotein I: Immunomodulation of experimental antiphospholipid syndrome. J. Immunol. 161: 5303-5312.

Brennan, F.R., Gilleland, L.B., Staczek, J., Bendig, M.M., Hamilton, W.D., and Gilleland, H.E., Jr. 1999a. A chimaeric plant virus vaccine protects mice against a bacterial infection. Microbiology. 145: 2061-2067.

Brennan, F.R., Jones, T.D., Longstaff, M., Chapman, S., Bellaby, T., Smith, H., Xu, F., Hamilton, W.D., and Flock, J.I. 1999b. Immunogenicity of peptides derived from a fibronectin-binding protein of *S. aureus* expressed on two different plant viruses. Vaccine. 17: 1846-1857.

Brod, S.A., al Sabbagh, A., Sobel, R.A., Hafler, D.A., and Weiner, H.L. 1991. Suppression of experimental autoimmune encephalomyelitis by oral administration of myelin antigens: IV. Suppression of chronic relapsing disease in the Lewis rat and strain 13 guinea pig. Ann. Neur. 29: 615-622.

Cabral, G.A., Marciano-Cabral, F., Funk, G.A., Sanchez, Y., Hollinger, F.B., Melnick, J.L., and Dreesman GR. 1978. Cellular and humoral immunity in guinea pigs to two major polypeptides derived from hepatitis B surface antigen. J. Gen. Virol. 38: 339-350.

Cainelli, F., Manzaroli, D., Renzini, C., Casali, F., Concia, E., and Vento, S. 2000. Coxsackie B virus-induced autoimmunity to GAD does not lead to type 1 diabetes. Diabetes Care. 23: 1021-1022.

Carrillo, C., Wigdorovitz, A., Oliveros, J.C., Zamorano, P.I., Sadir, A.M., Gomez, N., Salinas, J., Escribano, J.M., and Borca, M.V. 1998. Protective immune response to foot-and-mouth disease virus with VP1 expressed in transgenic plants. J. Virol. 72: 1688-1690.

Castanon, S., Marin, M.S., Martin-Alonso, J.M., Boga, J.A., Casais, R., Humara, J.M., Ordas, R.K., and Parra, F. 1999. Immunization with potato plants expressing VP 60 protein protects against rabbit hemorrhagic disease virus. J. Virol. 73: 4452-4455.

Castro, G.A., and Arntzen, C.J. 1993. Immunophysiology of the gut: A research frontier for integrative studies of the common mucosal immune system. Am. J. Physiol. 265: 599-610.

Chase, A. 1982. Magic Shots. William Morrow and Company Inc., New York, NY. p. 234-235.

Clements, J.D., Hartzog, N.M., and Lyon, F.L. 1988. Adjuvant activity of *Escherichia coli* heat-labile enterotoxin and effect on the induction of oral tolerance in mice to unrelated protein antigens. Vaccine. 6: 269-277.

Cruz, S.S., Chapman, S., Roberts, A.G., Roberts, I.M., and Prior, D.A. 1996. Assembly and movement of a plant virus carrying a green fluorescent protein overcoat. Proc. Natl. Acad. Sci. USA. 93: 6286-6290.

Curtiss, R., and Cardineau, G.A. 1990. World Intellectual Property Organization PCT/US89/03799.

Czerkinsky, C., Sun, J.B., Lebens, M., Li, B.L., Rask, C., Lindblad, M., and Holmgren, J. 1996. Cholera toxin B subunit as transmucosal carrier-delivery and immunomodulating system for induction of antiinfectious and antipathological immunity. Ann. NY Acad. Sci. 778: 185-193.

Dalsgaard, K., Uttenthal, A., Jones, T.D., Xu, F., Merryweather, A., Hamilton, W.D., Langeveld, J.P., Boshuizen, R.S., Kamstrup, S., Lomonossoff, G.P., Porta, C., Vela, C., Casal, J.I., Meloen, R.H., and Rodgers, P.B. 1997. Plant-derived vaccine protects target animals against a viral disease. Nat. Biotechnol. 15: 248-252.

Daniell, H., Lee, S.B., Panchal, T., and Wiebe, P.O. 2001. Expression of the native cholera toxin B subunit gene and assembly as functional oligomers in transgenic tobacco chloroplasts. J. Mol. Biol. 311: 1001-1009.

Daniell, H., Vivekananda, J., Nielsen, B.L., Ye, G.N., Tewari, K.K., and Sanford, J.C. 1990. Transient foreign gene expression in chloroplasts of cultured tobacco cells after biolistic delivery of chloroplast vectors. Proc. Natl. Acad. Sci. USA. 87: 88-92.

Danner, K. 1997. Acceptability of bio-engineered vaccines. Comp. Immunol. Microbiol. Infect. Dis. 20: 3-12.

Dasgupta, A., Ramaswamy, K., Giraldo, J., Taniguchi, M., Amenta, P.S., and Das, K.M. 2001. Colon epithelial cellular protein induces oral tolerance in the experimental model of colitis by trinitrobenzene sulfonic acid. J. Lab Clin. Med. 138: 257-269.

Davey, M.R., Rech, E.L., and Mulligan, B.J. 1989. Direct DNA transfer to plant cells. Plant Mol. Biol. 13: 273-285.

Davies, J.M. 1997. Molecular mimicry: Can epitope mimicry induce autoimmune disease? Immunol. Cell Biol. 75: 113-126.

de Costa, B., Moar, W., Lee, S.B., Miller, M., and Daniell, H. 2001. Overexpression of the Bt *cry2Aa2* operon in chloroplasts leads to formation of insecticidal crystals. Nat. Biotechnol. 19: 71-74.

Dickinson, B.L. and Clements, J.D. 1996. Use of *Escherichia coli* heat labile enterotoxin as an oral adjuvant. In: Mucosal Vaccines. H. Kiyono, P.L. Orga, and J.R. McGhee, eds. Academic. San Diego, p. 73-101.

Douglas, C.J., Staneloni, R.J., Rubin, R.A., and Nester, E.W. 1985. Identification and genetic analysis of an *Agrobacterium tumefaciens* chromosomal virulence region. J. Bacteriol. 161: 850-860.

Fernandez-Fernandez, M.R., Martinez-Torrecuadrada, J.L., Casal, J.I., and Garcia, J.A. 1998. Development of an antigen presentation system based on plum pox potyvirus. FEBS Lett. 427: 229-235.

Fernandez-Fernandez, M.R., Mourino, M., Rivera, J., Rodriguez, F., Plana-Duran, J., and Garcia, J.A. 2001. Protection of rabbits against rabbit hemorrhagic disease virus by immunization with the VP60 protein expressed in plants with a potyvirus-based vector. Virology. 280: 283-291.

Friedman, A., and Weiner, H.L. 1994. Induction of anergy or active suppression following oral tolerance is determined by antigen dosage. Proc. Natl. Acad. Sci. USA. 91: 6688-6692.

Gavilondo, J.V. and Larrick, J.W. 2000. Antibody engineering at the millennium. Biotechniques. 29: 128-138.

Genain, C.P., Abel, K., Belmar, N., Villinger, F., Rosenberg, D.P., Linington, C., Raine, C.S., and Hauser, S.L. 1996. Late complications of immune deviation therapy in a nonhuman primate. Science. 274: 2054-2057.

Gil, F., Brun, A., Wigdorovitz, A., Catala, R., Martinez-Torrecuadrada, J.L., Casal, I., Salinas, J., Borca, M.V., and Escribano, J.M. 2001. High-yield expression of a viral peptide vaccine in transgenic plants. FEBS Lett. 488: 13-17.

Gomez, N., Carrillo, C., Salinas, J., Parra, F., Borca, M.V., and Escribano, J.M. 1998. Expression of immunogenic glycoprotein S polypeptides from transmissible gastroenteritis coronavirus in transgenic plants. Virology. 249: 352-358.

Gomez, N., Wigdorovitz, A., Castanon, S., Gil, F., Ordas, R., Borca, M.V., and Escribano, J.M. 2000. Oral immunogenicity of the plant derived spike protein from swine-transmissible gastroenteritis coronavirus. Arch. Virol. 145: 1725-1732.

Haq, T.A., Mason, H.S., Clements, J.D., and Arntzen, C.J. 1995. Oral immunization with a recombinant bacterial antigen produced in transgenic plants. Science. 268: 714-716.

Hartmann, B., Bellmann, K., Ghiea, I., Kleemann, R., and Kolb, H. 1997. Oral insulin for diabetes prevention in NOD mice: Potentiation by enhancing Th2 cytokine expression in the gut through bacterial adjuvant. Diabetologia. 40: 902-909.

Higgins, P.J., and Weiner, H.L. 1988. Suppression of experimental autoimmune encephalomyelitis by oral administration of myelin basic protein and its fragments. J. Immunol. 140: 440-445.

Hoffman, R.W. 2001. T cells in the pathogenesis of systemic lupus erythematosus. Front. Biosci. 6: 1369-1378.

Holmgren, J., Lycke, N., and Czerkinsky, C. 1993. Cholera toxin and cholera B subunit as oral-mucosal adjuvant and antigen vector systems. Vaccine. 11: 1179-1184.

Huang, Z., Dry, I., Webster, D., Strugnell, R., and Wesselingh, S. 2001. Plant-derived measles virus hemagglutinin protein induces neutralizing antibodies in mice. Vaccine. 19: 2163-2171.

Joelson, T., Akerblom, L., Oxelfelt, P., Strandberg, B., Tomenius, K., and Morris, T.J. 1997. Presentation of a foreign peptide on the surface of tomato bushy stunt virus. J. Gen. Virol. 78: 1213-1217.

Joos, H., Inze, D., Caplan, A., Sormann, M., Van Montagu, M., and Schell, J. 1983. Genetic analysis of T-DNA transcripts in nopaline crown galls. Cell. 32: 1057-1067.

Jun, H.S., and Yoon, J.W. 2001. The role of viruses in type I diabetes: Two distinct cellular and molecular pathogenic mechanisms of virus-induced diabetes in animals. Diabetologia. 44: 271-285.

Kapusta, J., Modelska, A., Figlerowicz, M., Pniewski, T., Letellier, M., Lisowa, O., Yusibov, V., Koprowski, H., Plucienniczak, A., and Legocki, A.B. 1999. A plant-derived edible vaccine against hepatitis B virus. FASEB J. 13: 1796-1799.

Kim, N., Cheng, K.C., Kwon, S.S., Mora, R., Barbieri, M., and Yoo, T.J. 2001. Oral administration of collagen conjugated with cholera toxin induces tolerance to type II collagen and suppresses chondritis in an animal model of autoimmune ear disease. Ann. Otol. Rhinol. Laryngol. 110: 646-654.

Klein, T.M., Wolf, E.D., Wu, R., and Sanford, J.C. 1987. High-velocity microprojectiles for delivering nucleic acids into living cells. Nature. 327: 70-73.

Koo, M., Bendahmane, M., Lettieri, G.A., Paoletti, A.D., Lane, T.E., Fitchen, J.H., Buchmeier, M.J., and Beachy, R.N. 1999. Protective immunity against murine hepatitis virus (MHV) induced by intranasal or subcutaneous administration of hybrids of tobacco mosaic virus that carries an MHV epitope. Proc. Natl. Acad. Sci. USA. 96: 7774-7779.

Kusnadi, A.R., Nikolov, Z.L., and Howard, J.A. 1997. Production of recombinant proteins in transgenic plants: Practical considerations. Biotechnol. Bioeng. 56: 473-484.

Laude, H., Rasschaert, D., Delmas, B., Godet, M., Gelfi, J., and Charley, B. 1990. Molecular biology of transmissible gastroenteritis virus. Vet. Microbiol. 23: 147-154.

Lauterslager, T.G., Florack, D.E., van der Wal, T.J., Molthoff, J.W., Langeveld, J.P., Bosch, D., Boersma, W.J., and Hilgers, L.A. 2001. Oral immunisation of naive and primed animals with transgenic potato tubers expressing LT-B. Vaccine. 19: 2749-2755.

Lee, R.W., Strommer, J., Hodgins, D., Shewen, P.E., Niu, Y., and Lo, R.Y. 2001. Towards development of an edible vaccine against bovine pneumonic pasteurellosis using transgenic white clover expressing a *Mannheimia haemolytica* A1 leukotoxin 50 fusion protein. Infect. Immun. 69: 5786-5793.

Lee, S., Scherberg, N., and DeGroot, L.J. 1998. Induction of oral tolerance in human autoimmune thyroid disease. Thyroid. 8: 229-234.

Lider, O., Santos, L.M., Lee, C.S., Higgins, P.J., and Weiner, H.L. 1989. Suppression of experimental autoimmune encephalomyelitis by oral administration of myelin basic protein. J. Immunol. 142: 748-752.

Ma, J.K., Hikmat, B.Y., Wycoff, K., Vine, N.D., Chargelegue, D., Yu, L., Hein, M.B., and Lehner, T. 1998. Characterization of a recombinant plant monoclonal secretory antibody and preventive immunotherapy in humans. Nat. Med. 4: 601-606.

Ma, J.K. and Vine, N.D. 1999. Plant expression systems for the production of vaccines. Curr. Top. Microbiol. Immunol. 236: 275-292.

Ma, S.W., Zhao, D.L., Yin, Z.Q., Mukherjee, R., Singh, B., Qin, H.Y., Stiller, C.R., and Jevnikar, A.M. 1997. Transgenic plants expressing autoantigens fed to mice to induce oral immune tolerance. Nat. Med. 3: 793-796.

Mason, H.S., Ball, J.M., Shi, J.J., Jiang, X., Estes, M.K., and Arntzen, C.J. 1996. Expression of Norwalk virus capsid protein in transgenic tobacco and potato and its oral immunogenicity in mice. Proc. Natl. Acad. Sci. USA. 93: 5335-5340.

Mason, H.S., Haq, T.A., Clements, J.D., and Arntzen, C.J. 1998. Edible vaccine protects mice against *Escherichia coli* heat-labile enterotoxin (LT): Potatoes expressing a synthetic gene. Vaccine. 16: 1336-1343.

Mason, H.S., Lam, D.M., and Arntzen, C.J. 1992. Expression of hepatitis B surface antigen in transgenic plants. Proc. Natl. Acad. Sci. USA. 89: 11745-11749.

Marrack, P., Kappler, J., and Kotzin, B.L. 2001. Autoimmune disease: Why and where it occurs. Nat. Med. 7: 899-905.

McBride, K.E., Svab, Z., Schaaf, D.J., Hogan, P.S., Stalker, D.M., and Maliga, P. 1995. Amplification of a chimeric Bacillus gene in chloroplasts leads to an extraordinary level of an insecticidal protein in tobacco. Bio/Technology. 13: 362-365.

McCormick, A.A., Kumagai, M.H., Hanley, K., Turpen, T.H., Hakim, I., Grill, L.K., Tuse, D., Levy, S., and Levy, R. 1999. Rapid production of specific vaccines for lymphoma by expression of the tumor-derived single-chain Fv epitopes in tobacco plants. Proc. Natl. Acad. Sci. USA. 96: 703-708.

McGhee, J.R., and Kiyono, H. 1993. New perspectives in vaccine development: Mucosal immunity to infections. Infect. Agents. Dis. 2: 55-73.

McGarvey, P.B., Hammond, J., Dienelt, M.M., Hooper, D.C., Fu, Z.F., Dietzschold, B., Koprowski, H., Michaels, F.H., 1995. Expression of the rabies virus glycoprotein in transgenic tomatoes. Bio/Technology. 13: 1484-1487.

McHughen, A. 2000. Pandora's Picnic Basket: The Potential and Hazards of Genetically Modified Foods. Oxford University Press, New York.

McLain, L., Durrani, Z., Wisniewski, L.A., Porta, C., Lomonossoff, G.P., and Dimmock, N.J. 1996. Stimulation of neutralizing antibodies to human immunodeficiency virus type 1 in three strains of mice immunized with a 22 amino acid peptide of gp41 expressed on the surface of a plant virus. Vaccine. 14: 799-810.

Mitchell, V.S., Philipose, N.M., and Sanford J.P. 1993. The children's vaccine initiative. National Academy Press. 36: 57-59.

Moffat, A.S. 1995. Exploring transgenic plants as a new vaccine source. Science. 268: 658-660.

Modelska, A., Dietzschold, B., Sleysh, N., Fu, Z.F., Steplewski, K., Hooper, D.C., Koprowski, H., and Yusibov, V. 1998. Immunization against rabies with plant-derived antigen. Proc. Natl. Acad. Sci. USA. 95: 2481-2485.

Morse, S.S., 1995. Factors in the emergence of infectious diseases. Emerg. Infect. Dis. 1: 7-15. http://www.cdc.gov/ncidod/eid/vol1no1/downmors.htm

Mor, T.S., Gomez-Lim, M.A., and Palmer, K.E. 1998. Perspective: Edible vaccines—A concept coming of age. Trends Microbiol. 6: 449-453.

Nagler-Anderson, C., Bober, L.A., Robinson, M.E., Siskind, G.W., and Thorbecke, G.J. 1986. Suppression of type II collagen-induced arthritis by intragastric administration of soluble type II collagen. Proc. Natl. Acad. Sci. USA. 83: 7443-7446.

Nemchinov, L.G., Liang, T.J., Rifaat, M.M., Mazyad, H.M., Hadidi, A., and Keith, J.M. 2000. Development of a plant-derived subunit vaccine candidate against hepatitis C virus. Arch. Virol. 145: 2557-2573.

Newell, C.A. 2000. Plant transformation technology. Developments and applications. Mol. Biotechnol. 16: 53-65.

Nester, E.W., Gordon, M.P., Amasino, R.M., and Yanofsky, M.F. 1984. Crown gall: A molecular and physiological analysis. Annu. Rev. Plant Physiol. 35: 387.

Neurath, M.F., Fuss, I., Kelsall, B.L., Presky, D.H., Waegell, W., and Strober, W. 1996. Experimental granulomatous colitis in mice is abrogated by induction of TGF-beta-mediated oral tolerance. J. Exp. Med. 183: 2605-2616.

Nordlee, J.A., Taylor, S.L., Townsend, J.A., Thomas, L.A., and Bush, R.K. 1996. Identification of a Brazil-nut allergen in transgenic soybeans. N. Engl. J. Med. 334: 688-692.

Nussenblatt, R.B., Caspi, R.R., Mahdi, R., Chan, C.C., Roberge, F., Lider, O., and Weiner, H.L. 1990. Inhibition of S-antigen induced experimental autoimmune uveoretinitis by oral induction of tolerance with S-antigen. J. Immunol. 144: 1689-1695.

Nussenblatt, R.B., Whitcup, S.M., de Smet, M.D., Caspi, R.R., Kozhich, A.T., Weiner, H.L., Vistica, B., and Gery, I. 1996. Intraocular inflammatory disease (uveitis) and the use of oral tolerance: A status report. Ann. NY Acad. Sci. 778: 325-337.

Olson, J.K., Croxford, J.L., and Miller, S.D. 2001. Virus-induced autoimmunity: Potential role of viruses in initiation, perpetuation, and progression of T-cell-mediated autoimmune disease. Viral Immunol. 14: 227-250.

Peralta, E.G., and Ream, L.W. 1985. T-DNA border sequences required for crown gall tumorigenesis. Proc. Natl. Acad. Sci. USA. 82: 5112-5116.

Peterson, K.E., and Braley-Mullen, H. 1995. Suppression of murine experimental autoimmune thyroiditis by oral administration of porcine thyroglobulin. Cell Immunol. 166: 123-130.

Porta, C., Spall, V.E., Loveland, J., Johnson, J.E., Barker, P.J., and Lomonossoff, G.P. 1994. Development of cowpea mosaic virus as a high-yielding system for the presentation of foreign peptides. Virology. 202: 949-955.

Richter, L., and Kipp, P.B. 1999. Transgenic plants as edible vaccines. Curr. Top. Microbiol. Immunol. 240: 159-176.

Richter, L.J., Thanavala, Y., Arntzen, C.J., and Mason, H.S. 2000. Production of hepatitis B surface antigen in transgenic plants for oral immunization. Nat. Biotechnol. 18: 1167-1171.

Ruf, S., Hermann, M., Berger, I.J., Carrer, H., and Bock, R. 2001. Stable genetic transformation of tomato plastids and expression of a foreign protein in fruit. Nat. Biotechnol. 19: 870-875.

Sandhu, J.S., Krasnyanski, S.F., Domier, L.L., Korban, S.S., Osadjan, M.D., and Buetow, D.E. 2000. Oral immunization of mice with transgenic tomato fruit expressing respiratory syncytial virus-F protein induces a systemic immune response. Transgenic Res. 9: 127-135.

Sandhu, J.S., Osadjan, M.D., Krasnyanski, S.F., Domier, L.L., Korban, S.S., and Buetow, D.E. 1999. Enhanced expression of the human respiratory syncytial virus-F gene in apple leaf protoplasts. Plant Cell Rep. 18: 394-397.

Satcher, D. 1995. Emerging infections: Getting ahead of the curve. Emerg. Infect. Dis. 1: 1-6. http://www.cdc.gov/ncidod/eid/vol1no1/downsatc.htm

Schell, J.S. 1987. Transgenic plants as tools to study the molecular organization of plant genes. Science. 237: 1176.

Schloot, N.C., Willemen, S.J., Duinkerken, G., Drijfhout, J.W., de Vries, R.R., and Roep, B.O. 2001. Molecular mimicry in type 1 diabetes mellitus revisited: T-cell clones to GAD65 peptides with sequence homology to Coxsackie or proinsulin peptides do not crossreact with homologous counterpart. Hum. Immunol. 62: 299-309.

Shahin, E.A., and Simpson, R.B. 1986. Gene transfer system for potato. HortScience. 21: 1199-1201.

Stachel, S.E., and Nester, E.W. 1986. The genetic and transcriptional organization of the vir region of the A6 Ti plasmid of *Agrobacterium tumefaciens*. EMBO J. 5: 1445-1454.

Stachel, S.E., Timmerman, B., and Zambryski, P. 1987. Activation of *Agrobacterium tumefaciens* vir gene expression generates multiple single-stranded T-strand molecules from the pTiA6 T-region: requirement for 5' virD gene products. EMBO J. 6: 857-863.

Stoger, E., Vaquero, C., Torres, E., Sack, M., Nicholson, L., Drossard, J., Williams, S., Keen, D., Perrin, Y., Christou, P., and Fischer, R. 2000. Cereal crops as viable production and storage systems for pharmaceutical scFv antibodies. Plant Mol. *Biol*. 42: 583-590.

Streatfield, S.J., Jilka, J.M., Hood, E.E., Turner, D.D., Bailey, M.R., Mayor, J.M., Woodard, S.L., Beifuss, K.K., Horn, M.E., Delaney, D.E., Tizard, I.R., and Howard, J.A. 2001. Plant-based vaccines: Unique advantages. Vaccine. 19: 2742-2748.

Sugiyama, Y., Hamamoto, H., Takemoto, S., Watanabe, Y., and Okada, Y. 1995. Systemic production of foreign peptides on the particle surface of tobacco mosaic virus. FEBS Lett. 359: 247-250.

Sun, J.B., Xiao, B.G., Lindblad, M., Li, B.L., Link, H., Czerkinsky, C., and Holmgren, J. 2000. Oral administration of cholera toxin B subunit conjugated to myelin basic protein protects against experimental autoimmune encephalomyelitis by inducing transforming growth factor-beta-secreting cells and suppressing chemokine expression. Int. Immunol. 12: 1449-1457.

Tackaberry, E.S., Dudani, A.K., Prior, F., Tocchi, M., Sardana, R., Altosaar, I., and Ganz, P.R. 1999. Development of biopharmaceuticals in plant expression systems: Cloning, expression and immunological reactivity of human cytomegalovirus glycoprotein B (UL55) in seeds of transgenic tobacco. Vaccine. 17: 3020-3029.

Tacket, C.O., Mason, H.S., Losonsky, G., Clements, J.D., Levine, M.M., and Arntzen, C.J. 1998. Immunogenicity in humans of a recombinant bacterial antigen delivered in a transgenic potato. Nat. Med. 4: 607-609.

Tacket, C.O., Mason, H.S., Losonsky, G., Estes, M.K., Levine, M.M., and Arntzen, C.J. 2000. Human immune responses to a novel norwalk virus vaccine delivered in transgenic potatoes. J. Infect. Dis. 182: 302-305.

Thanavala, Y., Yang, Y.F., Lyons, P., Mason, H.S., and Arntzen, C. 1995. Immunogenicity of transgenic plant-derived hepatitis B surface antigen. Proc. Natl. Acad. Sci. USA. 92: 3358-3361.

Todd, J.A., Bell, J.I., and McDevitt, H.O. 1987. HLA-DQ beta gene contributes to susceptibility and resistance to insulin-dependent diabetes mellitus. Nature. 329: 599-604.

Tuboly, T., Yu, W., Bailey, A., Degrandis, S., Du, S., Erickson, L., and Nagy, E. 2000. Immunogenicity of porcine transmissible gastroenteritis virus spike protein expressed in plants. Vaccine. 18: 2023-2028.

Turpen, T.H., Reinl, S.J., Charoenvit, Y., Hoffman, S.L., Fallarme, V., and Grill, L.K. 1995. Malarial epitopes expressed on the surface of recombinant tobacco mosaic virus. Bio/Technology. 13: 53-57.

Usha, R., Rohll, J.B., Spall, V.E., Shanks, M., Maule, A.J., Johnson, J.E., and Lomonossoff, G.P. 1993. Expression of an animal virus antigenic site on the surface of a plant virus particle. Virology. 197: 366-374.

Vaquero, C., Sack, M., Chandler, J., Drossard, J., Schuster, F., Monecke, M., Schillberg, S., and Fischer, R. 1999. Transient expression of a tumor-specific single-chain fragment and a chimeric antibody in tobacco leaves. Proc. Natl. Acad. Sci. USA. 96: 11128-11133.

Verch, T., Yusibov, V., and Koprowski, H. 1998. Expression and assembly of a full-length monoclonal antibody in plants using a plant virus vector. J. Immunol. Methods. 220: 69-75.

Wang, K., Herrera-Estrella, L., Van Montagu, M., and Zambryski, P. 1984. Right 25 bp terminus sequence of the nopaline T-DNA is essential for and determines direction of DNA transfer from agrobacterium to the plant genome. Cell. 38: 455-462.

Wang, Z.Y., Qiao, J., and Link, H. 1993. Suppression of experimental autoimmune myasthenia gravis by oral administration of acetylcholine receptor. J. Neuroimmunol. 44: 209-214.

Wigdorovitz, A., Carrillo, C., Dus Santos, M.J., Trono, K., Peralta, A., Gomez, M.C., Rios, R.D., Franzone, P.M., Sadir, A.M., Escribano, J.M., and Borca, M.V. 1999a. Induction of a protective antibody response to foot and mouth disease virus in mice following oral or parenteral immunization with alfalfa transgenic plants expressing the viral structural protein VP1. Virology. 255: 347-353.

Wigdorovitz, A., Perez Filgueira, D.M., Robertson, N., Carrillo, C., Sadir, A.M., Morris, T.J., and Borca, M.V. 1999b. Protection of mice against challenge with foot and mouth disease virus (FMDV) by immunization with foliar extracts from plants infected with recombinant tobacco mosaic virus expressing the FMDV structural protein VP1. Virology. 264: 85-91.

Willmitzer, L., Simons, G., and Schell, J. 1982. The TL-DNA in octopine crown-gall tumours codes for seven well defined polyadenylated transcripts. EMBO J. 1: 139-146.

Yamamoto, A., Iwahashi, M., Yanofsky, M.F., Nester, E.W., Takebe, I., and Machida, Y. 1987. The promoter proximal region in the *virD* locus of *Agrobacterium tumefaciens* is necessary for the plant-inducible circularization of T-DNA. Mol. Gen. Genet. 206: 174-177.

Yanofsky, M.F., Porter, S.G., Young, C., Albright, L.M., Gordon, M.P., and Nester, E.W. 1986. The *virD* operon of *Agrobacterium tumefaciens* encodes a site-specific endonuclease. Cell. 47: 471-477.

Yu, J., and Langridge, W.H.R. 2001. A plant-based multicomponent vaccine protects mice from enteric diseases. Nat. Biotechnol. 19: 548-552.

Yusibov, V., Modelska, A., Steplewski, K., Agadjanyan, M., Weiner, D., Hooper, D.C., and Koprowski, H. 1997. Antigens produced in plants by infection with chimeric plant viruses immunize against rabies virus and HIV-1. Proc. Natl. Acad. Sci. USA. 94: 5784-5788.

Zambryski, P. C. 1992. Chronicles from the *Agrobacterium*-plant cell DNA transfer story. Ann. Rev. Plant Physiol. Plant Mol. Biol. 43: 465-490.

Zambryski, P., Depicker, A., Kruger, K., and Goodman, H.M. 1982. Tumor induction by *Agrobacterium tumefaciens*: analysis of the boundaries of T-DNA. J. Mol. Appl. Genet. 1: 361-370.

Zeitlin, L., Olmsted, S.S., Moench, T.R., Co, M.S., Martinell, B.J., Paradkar, V.M., Russell, D.R., Queen, C., Cone, R.A., and Whaley, K.J. 1998. A humanized monoclonal antibody produced in transgenic plants for immunoprotection of the vagina against genital herpes. Nat. Biotechnol. 16: 1361-1364.

Zhang, Z.J., Davidson, L., Eisenbarth, G., and Weiner, H.L. 1991. Suppression of diabetes in nonobese diabetic mice by oral administration of porcine insulin. Proc. Natl. Acad. Sci. USA. 88: 10252-10256.

Zhang, Z.Y., Lee, C.S., Lider, O., and Weiner, H.L. 1990. Suppression of adjuvant arthritis in Lewis rats by oral administration of type II collagen. J. Immunol. 145: 2489-2493.

From: *Transgenic Plants: Current Innovations and Future Trends*
Edited by: C. Neal Stewart, Jr.

Chapter 11

Genomics Using Transgenic Plants

Mentewab Ayalew

Abstract

Plant transformation technology is an established toolbox for the study of plant genomics; particularly so for model species such as *Arabidopsis thaliana* and rice. Progress in transformation methods for these species has allowed the use of several strategies for gene identification and characterization. T-DNA and transposons have been used for random insertional mutagenesis with gene constructs designed to create gene knockouts, promoter traps, enhancer traps, and activation tags. RNA silencing is widely used to create down-regulated mutants. The efficiency gained with a better understanding of the mechanisms of post-transcriptional gene silencing allows the use of the method for large-scale approach. Targeted gene inactivation by homologous recombination is an emerging method for plant transformation. These genomics methods already available to plant molecular biologists or under development are described.

Introduction

Plant transformation technology is not only crucial for the production of transgenic crops, but is also a powerful tool for the study of various aspects of plant genomics: the identification of genes, the study of their expression patterns, and their interactions to elucidate function. The possibility of generating transgenic plants in a variety of plant species has been of importance in the advance of several major areas of research. Analysis of gene regulation was facilitated by the use of reporter gene constructs that allow easy assessment of the transcriptional activity conferred by cis-acting elements. The functional analysis of physiological or developmental processes was aided by the possibility of overexpressing particular genes by a strong promoter. Finally it allowed the establishment of gene tagging systems to isolate genes only characterized by a mutant phenotype or by a characteristic expression pattern.

Plant genomics is entering a new phase with the rapid accumulation of DNA sequence data in public and private databases. The entire *Arabidopsis thaliana* genome sequence is already available (The Arabidopsis Genome Initiative, 2000) and the rice (*Oryza sativa*) genome is expected to be completed in 2003 (Sasaki and Burr, 2000; Pennisi, 2000). In addition to these two model species, large-scale EST (expressed sequence tag) sequencing projects are ongoing for another 21 plant species (http://www.ncbi.nlm.nih.gov/PMGifs/Genomes/PlantList.html). Gene function can sometimes be deduced from nucleotide sequence analysis by the comparison of the predicted gene products to those previously characterized. More often than not the sequence is novel and therefore provides little information as to gene function. For this reason, new high-throughput methods are being developed for expression analysis as well as the recovery and identification of mutants. Major developments of genome analysis have been in the field of transcript expression analysis using a variety of high-throughput methods, the most popular being microarray analysis. Such methods only provide correlative data and must be confirmed by other experimental methods. They also lack the resolution power to monitor the expression of specific genes within single or few cells. The different transgenic strategies coupled with the production of enormous amounts of DNA sequence information are opening up new experimental opportunities for the functional analysis of plant genes. An overview is provided here on current and future use of transgenic strategies in genome-wide approaches as well as targeted approaches in plant genomics.

Random Insertional Mutagenesis

Mutagenesis or the production of changes in the DNA sequence that affect the expression or structure of gene products, is one of the best methods for linking gene sequence and phenotype. The use of classical mutagenic agents such ethylmethane sulfonate, nitrosoguanidine, nitrosurea and radiation result in thousands of mutants. Saturation of the genome is relatively easy but gene characterization is based on positional cloning strategies. This is generally very laborious and time consuming but can be easier when more molecular markers are positioned on the map of the species of interest and the physical maps are more complete. The most effective current alternative strategy is random large-scale mutagenesis. Insertional mutagenesis using heterologous transposons or *Agrobacterium*-mediated T-DNA insertions are used for gene knockout, promoter trapping, enhancer trapping, and activation tagging. Since the sequence of the insert is known, the gene in which the T-DNA is inserted can be easily recovered using various cloning or PCR-based strategies.

T-DNA Tagging
T-DNA is a segment of the Ti (tumor inducing) plasmid DNA from *Agrobacterium tumefaciens* that becomes integrated into the plant genome upon transformation by the bacteria (Zambryski, 1988). A definitive model for T-DNA integration has not yet been established, but likely involves regions of microhomology between the T-DNA borders and the plant genome and possibly occurs by illegitimate recombination (Gheysen *et al.*, 1991). T-DNA insertions result in few insertions (1 to 3), are stable and easy to maintain, and insert in any chromosome (Azpiroz-Leehan and Feldmann, 1997). However integration preferentially occurs in gene-rich regions, making gene disruption more efficient than expected from random integration (Barakat *et al.*, 2000). The effect is that large-genomed species, such as tobacco, may be mutagenized with the same efficiency as small-genomed species like *Arabidopsis* and rice. Rearrangements of DNA including small deletions, duplications, translocations, base substitutions, and insertions of nucleotide sequences have been frequently reported (Herman *et al.*, 1990; Castle *et al.*, 1993; Ohba *et al.*, 1995; Nacry *et al.*, 1998; Laufs *et al.*, 1999; Tax and Vernon, 2001). A detailed characterization of the mutants is often needed before linking the tagged genes and the mutant phenotypes (Tax and Vernon, 2001).

Transposon-Tagging

T-DNA insertional mutagenesis has provided many useful mutations. However, many laboratories have not been able to generate large populations of transformants. Furthermore, the high ratio of non-tagged to tagged mutants in transformants generated from tissue (escapes) often does not make this approach an attractive option. These factors have provided the incentive for the development of transposon-tagging systems in *Arabidopsis*. The lack of well-characterized endogenous transposons in *Arabidopsis* stimulated a number of groups to introduce elements from other species. These include the maize Ac/Ds (Long *et al.*, 1993), En/Spm (Aarts *et al.*, 1995; Wisman *et al.*, 1998), and the *Antirrhinum* Tam3 element (De Greef and Jacobs, 1996). Most work and success to date has been with the maize Ac/Ds and the En/Spm elements. The main advantage of both elements lies in the fact that new transposition events can be generated in the presence of the transposase. The element can also be stabilized upon segregation of the active transposase source. In rice the most frequently used element is Tos 17, an endogenous retro-element which can be activated by tissue culture (Hirochika, 2001).

The behavior of Ac/Ds and En/Spm elements in *Arabidopsis* has shown that transposition events occur predominantly to linked loci. For example, in studies focusing on chromosome V of *Arabidopsis*, 14 to 20% of transposition events occurred within 1Mb surrounding the donor site (Ito *et al.*, 1999; Seki *et al.*, 1999). The propensity of transposons to transpose to linked sites can be exploited to study a specific region in the genome. With just a few plants, Speulman *et al.* (2000) obtained insertions in all of the 15 target genes from a single donor site on chromosome IV. When linked transpositions are not desirable, the vectors can be designed to select against linked transpositions (Sundaresan *et al.*, 1995; Tissier et al., 1999).

When using transposons, the criteria for identifying a tagged mutant could be reversion to a wild type phenotype after excision of the element rather than complementation experiments (Long *et al.*, 1993). Furthermore the presence of a footprint after the element has excised could be used to generate an allelic series at the locus of interest. Hot spots with a higher frequency of transposition have been reported (Parinov *et al.*, 1999; Meissner *et al.*, 1999). Despite this abnormal frequency, the tagging frequency for any given gene was higher than expected from random insertion, showing that transposons also integrate in gene-rich regions.

Gene Knockouts

Gene knockout constructs only need to contain a selection cassette, typically a kanamycin resistance gene (Figure 1A), but transformation with any other construct can create a mutant. The insertion of the foreign DNA creates a mutation and tags the gene for identification either in forward or reverse genetic approaches (Azpiroz-Leehan and Feldman, 1997; Krysan *et al.*, 1999).

Genome Saturation

The key parameter in the success of insertional mutagenesis is saturation of the genome. For example, given the size and gene density of the *Arabidopsis* genome, 50,000 T-DNA inserts give about 50% chance of having a mutation in a given gene of 1 kb and 200,000 T-DNA inserts give a 80% chance (Krysan *et al.*, 1999). In Arabidopis, the development of seed transformation method (Feldmann and Marks, 1987), and later the highly-efficient *in planta* transformation techniques (see Chapter 4; Bechtold *et al.*, 1993; Clough and Bent, 1998) have been instrumental in generating large populations of T-DNA insertion lines, while minimizing the effects of somaclonal variation due to tissue culture. Low copy transposons and T-DNA insertions require an impressive number of plants, while saturation can be easily achieved with high copy transposons lines.

Reverse Genetics

A number of groups have sequenced the regions flanking T-DNA and transposons to build an insertion sequence database for reverse genetics (Parinov *et al.*, 1999; Speulman *et al.*, 1999). Alternatively, the selection of defined gene mutations is done by PCR using a pair of primers, one of which anneals to the insertion sequence and the other to the specific target gene. Degenerate primers can also be used to screen for T-DNA insertions within members of a gene family (McKinney *et al.*, 1995). For an efficient screen, these populations are typically organized into pools of DNA from more than 2000 plants (Krysan *et al.*, 1996) or multidimensional pools (Azpiroz-Leehan and Feldman, 1997; Meissner *et al.*, 1999; Tissier *et al.*, 1999). Rearrangements and deletions associated with T-DNA insertions make it necessary to use primers from both right and left borders for PCR detection of genes of interest (Meissner *et al.*, 1999). In contrast to the often complex nature of T-DNA insertions, transposon inserts have a defined terminal sequence, rendering them all accessible by PCR selection.

Recently, a method combining TAIL-PCR and microarrays was assessed for its use in screening insertional mutants (Mahalingam and Fedoroff, 2001). They showed that a DNA pool from 100 plants could be amplified with transposon-specific and degenerate primers and then hybridized to a cDNA

A. Gene knockout

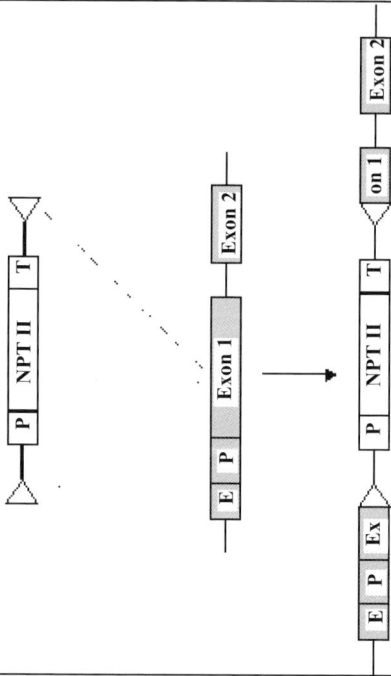

B. Promoter trapping

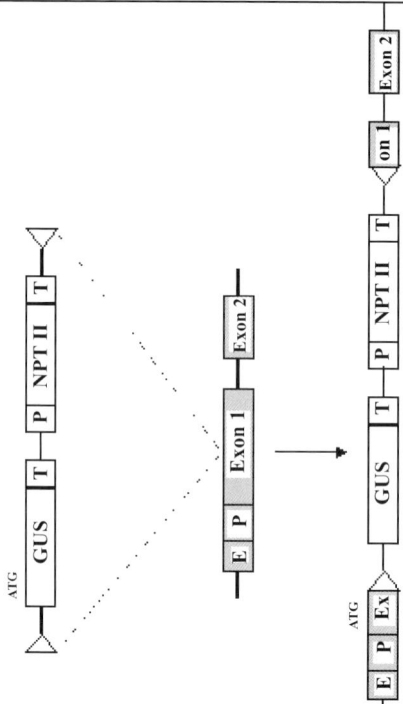

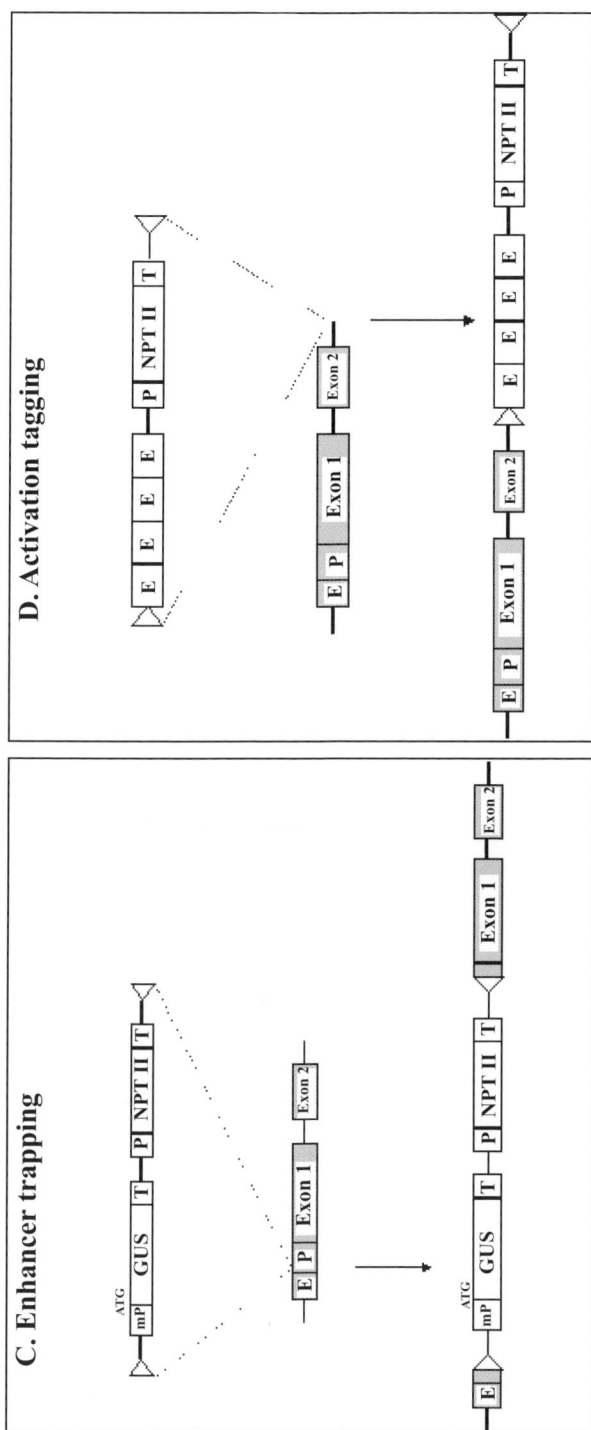

Figure 1. T-DNA insertion for trapping genes, promoters and enhancers. GUS: *uidA* reporter gene; P: promoter; mP: minimal promoter; E: enhancer; T: polyadenylation signal; NPT-II: neomycin phospo-transferase gene conferring kanamycin resistance. A: Gene knockout. Loss-of-function mutations are generated when the T-DNA inserts in exons, but also in introns and promoter regions of a native gene. B: promoter trapping. The promoterless reporter gene *uidA* is expressed following insertion downstream of a native promoter to generate a transcriptional (or at a lower frequency a translational) gene fusion with a native gene. Pattern of reporter gene expression would be similar or identical to that of the native gene. C: Enhancer trapping. The reporter gene with a minimal promoter can be expressed when integrated in the vicinity of a native enhancer. D: Activation tagging. A strong enhancer construct, such as quadruple copies of CaMV 35S enhancer sequences, activate genes in 5' or 3' directionsupon integration in the genome, creating a dominant gain-of-function mutant.

chip. The advantage of this method lies in its potential to identify all of the insertions in a single screen, thereby allowing an efficient creation of a database of insertion.

Examples abound in the literature where genes have been characterized via insertional mutants. Nevertheless the use of insertional mutagenesis with has some limitations. It is limited to genes whose disruption leads to mutant phenotypes. Many genes are functionally redundant so that their disruption doesn't show any recognizable phenotype. The completion of the *Arabidopsis* genome sequence has revealed pronounced redundancy with 37.4% proteins belonging to families of more than five members (The *Arabidopsis* Genome Initiative, 2000). Thus a systematic screen of mutants in the MYB transcription factor gene family has shown that a significant percentage of genes (20 out of 26) have no obvious phenotype under various experimental conditions (Meissner *et al.*, 1999). Another limitation of insertional mutagenesis is the cloning of genes leading to a lethal mutant such as those involved in early embryogenesis. Furthermore, a mutant phenotype leading to lethality may only reflect a partial role of the gene.

Promoter Trapping
To discover functions of genes that do not display phenotypes or give lethal mutants when mutated, insertion sequences have been designed to monitor changes of the expression patterns of adjacent genes (Sundaresan *et al.*, 1995; Martienssen, 1998). Promoter trap vectors contain reporter genes that lack promoters, and that are situated close to the T-DNA borders so that reporter gene expression can occur only when the reporter gene inserts within a transcribed chromosomal gene, creating a transcriptional fusion (Figure 1B). Enhancer trap vectors possess a minimal promoter 5' to the reporter gene and are expressed when inserted near cis-acting chromosomal enhancers (Figure 1C) (Fobert *et al.*, 1991; Topping *et al.*, 1991; Goldsbrough and Bevan, 1991). By virtue of the fact that they insert in the regulatory sequences, promoter and enhancer traps are likely to better reflect endogenous gene expression than promoter-reporter gene fusions that are extensively used for the study of the expression pattern and regulation of genes. The GUS (*uidA*) reporter gene is the most often used gene due to the absence of endogenous GUS activity in plants and the sensitivity of the histochemical detection method (Jefferson *et al.*, 1987).

Promoter and enhancer traps have been successfully used to identify genes specifically expressed in inflorescence and seedlings (Sundaresan *et al.*, 1995; Campisi *et al.*, 1999; Swaminathan *et al.*, 2000), in subsets of embryo cells (Lindsey *et al.*, 1998), in ovule (Grossniklaus and Schneitz,

1998) and during senescence (He *et al.*, 2001). The frequency of reporter gene expression is in the range of 30 to 50%, much higher than expected from random integration of the T-DNA (Koncz *et al.*, 1989; Kertbundit *et al.*, 1991; Fobert *et al.*, 1991; Topping *et al.*, 1991; Campisi *et al.*, 1999). These observations support the hypothesis that T-DNA integrates preferentially into transcriptionally active regions of chromatin. Another factor that may contribute to the high proportion of active integrated reporter gene is the tagging of cryptic promoters not associated with transcribed genes (Fobert *et al.*, 1994; Foster *et al.*, 1999).

Activation Tagging

Activation tagging is another avenue for addressing the function of redundant genes and lethal mutations by creating dominant gain of function phenotypes. It is based on the use of an insertion element carrying a strong enhancer such as a tetramer of the CaMV 35S enhancer region whose insertion creates an ectopic activation of the flanking gene (Figure 1D) (Weigel *et al.*, 2000). Some spectacular mutants have been obtained by activation tagging, proving the method is particularly suited for isolating genes implicated in plant growth and development (Walden *et al.*, 1994), and dissecting metabolic pathways. For example, knockout of a plastidial phosphoenolpyruvate transporter resulted in compromised aromatic amino acid synthesis, underexpression of nuclear encoded proteins, and mesophyll-specific cell defects causing reticulate leaves (Streatfield *et al.*, 1999). The activation of a conserved MYB transcription factor resulted in the upregulation of the phenylpropanoid biosynthesis pathway including anthocyanins responsible for the purple color observed in the mutant (Borevitz *et al.*, 2000). In contrast to T-DNA insertions designed to create a loss of function mutant, activation tagging results in less chromosomal rearrangements, making it easier to identify the flanking sequences and link the gene to the observed phenotype (Weigel *et al.*, 2000; Huang *et al.*, 2001).

RNA Silencing Mutants

RNA silencing methods exploit plants ability to degrade RNA in a sequence specific manner during post-transcriptional gene silencing (PTGS) and virus induced gene silencing (VIGS) as a defense against transgenes, endogenous genes and viruses (Ding, 2000). Gene silencing using antisense methods was first tested in plants by inhibiting the expression of chloramphenicol-acetyl-transferase (CAT) in transient experiments with carrot protoplasts (Ecker and Davis, 1986). Downregulation of endogenous genes was later demonstrated in transgenic petunia and tobacco plants expressing an antisense

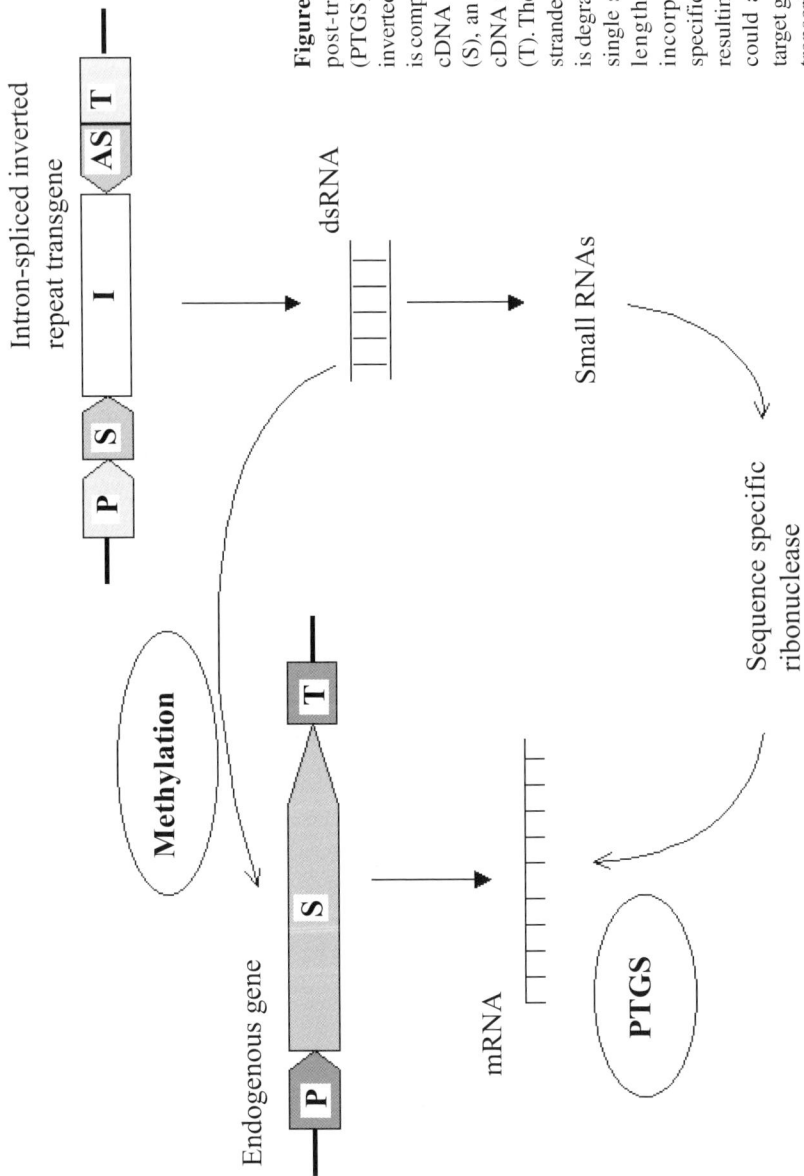

Figure 2. Proposed mechanisms of post-transcriptional gene silencing (PTGS) triggered by an intron-spliced inverted repeat construct. The construct is comprised of a promoter (P), a partial cDNA sequence in sense orientation (S), an intron (I), the antisense partial cDNA sequence (AS), and a terminator (T). The transcript is spliced and double stranded RNA (dsRNA) is produced. It is degraded by an RNAse to form small single strand RNAs of about 25 nt in length. The small RNAs are incorporated in ribonuclease that specifically degrades the target RNA resulting in PTGS. Double strand RNA could also cause methylation of the target gene, interfering with subsequent transcription.

chalcone synthase (CHS) gene (van der Krol *et al.*, 1988). Since then, antisense strategy has been a widely used tool for the creation of down-regulated mutants in many different plant species. Some early examples are the use of antisense polygalacturonase (Smith et al, 1990) and 1-aminocyclopropane-1-carboxylate (ACC) synthase (Theologis *et al.*, 1993) for dissection of fruit ripening in tomato. Gene silencing resulting from sense constructs was first reported when additional copies of CHS genes introduced in *Petunia* resulted in the loss of pigmentation due to the suppression of the endogenous genes (Napoli *et al.*, 1990). Silencing by viruses was also reported in which there was sequence similarity between the virus and either a transgene or an endogenous gene (Lindbo *et al.*, 1993; Kumagai *et al.*, 1995). The effectiveness of gene silencing obtained with sense/antisense constructs or with viruses was variable and not controllable, nevertheless the discovery of the phenomenon prompted the study of the mechanisms underlying post-transcriptional gene silencing.

Mechanism
PTGS also referred to as 'RNA interference' in *Caenorhabtidis elegans* and quelling in *Neurospora crassa* is now regarded as ubiquitous phenomenon sharing similarities across kingdoms (Cogoni and Macino, 2000). The molecular mechanisms of PTGS are still not fully understood, but the formation of a double stranded RNA seems to be critical for the degradation of homologous RNA and the specific methylation of nuclear DNA (Figure 2) (Matzke *et al.*, 2001; Vaucheret *et al.*, 2001). Theoretically, RNA transcribed from a gene inserted in the antisense direction will bind to the RNA of the same gene in its correct orientation. The double stranded RNA is consequently degraded by a double strand RNAse and no translation takes place. During silencing with sense constructs the formation of double stranded RNA is mediated by an RNA-dependent RNA polymerase (RdRp) since RdRp mutants are unable to trigger PTGS (Dalmay *et al.*, 2000). However, during VIGS, the double stranded RNA is formed during the process of viral genome replication and the RdRp not required. Small RNA molecules about 25 nucleotides in length result from the degradation of double stranded RNA (Hamilton and Baulcombe, 1999). They are thought to be incorporated into a ribonuclease, serving as a guide to find homologous target RNAs (Bass, 2000). Additionally, double stranded RNA was shown to causes DNA methylation in plants although it is not known whether it was the result of RNA-DNA interactions in the nucleus, or if it is mediated by the small RNAs resulting from the double strand degradation (Mette *et al.*, 2000). Grafting experiments Palauqui *et al.* (1997) have shown that silencing is systemically transmitted but the nature of the mediator of the signal is as yet unknown.

Application

The origin of the DNA for antisense RNA production has usually been a full-length cDNA for the endogenous target gene (van der Kroll *et al.*, 1988; Lee *et al.*, 1999; Wenderoth and von Schaewen, 2000; Papenbrock *et al.*, 2001). Efficient inhibition is also reported with the 5' end of the cDNA (Smith *et al.*, 1988) and 3' end of cDNA (van der Kroll *et al.*, 1990). Latter studies showed that short strands covering only 41 bases spanning the translation start codon of the mRNA were effective in silencing the GUS gene (Cannon *et al.*, 1990). The emergence of double stranded RNA as a potent inducer of silencing prompted researchers to design constructs producing duplex forming RNA (Waterhouse *et al.*, 1998). Chuang and Meyerowitz (2000) showed that such constructs lead to silencing of 87 to 99 % of transgenic *Arabidopsis* plants, depending on the targeted gene. The phenotypes obtained were similar to those resulting from loss-of-function and reduction-of-function mutants previously characterized. The silencing was also specific and heritable. In subsequent studies, constructs encoding intron-spliced, self-complementary RNA (Figure 2) gave efficient gene silencing, in the range of 90 to 100%, in several plant species (Smith *et al.*, 2000; Singh *et al.*, 2000; Wesley *et al.*, 2001). It is hypothesized that the excision of the intron promotes the formation of the duplex by bringing together the sense and antisense strands of the RNA.

A valuable feature of PTGS methods is their potential of recovering knockout phenotypes for redundant genes and gene families. For example, antisense expression of the cell wall-associated protein kinase WAK4 resulted in the reduction to undetectable levels of other 5 highly related members of the WAK gene family and allowed to establish their role in cell elongation and morphology (Lally *et al.*, 2001). From antisense experiments using full-length cDNA sequences it is estimated that a 60 to 70% homology in the coding-region is required for silencing of homologous genes (Elomaa *et al.*, 1996; Kunz *et al.*, 2001). However the length and sequence similarities necessary for degrading the mRNA from homologous genes is not yet established for inverted repeat constructs. In theory, a small conserved region in the construct should be sufficient as the hybridization to the different mRNAs is bypassed.

Silencing can be spacially or temporally controlled by using tissue-specific promoters or inducible promoters. This feature facilitates the study of genes essential for plant development, unlike for insertional mutants in which the knockout of genes required at early developmental stages precludes the study of their possible role at later stages. An example is the use of a glucocorticoid-inducible promoter to trigger the silencing of cell wall-associated kinase genes that are otherwise required for development (Lally *et al.*, 2001).

VIGS can be used for reverse genetics simply by infecting plants with virus vectors carrying fragments of the host gene (Baulcombe, 1999). The major advantage of the VIGS system is the speed with which the role of gene product can be defined compared with the slower antisense and sense approaches in transgenic plants. VIGS was successfully used for the first time in reverse genetics to suppress plant cellulase synthase genes by infecting *Nicotiana* plants with potato virus X harboring putative cellulose synthase cDNA fragments (Burton *et al.*, 2000). To eliminate virus-induced symptoms and obtain gene silencing throughout the plant an infectious cDNA clone has been designed. It consists of a tobacco rattle virus (TRV) cDNA modified to facilitate insertion of non-viral sequence (Ratcliff *et al.*, 2001). VIGS is limited to mature plants, and is, for the moment, restricted to *Nicotiana* species.

A major drawback of using PTGS is that mutants are leaky and it may not be possible to obtain complete silencing. Factors such as environmental conditions or developmental stages influence the extent of silencing (Hart *et al.*, 1992; Wenderoth and von Schaewen, 2000; Kunz *et al.*, 2001). Reduction of RNA levels is sometimes not enough to create a loss or reduction-of-function mutant. For example, Temple *et al.* (1998) observed that despite an 80% reduction of mRNA with the antisense construct of glutamine synthase, no reduction of GS activity was acheived. Similarly, transgenic tomato plants expressing antisense RNA to the key enzyme in the ethylene biosynthetic pathway, ACC synthase show a mutant phenotype only when ethylene production is suppressed by more than 99% (Theologis *et al.*, 1993). Therefore, downregulation in some cases may not lead to a phenotypical change making it difficult to assert the function of the gene. Conversely, reduction-of-function leaves the possibility of recovering mutants that would otherwise be lethal.

Silencing strategies are best suited to study gene function where no insertional mutants are available, or when dealing with redundant genes. Only a few transgenic plants are required to obtain the desired mutant as opposed to thousands in the case of random insertional mutagenesis, which makes the method particularly attractive for plant species not routinely amenable to transformation. The efficiency of silencing obtained by intron-spiced double hairpin forming sequences has prompted the design of vectors to generate silencing constructs from cDNA libraries (Wesley *et al.*, 2001). This breakthrough is expected to facilitate the large-scale discovery of plant gene functions in a similar fashion that RNAi has been instrumental for the genetics of *C. elegans* (Fire, 1999; Hunter, 1999).

Another method explored for degrading target RNA takes advantage of naturally occuring ribozymes, such as the RNAse P responsible for the

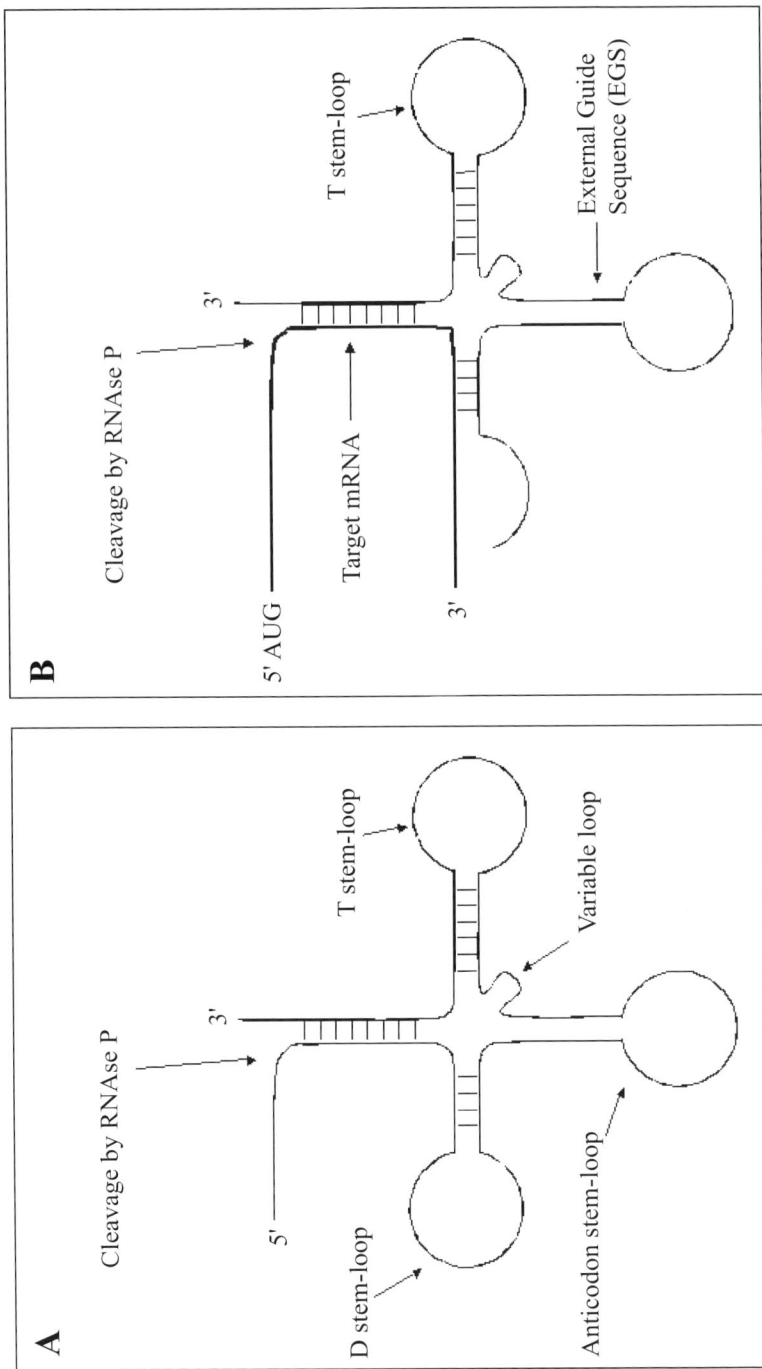

Figure 3. The concept underlying RNAse P-mediated cleavage of a target mRNA. A: structure of a ptRNA, substrate of RNAse P. B: Complex formed by the target mRNA and the External Guide Sequence (EGS) resembling that of a ptRNA.

5'maturation of precursor t-RNAs. RNAse P recognizes the structure of its substrate, rather than the sequence to remove the 5' leader sequence. Constructs can be designed to form a sequence specific complex with a given mRNA via an external guide sequence to mimic ptRNAs (Figure 3). They are then cleaved by RNAse P (Forster and Altman, 1990; Guerrier-Takada and Altman, 2000). This method first developed in mammalian systems, was shown to function with rice and maize RNAse P in vitro (Raj *et al.*, 2001). If the extent of silencing obtained with such constructs is proven as effective as double strand RNA silencing, this method could provide an alternative for targeted gene knockouts.

Gene Targeting

Gene targeting consists of altering a precise chromosomal locus by homologous recombination with foreign DNA. Targeted approaches of gene mutation have been successful in yeast in which the relatively small genome and a high recombination frequency allows the disruption of specific genes by introducing PCR products containing 30-45 bp sequence homology flanking the gene of interest. The high substitution efficiency, over 70%, has allowed the generation of a collection of more than 2000 mutants for the functional study of the yeast genome (Güldener *et al.*, 1996 ; Wach *et al.*, 1997). The method has been adopted in mammalian research but suffers from a low frequency of targeted gene disruption thus making it a laborious technique. In plants, homologous recombination was achieved after direct delivery of DNA (Paszkowski *et al.*, 1988; Halfter *et al.*, 1992) or by *Agrobacterium*-mediated infection (Lee *et al.*, 1990; Offringa *et al.*, 1990; Kempin *et al.*, 1997). Only a few targeted replacements were obtained, making it difficult to estimate actual homologous recombination frequencies and the much higher illegitimate integration frequency (10^4 to 10^5 fold), has limited the use of the method. Strategies to increase the recombination frequency by extending the length of homology to the target gene or the use of negative selection to enrich for targeted events did not improve significantly the targeting efficiency (Thykjaer *et al.*, 1997; Xiaohui Wang *et al.*, 2001). So far, the best efficiency for the targeting of a precise chromosomal locus, 1 in 750, was obtained when PCR selection was performed on pooled samples from transgenic plants using primers complementary to the inserted sequence and to the genomic region (Kempin *et al.*, 1997).

A major breakthrough in targeted mutagenesis was recently achieved by introducing tailor-made chimeric DNA/RNA oligonucleotides that form a heteroduplex with double hairpin caps at the end and contain a mismatch to the target sequence. This approach was initially developed for mammalian gene therapy and correction of the point mutation was accomplished with a frequency approaching 30% (Yoon *et al.*, 1996). The method was used to create a mutation in the acetolactate synthase (ALS) gene in tobacco (Beetham *et al.*, 1999) and a mutation in the acetohydroxyacid synthase (AHAS) in maize (Zhu *et al.*, 1999) albeit at a very low frequencies. It is envisioned that the method could provide an alternative gene targeting technology for plants (Hohn and Puchta, 1999). Mutations induced by chimeric oligonucleotides generally involve one or two base pairs in the target site, which make the technique suitable for generating allelic series. At present, the low frequency of gene repair needs to be increased before it can become a tool in plant genomics. Progress can be expected from studies focusing on the mechanisms and structures of the chimeric oligonucleotides required for an efficient targeting. It is hypothesized that the mismatch is recognized by DNA repair enzymes, which results in the stable integration of the oligo in the endogenous gene (Rice *et al.*, 2000; Gamper *et al.*, 2000). While undergoing the DNA pairing process the RNA helps in stabilizing the central reaction intermediate rendering it resistant to nuclease activity. Several chimeras were also tested to determine the key features of the chimeric oligos necessary for nucleotide conversion (Gamper *et al.* 2000). Their results show that the DNA strand of the chimera directs the gene repair activity and the length of homology with the targeted region affects gene-targeting frequency.

Several studies undertaken to better understand the mechanisms of homologous recombination could potentially open the way for a more effective gene targeting. Candidate genes encoding proteins involved in the reactions of strand transfer, required for nuclease activity, and helicases from yeast and bacteria are being examined for enhancing gene targeting in plants. For example, overexpression of bacterial *RuvC* led to a 56-fold increase in extrachromosomal homologous recombination frequencies between plasmids cotransformed into young leaves via particle bombardment (Shalev *et al.*, 1999). Conversely genes that could depress non-homologous recombination due to random integration and DNA damage signaling pathways also need to be examined for stimulating homologous recombination (Vasquez *et al.*, 2001). Recombinogenic plants can increase the frequency of targeted gene replacement and make the method useful for genomic studies.

Remarkably there is one plant species, *Physcomitrella patens*, for which a recombination frequency of up to 90% can be achieved (Schaefer and Zryd, 1997). It is postulated that the gametophytic nature of the tissue used for transformation, its synchronized division and its distinctive cell cycle arrest in G2/M for most part of the day could account for the high homologous recombination frequencies observed (Reski, 1998). Targeted mutagenesis has been used in this specie to characterize a novel delta 6-acyl-lipid desaturase (Girke *et al.*, 1998) and the *FtsZ* gene essential for chloroplast division, the first organellar division protein from a eukaryote (Strepp *et al.*, 1998).

Potentially, targeted mutagenesis could be used to specifically introduce pinpointed modification of the target sequence at the nucleotide level and create allelic variants. It could also be used to insert reporter genes within a given gene to examine its expression pattern in its original chromosomal location, thereby avoiding positional effects associated with promoter reporter constructs randomly inserted in the genome. For *Physcomitrella* gene-trap and enhancer-trap lines using the GUS reporter gene have already been obtained by homologous recombination (Hiwatashi *et al.*, 2001).

Conclusion

A transgenic approach remains a method of choice for validating functional genomics and in elucidating metabolic and physiological processes in plants. Although much information can be gained from physiological studies and expression analysis, the interaction of genes with other components of an intact organism provides a much more complete and physiologically relevant picture of gene function than could be achieved in any other way. Two plant species, *Arabidopsis* and rice, have relatively small genomes, and are the focus of sequencing projects. In addition, efficient transformation systems exist, and these species have benefited in the elucidation of their genomics. The high level of redundancy in their genomes will be addressed, at least in part, by a shift of genomic studies from gene function to gene regulation. Redundancy can be more apparent than real because even though gene families arise through duplications, their persistence through time relies on their ability to fulfill one or both of two functions: to provide a level of control of expression that a single gene is unable to provide or to provide proteins with differing functionality (Durbin *et al.*, 2000). We can exploit reporter genes, targeted insertions, and promoter traps to distinguish the role of cis-regulatory elements in the various spatial and developmental regulation of gene family members.

Although much information can be gained from *Arabidopsis* and rice, there is still a vast array of economically-important species for which genome sequence and large-scale approaches to functional analysis are lacking. One of the transformation-based techniques that is being developed is transposon arrayed gene knockouts (TAGKO), in which cosmid clones are first mutagenized with transposons to create an annotated collection of insertional gene disruption vectors. They are then used for targeted integration, which gives a much higher frequency than with conventional gene disruption vectors (Hamer *et al.*, 2001). This method has been tested in different species of filamentous fungi and can potentially be applied to flowering plants amenable to transformation and gene targeting. With the concomitant improvement of transformation methods, the versatility of the constructs used for transformation, significant progress can be expected for plant species with large genomes and for which there is no sequencing program.

References

Aarts, M.G., Corzaan, P., Stiekema, W.J., and Pereira, A. 1995. A two-element enhancer-inhibitor transposon system in *Arabidopsis thaliana*. Mol. Gen. Genet. 247: 555-564.

Azpiroz-Leehan, R., and Feldmann, K.A. 1997. T-DNA insertion mutagenesis in *Arabidopsis*: going back and forth. Trends Genet. 13: 152-156.

Barakat, A., Gallois, P., Raynal, M., Mestre-Ortega, D., Sallaud, C., Guiderdoni, E., Delseny, M., and Bernardi, G. 2000. The distribution of T-DNA in the genomes of transgenic *Arabidopsis* and rice. FEBS Lett. 471: 161-164.

Bass, B.L. 2000. Double-stranded RNA as a template for gene silencing. Cell. 101: 235-238.

Baulcombe, D.C. 1999. Fast forward genetics based on virus-induced gene silencing. Curr. Opin. Plant Biol. 2: 109-113.

Bechtold, N., Ellis, J., and Pelletier, G. 1993. *In planta Agrobacterium* mediated gene transfer by infiltration of adult *Arabidopsis thaliana* plants. C. R. Acad. Sci. 316: 1194-1199.

Beetham, P.R., Kipp, P.B., Sawycky, X.L., Arntzen, C.J., and May, G.D. 1999. A tool for functional plant genomics: chimeric RNA/DNA oligonucleotides cause *in vivo* gene-specific mutations. Proc. Natl. Acad. Sci. USA. 96: 8774-8778.

Borevitz, J.O., Xia, Y., Blount J., Dixon, R.A., and Lamb, C. 2000. Activation tagging identifies a conserved MYB regulator of phenylpropanoid biosynthesis. Plant Cell. 12:2383-2394.

Burton, R.A., Gibeaut, D.M., Bacic, A., Findlay, K., Roberts, K., Hamilton, A., Baulcombe, D.C., and Fincher, G.B. 2000. Virus-induced silencing of a plant cellulose synthase gene. Plant Cell. 12: 691-706.

Campisi, L., Yang, Y., Yi, Y., Heilig, E., Herman, B., Cassista, A.J., Allen, D.W., Xiang, H., and Jack, T. 1999. Generation of enhancer trap lines in *Arabidopsis* and characterization of expression patterns in the inflorescence. Plant J. 17: 699-707.

Cannon. M., Platz. J., O'Leary. M., Sookdeo. C.. and Cannon. F. 1990. Organ-specific modulation of gene expression in transgenic plants using antisense RNA. Plant Mol. Biol. 15: 39-47.

Castle, L.A., Errampalli, D., Atherton, T.L., Franzmann, L.H., Yoon, E.S., and Meinke D.W. 1993. Genetic and molecular characterization of embryonic mutants identified following seed transformation in *Arabidopsis*. Mol. Gen. Genet. 241: 504-14.

Chuang, C.F., and Meyerowitz, E.M. 2000. Specific and heritable genetic interference by double-stranded RNA in *Arabidopsis thaliana*. Proc. Natl. Acad. Sci. USA. 97: 4985-4990.

Clough, S.J., and Bent, A.F. 1998. Floral dip: a simplified method for *Agrobacterium*-mediated transformation of *Arabidopsis thaliana*. Plant J. 16: 735-743.

Cogoni, C., and Macino, G. 2000. Post-transcriptional gene silencing across kingdoms. Curr. Opin. Genet. Dev. 10: 638-643.

Dalmay, T., Hamilton, A., Rudd, S., Angell, S., and Baulcombe, D.C. 2000. An RNA-dependent RNA polymerase gene in *Arabidopsis* is required for post-transcriptional gene silencing mediated by a transgene but not by a virus. Cell. 101: 543-553.

De Greef, B., and Jacobs, M. 1996. Evidence for Tam3 activity in transgenic *Arabidopsis thaliana*. *In Vitro* Cell. Dev. Biol. Plant. 32: 241-248.

Ding, S.W. 2000. RNA silencing. Curr. Opin. Biotechnol. 11:152-156.

Durbin, M.L., McCaig, B., and Clegg, M.T. 2000. Molecular evolution of the chalcone synthase multigene family in the morning glory genome. Plant Mol. Biol. 42: 79-92.

Ecker, J.R., and Davis, R.W. 1986. Inhibition of gene expression in plant cells by expression of antisense RNA. Proc. Natl. Acad. Sci. USA. 83: 5372-5376.

Elomaa, P., Helariutta, Y., Kotilainen, M., and Teeri, T. H. 1996. Transformation of antisense constructs of the chalcone synthase gene superfamily into *Gerbera hybrida*: differential effect on the expression of family members. Mol. Biol. 2: 41-50.

Feldmann, K.A., and Marks, M.D. 1987. *Agrobacterium*-mediated transformation of germinating seeds of *Arabidopsis thaliana*: a non-tissue culture approach. Mol. Gen. Genet. 208: 1-9.

Fire, A. 1999. RNA-triggered gene silencing. Trends Genet. 15: 358-363.

Fobert, P.R., Labbe, H., Cosmopoulos, J., Gottlob-McHugh, S., Ouellet, T., Hattori, J., Sunohara, G., Iyer, V.N., and Miki B.L. 1994. T-DNA tagging of a seed coat-specific cryptic promoter in tobacco. Plant J. 6: 567-577.

Fobert, P.R., Miki, B.L., and Iyer, V.N. 1991. Detection of gene regulatory signals in plants revealed by T-DNA-mediated fusions. Plant Mol. Biol. 17: 837-851.

Forster, A.C., and Altman, S. 1990. External guide sequences for an RNA enzyme. Science. 249: 783-786.

Foster, E., Hattori, J., Labbe, H., Ouellet, T., Fobert, P.R., James, L.E., Iyer, V.N., and Miki, B.L. 1999. A tobacco cryptic constitutive promoter, tCUP, revealed by T-DNA tagging. Plant Mol. Biol. 41: 45-55.

Gamper, H.B., Parekh, H., Rice, M.C., Bruner, M., Youkey, H., and Kmiec, E.B. 2000 The DNA strand of chimeric RNA/DNA oligonucleotides can direct gene repair/conversion activity in mammalian and plant cell-free extracts. Nucleic Acids Res. 28: 4332-4339.

Gamper, H.B., Parekh, H., Rice, M.C., Bruner, M., Youkey, H., and Kmiec, E.B. 2000. The DNA strand of chimeric RNA/DNA oligonucleotides can direct gene repair/conversion activity in mammalian and plant cell-free extracts. Nucleic Acids Res. 28: 4332-4339.

Gheysen, G., Villarroel, R., and Van Montagu, M. 1991. Illegitimate recombination in plants: a model for T-DNA integration. Genes Dev. 5: 287-297.

Girke, T., Schmidt, H., Zahringer, U., Reski, R., and Heinz, E. 1998. Identification of a novel delta 6-acyl-group desaturase by targeted gene disruption in *Physcomitrella patens*. Plant J. 15: 39-48.

Goldsbrough, A., and Bevan, M. 1991. New patterns of gene activity in plants detected using an *Agrobacterium* vector. Plant Mol. Biol. 16: 263-269.

Grossniklaus, U., and Schneitz, K. 1998. The molecular and genetic basis of ovule and megagametophyte development. Semin. Cell. Dev. Biol. 9: 227-238.

Guerrier-Takada, C., and Altman, S. 2000. Inactivation of gene expression using ribonuclease P and external guide sequences. Methods Enzymol. 313: 442-456.

Güldener, U., Heck, S., Fiedler, T., Beinhauer, J., and Hegemann, J.H. 1996. A new efficient gene disruption cassette for repeated use in budding yeast. Nucleic Acids Res. 24: 2519-2524.

Halfter, U., Morris, P.C., and Willmitzer, L. 1992. Gene targeting in *Arabidopsis thaliana*. Mol. Gen. Genet. 231: 186-193.

Hamer, L., Adachi, K., Montenegro-Chamorro, M.V., Tanzer, M.M., Mahanty, S.K., Lo, C., Tarpey, R.W., Skalchunes, A.R., Heiniger, R.W., Frank, S.A., Darveaux, B.A., Lampe, D.J., Slater, T.M., Ramamurthy, L., DeZwaan, T.M., Nelson, G.H., Shuster, J.R., Woessner, J., and Hamer, J.E. 2001. Gene discovery and gene function assignment in filamentous fungi. Proc. Natl. Acad. Sci. USA. 98: 5110-5115.

Hamilton, A.J., and Baulcombe, D.C. 1999. A species of small antisense RNA in posttranscriptional gene silencing in plants. Science. 286: 950-952.

Hart, C.M., Fischer, B., Neuhaus, J.M., and Meins, F., Jr. 1992. Regulated inactivation of homologous gene expression in transgenic *Nicotiana sylvestris* plants containing a defense-related tobacco chitinase gene. Mol. Gen. Genet. 235: 179-188.

He, Y., Tang, W., Swain, J.D., Green, A.L., Jack, T.P., and Gan, S. 2001. Networking senescence-regulating pathways by using *Arabidopsis* enhancer trap lines. Plant Physiol. 126: 707-716.

Herman, L., Jacobs, A., Van Montagu, M., and Depicker, A. 1990. Plant chromosome/marker gene fusion assay for study of normal and truncated T-DNA integration events. Mol. Gen. Genet. 224: 248-256.

Hirochika, H. 2001. Contribution of the Tos17 retrotransposon to rice functional genomics. Curr. Opin. Plant Biol. 4: 118-122.

Hiwatashi, Y., Nishiyama, T., Fujita, T., and Hasebe, M. 2001. Establishment of gene-trap and enhancer-trap systems in the moss *Physcomitrella patens*. Plant J. 28: 105-116.

Hohn, B., and Puchta, H. 1999. Gene therapy in plants. Proc. Natl. Acad. Sci. USA. 96: 8321-8323.

Huang, S., Cerny, R.E., Bhat, D.S., and Brown, S.M. 2001. Cloning of an *Arabidopsis* patatin-like gene, *STURDY*, by activation T-DNA tagging. Plant Physiol. 125: 573-584.

Hunter, C.P. 1999. Genetics: a touch of elegance with RNAi. Curr. Biol. 9: R440-442.

Ito, T., Seki, M., Hayashida, N., Shibata, D., and Shinozaki, K. 1999. Regional insertional mutagenesis of genes on *Arabidopsis thaliana* chromosome V using the Ac/Ds transposon in combination with a cDNA scanning method. Plant J. 17: 433-444.

Jefferson, R.A., Kavanagh, T.A., and Bevan, M.W. 1987. GUS fusions: beta-glucuronidase as a sensitive and versatile gene fusion marker in higher plants. EMBO J. 6: 3901-3907.

Kempin, S.A., Liljegren, S.J., Block, L.M., Rounsley, S.D., Yanofsky, M.F., and Lam, E. 1997. Targeted disruption in *Arabidopsis*. Nature. 389: 802-803.

Kertbundit, S., De Greve, H., Deboeck, F., Van Montagu, M., and Hernalsteens, J.P. 1991. *In vivo* random beta-glucuronidase gene fusions in *Arabidopsis thaliana*. Proc. Natl. Acad. Sci. USA. 88: 5212-5216.

Koncz, C., Martini, N., Mayerhofer, R., Koncz-Kalman, Z., Korber, H., Redei, G.P., and Schell, J. 1989. High-frequency T-DNA-mediated gene tagging in plants. Proc. Natl. Acad. Sci. USA. 86: 8467-8471.

Krysan, P.J., Young, J.C., and Sussman, M.R. 1999. T-DNA as an insertional mutagen in *Arabidopsis*. Plant Cell. 11: 2283-2290.

Krysan, P.J., Young, JC., Tax, F., and Sussman, M.R. 1996. Identification of transferred DNA insertions within *Arabidopsis* genes involved in signal transduction and ion transport. Proc. Natl. Acad. Sci. USA. 93: 8145-8150.

Kumagai, M.H., Donson, J., della-Cioppa, G., Harvey, D., Hanley, K., and Grill, L.K. 1995. Cytoplasmic inhibition of carotenoid biosynthesis with virus-derived RNA. Proc. Natl, Acad. Sci. USA. 92: 1679-1683.

Kunz, C., Schob, H., Leubner-Metzger, G., Glazov, E., and Meins, F., Jr. 2001. Beta-1, 3-glucanase and chitinase transgenes in hybrids show distinctive and independent patterns of posttranscriptional gene silencing. Planta. 212: 243-249.

Lally, D., Ingmire, P., Tong, H.Y., and He, Z.H. 2001. Antisense expression of a cell wall-associated protein kinase, WAK4, inhibits cell elongation and alters morphology. Plant Cell. 13: 1317-1331.

Laufs, P., Autran, D., and Traas, J. 1999. A chromosomal paracentric inversion associated with T-DNA integration in *Arabidopsis*. Plant J. 18: 131-139.

Lee, K.Y., Lund, P., Lowe, K., and Dunsmuir, P. 1990. Homologous recombination in plant cells after *Agrobacterium*-mediated transformation. Plant Cell. 2: 415-425.

Lee, Y., Lloyd, A.M., and Roux, S.J. 1999. Antisense expression of the CK2 alpha-subunit gene in *Arabidopsis*. Effects on light-regulated gene expression and plant growth. Plant Physiol. 119: 989-1000.

Lindbo, J.A., Silva-Rosales, L., Proebsting, W.M., and Dougherty, W.G. 1993. Induction of a highly specific antiviral state in transgenic plants: implications for regulation of gene expression and viral resistance. Plant Cell. 5: 1749-1759.

Lindsey, K., Topping, J.F., Muskett, P.R., Wei, W., and Horne, K.L. 1998. Dissecting embryonic and seedling morphogenesis in *Arabidopsis* by promoter trap insertional mutagenesis. Symp. Soc. Exp. Biol. 51: 1-10.

Long, D., Martin, M., Sundberg, E., Swinburne, J., Puangsomlee, P., and Coupland, G. 1993. The maize transposable element system Ac/Ds as a mutagen in *Arabidopsis*: identification of an albino mutation induced by Ds insertion. Proc. Natl. Acad. Sci. USA. 90: 10370-10374.

Mahalingam, R., and Fedoroff, N. 2001. Screening insertion libraries for mutations in many genes simultaneously using DNA microarrays. Proc. Natl. Acad. Sci. USA. 98: 7420-7425.

Martienssen, R.A. 1998. Functional genomics: probing plant gene function and expression with transposons. Proc. Natl. Acad. Sci. USA. 95: 2021-2026.

Matzke, M.A., Matzke, A.J., Pruss, G.J., and Vance, V.B. 2001. RNA-based silencing strategies in plants. Curr. Opin. Genet. Dev. 11: 221-227.

McKinney, E.C., Ali, N., Traut, A., Feldmann, K.A., Belostotsky, D.A., McDowell, J.M., and Meagher, R.B. 1995. Sequence-based identification of T-DNA insertion mutations in *Arabidopsis*: actin mutants act2-1and act4-1. Plant J. 8: 613-622.

Meissner, R.C., Jin, H., Cominelli, E., Denekamp, M., Fuertes, A., Greco, R., Kranz, H.D., Penfield, S., Petroni, K., Urzainqui, A., Martin, C., Paz-Ares, J., Smeekens, S., Tonelli, C., Weisshaar, B., Baumann, E., Klimyuk, V., Marillonnet, S., Patel, K., Speulman, E., Tissier, AF., Bouchez, D., Jones, J.J., Pereira, A., Wisman, E., and Bevan, M. 1999. Function search in a large transcription factor gene family in *Arabidopsis*: assessing the potential of reverse genetics to identify insertional mutations in *R2R3 MYB* genes. Plant Cell. 11: 1827-1840.

Mette, M.F., Aufsatz, W., van der Winden, J., Matzke, M.A., and Matzke A.J. 2000. Transcriptional silencing and promoter methylation triggered by double-stranded RNA. EMBO J.19: 5194-5201.

Nacry, P., Camilleri, C., Courtial, B., Caboche, M., and Bouchez, D. 1998. Major chromosomal rearrangements induced by T-DNA transformation in *Arabidopsis*. Genetics. 149: 641-650.

Napoli, C., Lemieux, C., and Jorgensen, R. 1990. Introduction of a chimeric chalcone synthase gene into Petunia results in reversible co-suppression of homologous genes *in trans*. Plant Cell. 2: 279-289.

Offringa, R., de Groot, M.J., Haagsman, H.J., Does, M.P., van den Elzen, P.J., and Hooykaas, P.J. 1990. Extrachromosomal homologous recombination and gene targeting in plant cells after *Agrobacterium* mediated transformation. EMBO J. 9: 3077-3084.

Ohba, T., Yoshioka, Y., Machida, C., and Machida, Y. 1995. DNA rearrangement associated with the integration of T-DNA in tobacco: an example for multiple duplications of DNA around the integration target. Plant J. 7: 157-164.

Palauqui, J.C., Elmayan, T., Pollien, J.M., and Vaucheret, H. 1997. Systemic acquired silencing: transgene-specific post-transcriptional silencing is transmitted by grafting from silenced stocks to non-silenced scions. EMBO J. 16: 4738-4745.

Papenbrock, J., Mishra, S., Mock,,H.P., Kruse, E., Schmidt, E.K., Petersmann, A., Braun, H.P., and Grimm, B. 2001. Impaired expression of the plastidic ferrochelatase by antisense RNA synthesis leads to a necrotic phenotype of transformed tobacco plants. Plant J. 28: 41-50.

Parinov, S., Sevugan, M., De, Y., Yang, W.C., Kumaran, M., and Sundaresan, V. 1999. Analysis of flanking sequences from dissociation insertion lines: a database for reverse genetics in *Arabidopsis*. Plant Cell. 11: 2263-2270.

Paszkowski, J., Baur, M., Bogucki, A., and Potrykus, I. 1988. Gene targeting in plants. EMBO J. 7: 4021-4026.

Pennisi, E. 2000. Stealth genome rocks rice researchers. Science. 288: 239-241.

Raj, M.L., Pulukkunat, D.K., Reckard, J.F., III, Thomas, G., and Gopalan, V. 2001. Cleavage of bipartite substrates by rice and maize ribonuclease P. Application to degradation of target mRNAs in plants. Plant Physiol. 125: 1187-1190.

Ratcliff, F., Martin-Hernandez, A.M., and Baulcombe, D.C. 2001. Tobacco rattle virus as a vector for analysis of gene function by silencing. Plant J. 25: 237-245.

Reski, R. 1998. *Physcomitrella* and *Arabidopsis*: the David and Goliath of reverse genetics. Trends. Plant. Sci. 3: 209-210.

Rice, M.C., May, G.D., Kipp, P.B., Parekh, H., and Kmiec, E.B. 2000. Genetic repair of mutations in plant cell-free extracts directed by specific chimeric oligonucleotides. Plant Physiol.123: 427-438.

Sasaki, T., and Burr, B. 2000. International Rice Genome Sequencing Project: the effort to completely sequence the rice genome. Curr. Opin. Plant Biol. 3: 138-141.

Schaefer, D.G., and Zryd, J.P. 1997. Efficient gene targeting in the moss *Physcomitrella patens*. Plant J. 11: 1195-1206.

Seki, M., Ito, T., Shibata, D., and Shinozaki, K. 1999. Regional insertional mutagenesis of specific genes on the CIC5F11/CIC2B9 locus of *Arabidopsis thaliana* chromosome 5 using the Ac/Ds transposon in combination with the cDNA scanning method. Plant Cell Physiol. 40: 624-639.

Shalev, G., Sitrit, Y., Avivi-Ragolski, N., Lichtenstein, C., and Levy, A.A. 1999. Stimulation of homologous recombination in plants by expression of the bacterial resolvase RuvC Proc. Natl. Acad. Sci. USA. 96: 7398-7402.

Singh, S., Green, A., Stoutjesdijk, P., and Liu, Q. 2000. Inverted-repeat DNA: a new gene-silencing tool for seed lipid modification. Biochem. Soc. Trans. 28: 925-927.

Smith, C.J., Watson, C.F., Morris, P.C., Bird, C.R., Seymour, G.B., Gray, J.E., Arnold, C., Tucker, G.A., Schuch, W., Harding, S., and Grierson, D. 1990. Inheritance and effect on ripening of antisense polygalacturonase genes in transgenic tomatoes. Plant Mol. Biol. 14: 369-379.

Smith, C.J., Watson, C.F., Ray, J., Bird, C.R., Morris, P.C., Schuch, W., and Grierson, D. 1988. Antisense RNA inhibition of polygalacturonase gene expression in transgenic tomatoes. Nature. 334: 724-726.

Smith, N.A., Singh, S.P., Wang, M.B., Stoutjesdijk, P.A., Green, A.G., and

Waterhouse P.M. 2000. Total silencing by intron-spliced hairpin RNAs. Nature. 407: 319-320.

Speulman, E., Metz, P.L., van Arkel, G., te Lintel Hekkert, B., Stiekema, W.J., and Pereira, A. 1999. A two-component enhancer-inhibitor transposon mutagenesis system for functional analysis of the *Arabidopsis* genome. Plant Cell. 11: 1853-1866.

Speulman, E., van Asperen, R., van der Laak, J., Stiekema, W.J., and Pereira, A. 2000. Target selected insertional mutagenesis on chromosome IV of *Arabidopsis* using the En-I transposon system. J. Biotechnol. 78: 301-312.

Streatfield, S.J., Weber, A., Kinsman, E.A., Hausler, R.E., Li, J., Post-Beittenmiller, D., Kaiser, W.M., Pyke, K.A., Flugge, U.I., and Chory, J. 1999. The phosphoenolpyruvate/phosphate translocator is required for phenolic metabolism, palisade cell development, and plastid-dependent nuclear gene expression. Plant Cell. 11: 1609-1622.

Strepp, R., Scholz, S., Kruse, S., Speth, V., and Reski, R. 1998. Plant nuclear gene knockout reveals a role in plastid division for the homolog of the bacterial cell division protein FtsZ, an ancestral tubulin. Proc. Natl. Acad. Sci. USA. 95: 4368-4373.

Sundaresan, V., Springer, P., Volpe, T., Haward, S., Jones, J.D., Dean, C., Ma, H., and Martienssen R. 1995. Patterns of gene action in plant development revealed by enhancer trap and gene trap transposable elements. Genes Dev. 9: 1797-1810.

Swaminathan K., Yang, Y., Grotz, N., Campisi, L., and Jack, T. 2000. An enhancer trap line associated with a D-class cyclin gene in *Arabidopsis*. Plant Physiol. 124: 1658-1667.

Tax, F.E., and Vernon, D.M. 2001. T-DNA-associated duplication/translocations in *Arabidopsis*: implications for mutant analysis and functional genomics. Plant Physiol. 126: 1527-1538.

Temple, S.J., Bagga, S., and Sengupta-Gopalan, C. 1998. Down-regulation of specific members of the glutamine synthetase gene family in alfalfa by antisense RNA technology. Plant Mol. Biol. 37: 535-547.

The *Arabidopsis* Genome Initiative. 2000. Analysis of the genome sequence of the flowering plant *Arabidopsis thaliana*. Nature. 408: 796-815.

Theologis, A., Oeller, P.W., Wong, L.M., Rottmann, W.H., and Gantz, D.M. 1993. Use of a tomato mutant constructed with reverse genetics to study fruit ripening, a complex developmental process. Dev. Genet. 14: 282-295.

Thykjaer, T., Finnemann, J., Schauser, L., Christensen, L., Poulsen, C., and Stougaard, J. 1997. Gene targeting approaches using positive-negative selection and large flanking regions. Plant Mol. Biol. 35: 523-530.

Tissier, A.F., Marillonnet, S., Klimyuk, V., Patel, K., Torres, M.A., Murphy, G., and Jones, J.D. 1999. Multiple independent defective suppressor-mutator transposon insertions in *Arabidopsis*: a tool for functional genomics. Plant Cell. 11: 1841-1852.

Topping, J.F., Wei, W., and Lindsey, K. 1991. Functional tagging of regulatory elements in the plant genome. Development 112: 1009-1019.

van der Krol, A.R., Mur, L.A., de Lange, P., Mol, J.N., and Stuitje, A.R. 1990. Inhibition of flower pigmentation by antisense *CHS* genes: promoter and minimal sequence requirements for the antisense effect. Plant Mol. Biol. 14: 457-466.

van der Krol, R.A., Lenting, P.E., Veenstra, J., van der Meer, I.M., Koes, R.E., Gerats, A.G.M., Mol, J.N.M., and Stuitje, A.R. 1988. An antisense chalcone synthase gene in transgenic plants inhibit flower pigmentation. Nature. 333: 866-869.

Vasquez, K.M., Marburger, K., Intody, Z., and Wilson, J.H. 2001. Manipulating the mammalian genome by homologous recombination. Proc. Natl. Acad. Sci. USA. 98: 8403-8410.

Vaucheret, H., Beclin, C., and Fagard, M. 2001. Post-transcriptional gene silencing in plants. J. Cell Sci. 114: 3083-3091.

Wach. A, Brachat. A., Alberti-Segui, C., Rebischung, C., and Philippsen, P. 1997. Heterologous HIS3 marker and GFP reporter modules for PCR-targeting in *Saccharomyces cerevisiae*. Yeast. 13: 1065-1075.

Walden, R., Fritze, K., Hayashi, H., Miklashevichs, E., Harling, H., and Schell, J. 1994. Activation tagging: a means of isolating genes implicated as playing a role in plant growth and development. Plant Mol. Biol. 26: 1521-1528.

Waterhouse, P.M., Graham, M.W., and Wang, M.B. 1998. Virus resistance and gene silencing in plants can be induced by simultaneous expression of sense and antisense RNA. Proc. Natl. Acad. Sci. USA. 95: 13959-13964.

Weigel, D., Ahn, J.H., Blazquez, M.A., Borevitz, J.O., Christensen, S.K., Fankhauser, C., Ferrandiz, C., Kardailsky, I., Malancharuvil, E.J., Neff, M.M., Nguyen, J.T., Sato, S., Wang, Z.Y., Xia, Y., Dixon, R.A., Harrison, M.J., Lamb, C.J., Yanofsky, M.F., and Chory, J. 2000. Activation tagging in *Arabidopsis*. Plant Physiol. 122: 1003-1013.

Wenderoth, I., and von Schaewen, A. 2000. Isolation and characterization of plant N-acetyl glucosaminyltransferase I (GntI) cDNA sequences. Functional analyses in the *Arabidopsis* cgl mutant and in antisense plants. Plant Physiol. 123: 1097-1108.

Wesley, S.V., Helliwell, C.A., Smith, N.A., Wang, M., Rouse, D.T., Liu, Q., Gooding, P.S., Singh, S.P., Abbott, D., Stoutjesdijk, P.A., Robinson, S.P.,

Gleave, A.P., Green, A.G., and Waterhouse, P.M. 2001. Construct design for efficient, effective and high-throughput gene silencing in plants. Plant J. 27: 581-590.

Wisman, E., Cardon, G.H., Fransz, P., and Saedler, H. 1998. The behaviour of the autonomous maize transposable element En/Spm in *Arabidopsis thaliana* allows efficient mutagenesis. Plant Mol. Biol. 37: 989-999.

Xiaohui W-H., Viret, J.F., Eldridge, A., Perera, R., Signer, E.R., and Chiurazzi, M. 2001 Positive-negative selection for homologous recombination in *Arabidopsis*. Gene. 272: 249-255.

Yoon, K., Cole-Strauss, A., and Kmiec, E.B. 1996 Targeted gene correction of episomal DNA in mammalian cells mediated by a chimeric RNA.DNA oligonucleotide. Proc. Natl. Acad. Sci. USA. 93: 2071-2076.

Zambryski, P. 1988. Basic processes underlying *Agrobacterium*-mediated DNA transfer to plant cells. Annu. Rev. Genet. 22: 1-30.

Zhu, T., Peterson, D.J., Tagliani, L., St. Clair, G., Baszczynski, C.L., and Bowen, B. 1999. Targeted manipulation of maize genes in vivo using chimeric RNA/DNA oligonucleotides. Proc. Natl. Acad. Sci. USA. 96: 8768-8773.

Index

2,4-D 32, 48.

A

AadA See Antibiotic resistance genes.
Abies sp. See Conifers.
Activation tagging 17, 273.
Adventitious rooting 29-32.
Antibiotic resistance 5, 89-94 112-156.
Antibiotic resistance genes
    ampicillin (*bla*(TEM1)) 111-115,
      120-121, 126-127.
    bleomycin (*bleo*) 119.
    hygromycin (hpt) 53,75, 89 161,
      171.
    kanamycin (*nptII*, geneticin) 51-
      54, 68-70, 72, 77, 89, 115, 119-
      121, 124, 126-128, 133-135,
      161, 268.
    methotrexate (*DHFR*) 119.
    spectinomycin/streptomycin
      (*aadA*) 89-94, 115, 119-121,
      127, 142, 242.
Agrobacterium tumefaciens-mediated
    transformation 16, 46-48, 52-54,
    66, 70-77, 115-116, 130, 135-137,
    143, 163-165, 238-239, 267, 279.
Arabidopsis thaliana 2, 4, 11-15, 17,
    19-28, 32, 65-67, 70-77, 161- 164,
    169-171, 187-197, 265- 272.
Auxin *See* 2,4-D.
Avirulence (AVR) genes 181-183,
    196.

B

BA (benzylaminopurine) 48.
Bacillus thuringiensis (Bt) 51, 84, 92,
    94-96, 98, 222, 226-229, 239.
BAR *See* Herbicide tolerance.
Beta glucuronidase *See* GUS.
Biolistics *See* Gene gun-mediated
    transformation.
Biosafety 2-6, 120.
Bla(TEM1) *See* Antibiotic resistance
    genes.
Bleo See Antibiotic resistance genes.
Brassica sp. 66, 73, 77, 193, 196.
Bt *See Bacillus thuringiensis.*

C

CaMV 35S promoter *See* Promoters.
Chimeroplasty 279-280 *See also*
    Gene targeting.
Chloroplast transformation
    gene expression 87-89.
    horizontal gene transfer risks 123,
      127.
    insect resistance 94-95.
    pathogen resistance 96-97.
    potato 101.
    selection and selectable markers
      89-90.
    tobacco 84-85, 87-102.
    tomato 101.
    vertical gene transfer benefits 84,
      97-98.
Chromosome engineering 171.
Cocoa (*Theobroma cacao*) 50.
Coffee (*Coffea arabica*) 48.

S

Salicylic acid (SA) 189-191, 199.
SAR *See* Systemic acquired
    resistance.
Site-specific recombination *See*
    Marker-free transgenic plants.
Somatic embryogenesis 17, 45, 48-
    51.
Spruce *See* Conifers.
Systemic acquired resistance (SAR)
    188-192.

T

T-DNA tagging 267-268.
Taxus See Conifers.
Theombroma cacao See Cocoa.
Tobacco (*Nicotiana sp.*) 46, 69, 70,
    84-88, 98-102, 117, 133, 142,
    160-163, 171, 187-199, 202,
    238, 239, 243-245, 247, 251, 267,
    272-273, 277, 280.
Transgene stacking 169-170.
Transposon tagging 268.
Tulip poplar (*Liriodendron tulipifera*)
    56.

V

Vaccines *See* Edible vaccines.
Vacuum infiltration *See In planta*
    transformation.
Vir genes 52 *See also Agrobacterium*
    tumefaciens-mediated
        transformation.
Virus resistant transgenic plants
    198-200.

Y

Yew *See* Conifers.